Tackling Long-Term Global Energy Problems

ENVIRONMENT & POLICY

VOLUME 52

For further volumes:
http://www.springer.com/series/5921

Tackling Long-Term Global Energy Problems

The Contribution of Social Science

Edited by

Daniel Spreng
ETH Zurich, Switzerland

Thomas Flüeler
ETH Zurich, Switzerland

David L. Goldblatt
ETH Zurich, Switzerland

and

Jürg Minsch
minsch sustainability affairs, Zurich, Switzerland

Editors
Prof. Dr. Daniel Spreng
Energy Science Center (ESC)
ETH Zurich
Zürichbergstrasse 18
8032 Zurich
Switzerland
dspreng@ethz.ch

Dr. Thomas Flüeler
Institute for Environmental Decisions
  (IED)
ETH Zurich
Universitätsstrasse 22
8092 Zurich
Switzerland
thomas.flueeler@env.ethz.ch

Dr. David L. Goldblatt
Department of Management, Technology,
  and Economics (D-MTEC)
Centre for Energy Policy and
  Economics (CEPE)
ETH Zurich
Zürichbergstrasse 18
8032 Zurich
Switzerland
dgoldblatt@mtec.ethz.ch

Dr. Jürg Minsch
minsch sustainability affairs
Wehntalerstrasse 3
8057 Zurich
Switzerland
juerg.minsch@bluewin.ch

ISSN 1383-5130
ISBN 978-94-007-2332-0   e-ISBN 978-94-007-2333-7
DOI 10.1007/978-94-007-2333-7
Springer Dordrecht Heidelberg London New York

Library of Congress Control Number: 2011943627

© Springer Science+Business Media B.V. 2012
No part of this work may be reproduced, stored in a retrieval system, or transmitted in any form or by
any means, electronic, mechanical, photocopying, microfilming, recording or otherwise, without written
permission from the Publisher, with the exception of any material supplied specifically for the purpose
of being entered and executed on a computer system, for exclusive use by the purchaser of the work.

Printed on acid-free paper

Springer is part of Springer Science+Business Media (www.springer.com)

Foreword

This book makes the case for placing the social sciences
on an equal footing with engineering and natural sciences
within energy research.
(Chapter 1, this volume)

'Yet another book on energy!' So may a potential reader say to himself, bearing in mind the surge of articles and books, most of them intelligent and convincingly written, which for decades have presented doomsday scenarios on energy, resources, environment and climate. But the general public and most politicians seem to take little notice. We have become used to the coexistence of parallel worlds, one where scientists and engineers strive to develop accurate analyses based on sophisticated mathematical models, and the other of the majority of citizens and consumers led by short-sighted politicians, though the latter like to pretend they are carrying out the wishes of the former.

Why is that so? This book attempts to find answers from two different angles. The first refers to the research system itself, i.e., to the question of how scientific knowledge is traditionally produced and why this system is partially blind and thus bothered by black spots where problems resist proper analysis and solution. The central message of this book is that in energy research the social sciences belong to a black area. Social studies on energy per se are not necessarily better than the studies of engineers. What is missing is an interdisciplinary approach in which scientists of different disciplines engage in a true mutual cooperation that goes beyond the usual side-by-side pseudo-multidisciplinarity and requires that each discipline try to understand the approach and arguments of the others. The contributing authors present various case studies that demonstrate the failure of solutions in which just this proper engagement is missing.

In the citation that started this foreword the reader will notice however a remarkable asymmetry between natural sciences and engineering on one side and the social sciences on the other. Talking about the lack of true cooperation between different hemispheres of the scientific world is one thing, but pointing out that one hemisphere is not (yet) on equal footing with the other may suggest the arrogance of natural sciences and engineering vis-à-vis the social sciences. Perhaps arrogance really does play a part, as the physicist in me is tempted to admit, but self-criticism seems to be involved as well. Indeed, the asymmetry in the track records of problem solving between the two areas is salient. In a superficial sense engineers seem to be much more reliable in their accomplishments than economists, sociologists, political scientists or psychologists. If economists were able to give advice to national governments with the same certitude as engineers calculate and design bridges, we would live in paradise. It is true that bridges, buildings and machines sometimes fail as well, but these failures can usually be traced back to some violation of good engineering practice. Unfortunately, it is not so simple for the social sciences.[1]

Are the social sciences just lagging behind the natural sciences? Yes and no. Yes, because social scientists do not dispose of an undisputed set of laws and rules like physicists and engineers, the laws of classical mechanics, the Maxwell Equations, etc. No, because they never will, not because they are lazy and less gifted than physicists and engineers, but because they deal with people rather than with concrete, brick, stones and steel pipes, i.e., with well-defined materials and precisely designed pieces of machinery like integrated circuits that (almost) completely lack individual properties. Since the industrial revolution we have become so used to the reliability of objects designed by engineers that we have forgotten that this has come at a considerable cost, i.e., the cost of separating the world of engineers from the world of humans composed of individuals, families clans and states, which interact and which lack the unique properties of the lifeless world.

A little over a decade ago, when I was in charge of a project of the Swiss Federal Institutes of Technology called the *2000 Watt-Society*, together with Caroline Roggo Voegelin I dubbed the project *the moon flight of the 21st century*.[2] We argued that while the grand challenges of the 20th century were mainly of the type of the moon mission, that is, predominantly a challenge for the natural sciences and engineering – like so many other great achievements of that century – the great challenges of the 21st century will no longer allow us the luxury to simply ignore the much more complex issues of the social sciences. The establishment of a sustainable energy system, for which the *2000 Watt-Society* stands, is not a task for energy engineers alone (in fact, most of the necessary technical tools already exist) but primarily for social scientists who have to develop new societal and economic systems in which people can enjoy the achievements of modern society such as freedom, social justice and security at a significantly lower level of consumption of material and energy resources. We expressed the view that the new type of complexity holds

---

[1] Three epochal events, either recent or still unfolding as this book goes to print, demonstrate the fragility of 'engineered' solutions – in terms of classical engineering, risk management, as well as geopolitical strategy – to problems of energy supply, access, supply security and environmental protection: the 2010 Deepwater Horizon drilling rig explosion and oil spill in the Gulf of Mexico, the 2011 post-earthquake Japanese nuclear plant catastrophe and popular revolutions and rebellions against dictatorial regimes across the Arab world in 2011. All three challenge conventional thinking and planning for energy scientists in both the hard and soft sciences.

[2] ETH Bulletin Nr. 276 (2000), pp. 24–27.

Foreword vii

not only for the energy problem but for all other great challenges of the future – including poverty, political stability and access to essential resources such as water and land – where any scientific discipline alone would be hopelessly lost. This, it seems to me, is exactly the message of this book.

There is yet another angle from which the authors are looking at the global energy problem, the metaphor of the invisible. This book could not start with a more apt tale than that of *Gyges and the ring of invisibility* from Plato's dialogue *The Republic*, which tells about the power of the invisible and its use and misuse. In Plato's tale, Gyges uses the ring of invisibility to overthrow the king and usurp power. The lesson of the story is that invisibility produces secret power and thus threatens the freedom of those who are not aware of this power. In essence, the idea and value of scientific research is to make the invisible visible and thus to make it part of generally available knowledge. Academic research and democracy are correlative: this is the lesson to draw from Plato's story.

For most people a significant portion of the energy problem has remained invisible so far, either because it is hidden in the future or because it only manifests itself in other parts of the world or among other societal groups. Only a few seem to know of the invisible, of the hidden part of the swimming iceberg, and among those who know are clever politicians and businesspeople who use the invisibility of the problem to their own advantage. Again, the social sciences play a crucial role in making the invisible visible, since their occupation with human nature and with the history of man provides them with a kind of X-ray that can bring out structure and shape that would otherwise remain invisible.

It is to be hoped that the message of this book will be heard by all of us, natural scientists and engineers, social scientists, politicians and, most importantly, responsible citizens.

Mission to the Moon

Four years after the Sputnik shock of 1957, the cosmonaut Yuri Gagarin became the first human being to travel into space on *April 12, 1961*, greatly embarrassing the United States. In an address to a joint session of Congress on *May 25, 1961* intended to deliver a special message on urgent national needs, President Kennedy asked for an additional $7 to 9 billion over the next five years for the space programme, proclaiming that 'this nation should commit itself to achieving the goal, before the decade is out, of landing a man on the moon and returning him safely to the earth'. President Kennedy did not justify the needed expenditure on the basis of science and exploration. 'I believe we possess all the resources and talents necessary. But the facts of the matter are that we have never made the national decisions or marshalled the national resources required for such leadership. We have never specified long-range

viii   Foreword

> goals on an urgent time schedule, or managed our resources and our time so as to insure their fulfillment.'
> On *July 20, 1969* the Apollo 11 mission saw the realisation of this goal.[3]

Zurich, November 2011   Dieter M. Imboden
                                    President of the Research Council of the
                                    Swiss National Science Foundation, Berne

---

[3] Sources: http://history.nasa.gov/moondec.html, http://www.jfklibrary.org (> JFK > Historic Speeches > Address to Joint Session of Congress 25 May, 1961). Web links accessed November 16, 2011.

Contents

Part I

1 Introduction 3
David L. Goldblatt, Jürg Minsch, Thomas Flüeler, and Daniel Spreng

2 Energy-Related Challenges 11
Thomas Flüeler, David L. Goldblatt, Jürg Minsch, and Daniel Spreng

3 The Indispensable Role of Social Science in Energy Research . . . 23
Jürg Minsch, David L. Goldblatt, Thomas Flüeler, and Daniel Spreng

Part II

Invited Contributions 45
David L. Goldblatt, Thomas Flüeler, Jürg Minsch, and Daniel Spreng

4 What About Social Science and Interdisciplinarity?
A 10-Year Content Analysis of *Energy Policy* 47
Benjamin K. Sovacool, Saleena Saleem,
Anthony Louis D'Agostino, Catherine Regalado Ramos,
Kirsten Trott, and Yanchun Ong

5 Towards an Integrative Framework for Energy Transitions
of Households in Developing Countries 73
Shonali Pachauri and Daniel Spreng

6 A Socio-Cultural Analysis of Changing Household
Electricity Consumption in India 97
Harold Wilhite

7 The Changing Context for Efforts to Avoid the 'Curse of Oil' . . . 115
Jill Shankleman

8 Contributions of Economics and Ethics to an Assessment
of Emissions Trading 133
Adrian Muller

Contents

9   No Smooth, Managed Pathway to Sustainable Energy Systems – Politics, Materiality and Visions for Wind Turbine and Biogas Technology 167
Ulrik Jørgensen

10   Technical Fixes Under Surveillance – CCS and Lessons Learned from the Governance of Long-Term Radioactive Waste Management 191
Thomas Flüeler

11   Learning from the Transdisciplinary Case Study Approach: A Functional-Dynamic Approach to Collaboration Among Diverse Actors in Applied Energy Settings 227
Michael Stauffacher, Pius Krütli, Thomas Flüeler, and Roland W. Scholz

12   Lessons from the Invited Contributions 247
Daniel Spreng and David L. Goldblatt

13   Synthesis: Research Perspectives 263
David L. Goldblatt, Daniel Spreng, Thomas Flüeler, and Jürg Minsch

Part III

14   Lessons for Problem-Solving Energy Research in the Social Sciences 273
Jürg Minsch, Thomas Flüeler, David L. Goldblatt, and Daniel Spreng

Name Index 321

Subject Index 327

Contributors

Anthony Louis D'Agostino School of International and Public Affairs, Columbia University, New York, NY 10027, USA, ald2187@columbia.edu

Thomas Flüeler Institute for Environmental Decisions (IED), ETH Zurich, 8092 Zurich, Switzerland, thomas.flueeler@env.ethz.ch

David L. Goldblatt Department of Management, Technology, and Economics (D-MTEC), Centre for Energy Policy and Economics (CEPE), ETH Zurich, 8032 Zurich, Switzerland, dgoldblatt@mtec.ethz.ch

Ulrik Jørgensen Department of Management Engineering, Innovation and Sustainability, Technical University of Denmark, Lyngby, Denmark, uj@man.dtu.dk

Pius Krütli Institute for Environmental Decisions (IED), ETH Zurich, Natural and Social Science Interface (NSSI), CH-8092, Zurich, Switzerland, pius.kruetli@env.ethz.ch

Jürg Minsch minsch sustainability affairs, 8057 Zurich, Switzerland, juerg.minsch@bluewin.ch

Adrian Muller University Research Priority Programme in Ethics (UFSPE), University of Zurich, 8008 Zurich, Switzerland; Institute for Environmental Decisions (IED), Chair of Environmental Policy and Economics (PEPE), ETH Zurich, 8092 Zurich, Switzerland, amueller@env.ethz.ch

Yanchun Ong Formerly with Lee Kuan Yew School of Public Policy, National University of Singapore, Singapore, episcia3@yahoo.com

Shonali Pachauri International Institute for Applied Systems Analysis (IIASA), 2361 Laxenburg, Austria, pachauri@iiasa.ac.at

Catherine Regalado Ramos Centre for Skills, Performance and Productivity Research, Institute for Adult Learning, Singapore 248922, Singapore, catherine_ramos@ial.edu.sg

Saleena Saleem Formerly with Lee Kuan Yew School of Public Policy, National University of Singapore, Singapore, ssaleem@alum.mit.edu

Roland W. Scholz Institute for Environmental Decisions (IED), ETH Zurich, Natural and Social Science Interface (NSSI), CH-8092, Zurich, Switzerland, roland.scholz@env.ethz.ch

Jill Shankleman Woodrow Wilson International Center for Scholars, Washington, DC 20004-3027, USA, js@shankleman.com

Benjamin K. Sovacool Formerly with Lee Kuan Yew School of Public Policy, National University of Singapore, Singapore 259772, Singapore; current affiliation: Institute for Energy and the Environment, Vermont Law School, South Royalton, VT 05068-0444, USA, sovacool@vt.edu

Daniel Spreng Energy Science Center (ESC), ETH Zurich, 8032 Zurich, Switzerland, dspreng@ethz.ch

Michael Stauffacher Institute for Environmental Decisions (IED), ETH Zurich, Natural and Social Science Interface (NSSI), CH-8092 Zurich, Switzerland, michael.stauffacher@env.ethz.ch

Kirsten Trott Lee Kuan Yew School of Public Policy, National University of Singapore, Singapore 259772, Singapore, ktrott@nus.edu.sg

Harold Wilhite Centre for Development and the Environment (SUM), University of Oslo, 0317 Oslo, Norway, h.l.wilhite@sum.uio.no

Abbreviations

°C   Degree Celsius

A
ASRELEO   Agenda for Social-Science Research on Long-Term Energy Options

B
BP   British Petroleum

C
CEC   Commission of European Communities
CEO   Chief executive officer
CO_2 Carbon dioxide
CoRWM   Committee on Radioactive Waste Management (UK)
CCS   Carbon capture and storage
CDM   Clean Development Mechanism
CNPC   China National Petroleum Corporation

D
DMP   Decision-making process

E
EBS   Engineered barrier systems
EC   European Commission
EEA   European Environment Agency
EFDA   European Fusion Development Agreement
EITI   Extractive Industries Transparency Initiative
EMF   Energy Modelling Forum, Univ. Stanford
ETH   Swiss Federal Institute of Technology
ET   Emissions trading
ETS   EU Emissions Trading System
EU   European Union
EURc   Euro, Euro cent

F

FEPs	Features, events and processes
FSA	Formative Scenario Analysis

G

GHG	Greenhouse gas
GCM	Global climate modelling
GDP	Gross domestic product
GNP	Gross national product
Gt	Gigatonne (billion metric tonne)

H

HLW	High-level radioactive waste

I

IAEA	International Atomic Energy Agency
IBRD	The International Bank for Reconstruction and Development
IEA	International Energy Agency
IHDP	International Human Dimensions Programme on Global Environmental Change
ILW	(long-lived) Intermediate-level waste
IMF	International Monetary Fund
IPCC	Intergovernmental Panel on Climate Change
IRGC	International Risk Governance Council

K

kW	Kilowatt
kWh	Kilowatt hour
kyr	Kilo year (1000 years)

L

LCA	Life cycle analysis/assessment
LLW	Low-level radioactive waste
LNG	Liquefied natural gas
LPG	Liquefied petroleum gas
LTS	Large technical system

M

MCA	Multi-criteria analysis
MDG	Millennium Development Goal
MSc	Master of science

N
NEA	Nuclear Energy Agency
NGO	Non-governmental organisation
NWMO	Nuclear Waste Management Organization (Canada)
NIMBY	Not in my back yard

O
OECD	Organization for Economic Cooperation and Development

R
R&D	Research and development

S
SF	Spent fuel
STS	Science and technology studies

T
TA	Technology assessment
TD	Transdisciplinarity
TdCS	Transdisciplinary case study

U
UK	United Kingdom
UN	United Nations
UN AGECC	UN Advisory Group on Energy and Climate Change
UNDP	United Nations Development Programme
UNFCCC	UN Framework Convention on Climate Change
USA	United States of America
USD	US dollar

W
WB	World Bank
WBGU	German Advisory Council on Global Change
WEC	World Energy Council
WHO	World Health Organization
WRI	World Resources Institute
WWII	World War II (1939–1945)

List of Figures

Fig. 3.1	Interactions between hardware and software components form the current and future energy systems	30
Fig. 3.2	The weak Max-Neef transdisciplinarity pyramid of energy-relevant disciplines, with the field of energy analysis indicated by the shaded ellipse	41
Fig. 4.1	Disciplinary affiliations for energy policy articles. Total 1999–2008	57
Fig. 4.2	Disciplinary affiliations for *Energy Policy* articles, by year	58
Fig. 4.3	Regional affiliations for *Energy Policy* articles, 1999–2008	59
Fig. 4.4	Methodological approaches for *Energy Policy* articles, 1999–2008	62
Fig. 4.5	Case study focus for *Energy Policy* articles by region and top 10 countries, 1999–2008	64
Fig. 4.6	Country case study focus for *Energy Policy* articles, by year	65
Fig. 4.7	Specific technology focus for electricity articles	66
Fig. 4.8	Specific technology focus for transport fuel articles	67
Fig. 4.9	Specific policy mechanism focus for public policy articles	68
Fig. 4.10	Specific policy focus for carbon tax and cap-and-trade articles	68
Fig. 4.11	Cited references for *Energy Policy* articles by year, 1999–2008	68
Fig. 4.12	Total cited references for *Energy Policy* articles, 1999 to 2008	69
Fig. 5.1	Weak transdisciplinarity concept of Max-Neef applied to the study of energy transitions	76
Fig. 5.2	Different measures of electricity access in India	80
Fig. 5.3	Concept for a model. Household energy consumption is assumed to be linearly dependant on household expenditure	86
Fig. 5.4	The Consumption-Access matrix furnishes a two-dimensional view of poverty	89
Fig. 5.5	Energy consumption per capita – inequities between and within countries	92
Fig. 6.1	Ownership of household appliances in households with female heads of household working outside the home	

	and households with female heads of household not working outside the home	103
Fig. 6.2	Ownership of selected appliances, families with a member outside India (NRI) versus families with no family member working outside India	109
Fig. 9.1	The transition from central power utilities in 1980 (a) to distributed energy production today (b)	170
Fig. 9.2	Energy efficiency by production per m^2 and average size of wind turbines	175
Fig. 9.3	The quantitative change in electricity production by type of producer	177
Fig. 9.4	Production share covered by wind turbines	178
Fig. 9.5	Number of turbines installed and dismantled based on their year of installation	179
Fig. 9.6	Farm-scale and centralised biogas plants in Denmark	183
Fig. 10.1	Total contribution of CO_2 reduction options according to various scenario models, concentration targets and timelines .	192
Fig. 10.2	Contribution of reduction options over time (by 2100) and according to climate targets (CO_2 concentrations in ppm) .	193
Fig. 10.3	Contribution of CO_2 reduction options over time, as forecasted by the World Energy Outlook 2007 (2030) and ETP	193
Fig. 10.4	Main components of the CCS system	194
Fig. 10.5	Process chains associated with CO_2 capture	195
Fig. 10.6	Current (2005) maturity of CCS system components	199
Fig. 10.7	System components, drivers, actors and their interrelationships .	201
Fig. 10.8	Falloff of concentrations expected for various greenhouse gases, from a peak value following cessation of emissions	203
Fig. 10.9	Comparison of radioactive waste streams with respect to their radioactivity	203
Fig. 10.10	Multiple-barrier concept in radioactive waste management . . .	206
Fig. 10.11	Safety concept of CO_2 storage: CO_2 gas is injected into a 'storage formation' as the only barrier	206
Fig. 10.12	Phases and permits typically associated with CO_2 storage projects	211
Fig. 10.13	Swiss site selection procedure for radioactive waste disposal . .	212
Fig. 10.14	Global greenhouse gas abatement cost curve for 2030 and beyond business as usual	214
Fig. 11.1	Six steps in the transdisciplinary case study on landscape development in Appenzell Ausserrhoden	234
Fig. 11.2	Varying degrees of involvement and selection of involvement techniques and research methods applied in the case study on landscape development in Appenzell Ausserrhoden	239

| Fig. 13.1 | The 'perspectives cube' – Proposed scheme for characterising social-science energy research approaches | 264 |
| Fig. 13.2 | Studies in Chapters 5–11 positioned in the 'perspectives cube' | 266 |

List of Tables

Table 3.1	Examples of technical and social systems knowledge involved in global climate modelling (GCM)	33
Table 4.1	Sample of articles included in our content analysis	52
Table 4.2	Twenty categories for author training	53
Table 4.3	Intercoder reliability results	56
Table 4.4	Country affiliations for *Energy Policy* articles, 1999–2008	59
Table 4.5	Institutional affiliations for *Energy Policy* articles, 1999–2008	60
Table 4.6	Gender of *Energy Policy* authors, 1999–2008	60
Table 4.7	Interdisciplinary affiliations for *Energy Policy* articles, 1999–2008	61
Table 4.8	Country focus for *Energy Policy* articles, 1999–2008	63
Table 4.9	Comparative case study focus of *Energy Policy* articles, 1999–2008	65
Table 4.10	Technology focus for *Energy Policy* articles, 1999–2008	66
Table 4.11	Topic focus for *Energy Policy* articles, 1999–2008	67
Table 8.1	Basic characteristics of the EU-ETS	136
Table 8.2	The main ethics-based arguments against carbon emissions trading (ET) systems	156
Table 10.1	Climate stabilisation scenarios and resulting long-term global average temperature	192
Table 10.2	System and institutional properties of radioactive waste disposal and fossil CO_2 storage	204
Table 10.3	Risk-related properties of radioactive waste disposal and fossil CO_2 storage	205
Table 11.1	Levels of involvement in the case study on landscape development in Appenzell Ausserrhoden	238
Table 14.1	Summary of science and technology studies (STS) methods and concepts	294
Table 14.2	A recent agenda of priority areas or central challenges of sustainable development constructed from major documents demonstrates the relevance of energy as a persistent and pivotal research area in moving 'knowledge into action'	300

List of Boxes

Box 1.1	Gyges and the Ring of Invisibility	3
Box 3.1	Social Systems Knowledge in the Climate Knowledge Infrastructure	31
Box 14.1	Competencies for Sustainable Development	275
Box 14.2	The 2000 Watt Society – A Research- and Action-Guiding Vision of Prosperity Without Energy Consumption Growth	304

About the Authors

Anthony Louis D'Agostino is a PhD student in Columbia University's Sustainable Development program. He was formerly a Research Associate at the National University of Singapore's Centre on Asia and Globalisation. As part of the Rockefeller Foundation-funded Asian Trends Monitoring Bulletin project, his work focused on the pro-poor dimensions of energy security in Southeast Asia. His research interests encompass climate change adaptation policy in least developed countries, environmental decision analysis and impact evaluation.

Thomas Flüeler, Dr. sc. nat. ETH, is a Senior Research Associate and lecturer at ETH Zurich, Institute for Environmental Decisions, Switzerland. In addition to research and consulting, he has extensive experience in regulatory activities: He is Radioactive Waste/Power Plant Technology Unit Head at the Cantonal Directorate of Public Works, Zurich; and he was a member of the Swiss Federal Nuclear Safety Commission (KSA), an advisory committee to the Federal Government (1992–2004). His research foci are: decision making in complex socio-technical systems, especially institutional, regulatory and long-term aspects, concepts of robustness, resilience, and vulnerability in the context of long-term governance of environmental issues, stakeholder involvement and inclusive expertise in complex systems.

David L. Goldblatt, Dr. sc. nat. ETH, is a researcher and lecturer at the Centre for Energy Policy and Economics at ETH Zurich. He also holds a master's degree from Yale University's School of Forestry and Environmental Studies and a bachelor's degree in economics from Brown University. He has worked on energy, environment and public health issues in NGOs, government and the private sector in the US and Switzerland. From 2003 to 2005 he served as Risk Policy fellow in the AAAS Science & Technology Policy Fellowship Program in Washington DC. His past and present research interests include energy conservation and sustainable consumption, natural resources and political economy, renewable energy technologies, China and the environment and ethics and environmentalism.

Ulrik Jørgensen is a professor at the Department of Management Engineering at the Technical University of Denmark. He holds an M.Sc. in Engineering and a PhD in Innovation Economics from the Technical University of Denmark. Drawing on the theories and approaches from science and technology studies he is currently involved in strategic research on user involvement in design and enabling sustainable transitions. Earlier research comprised studies of waste handling, energy technologies, innovation in the health care sector, the role of experts in foresight and public advice, engineering and design competences as well as developments in the transport sector.

Pius Krütli is a senior scientist at the Chair of Natural and Social Science Interface (NSSI) at ETH Zurich. He graduated in Environmental Sciences (MSc, Dr. sc.) at the ETH. His research foci are: radioactive waste management, stakeholder involvement/public participation, justice issues, and transdisciplinary processes.

Jürg Minsch studied at the University of St. Gallen (PhD in ecological economics). He was a member of the Swiss Federal Board on Sustainable Development (1998–2000). Professor of Sustainable Development at the University of Natural Resources and Life Sciences (BOKU), Vienna, Austria (2000–2006). He is the founder of the Zurich-based independent think tank "minsch sustainability affairs – Strategien für Demokratie, Marktwirtschaft und Ökologie" (2007) and a lecturer at ETH Zurich and BOKU. He is a member of the Energy Committee of the Swiss Academy of Engineering Sciences SATW. His main research interests are: societal transformation, syndromes of non-sustainability, institutional innovations for sustainable development, strategies for a post-fossil and post-nuclear global economy, science and education for sustainable development.

Adrian Muller, PhD in theoretical physics, worked as a senior researcher at ETH Zurich, the University of Gothenburg and the University of Zurich. Currently he works at the Research Institute of Organic Agriculture FiBL and ETH Zurich. From 2007 to 2010, he followed a postgraduate programme in ethics. He works mainly in the area of environmental and resource economics, with a focus on climate change and policy instruments. His ethical work focuses on the possibility of consumption reduction as a policy strategy in liberal societies. His sectoral focus is energy and agriculture.

Shonali Pachauri is a Senior Research Scholar at the International Institute for Applied Systems Analysis and serves on the Executive Committee of the Global Energy Assessment. She received her PhD from ETH Zurich in 2002, following which she worked as a Post Doc with the Centre for Energy Policy and Economics at ETH. Her research foci include the analysis of the socio-economics of energy access, use and choice, resource use and access in relationship to lifestyles, poverty and development, energy demand and fuel choice modelling as well as the analysis of embodied energy of household consumption in developing countries.

Catherine Regalado Ramos holds a BA in Public Administration and a master's in Development Economics from the University of the Philippines. She is currently Senior Research Officer at the Institute for Adult Learning (IAL) in Singapore. Previously, she worked with Benjamin Sovacool as a research assistant on energy issues at the Lee Kuan Yew School of Public Policy. She has worked at the National

Institute of Education in Singapore on education research, in academia and in private organisations. Her 12 years of professional experience has involved social science research, human resources and training and legislative research.

Saleena Saleem was a research associate at the Centre on Asia and Globalisation (CAG), National University of Singapore, from 2008 to 2011, where she researched global energy governance and transparency and access to information.

Roland W. Scholz holds the Chair of Environmental Sciences: Natural and Social Science Interface (NSSI) at ETH Zurich. He is an adjunct professor of Psychology at the University of Zurich (Privatdozent), and was elected as the fifth holder of the King Carl XVI Gustaf's Professorship 2001/2002 hosted at the Center of Environment and Sustainability of Chalmers University of Technology and Gothenborg University (Sweden). Since 2002, he has been the speaker of the International Transdisciplinarity Network on Case Study Teaching (ITdNet).

Jill Shankleman, PhD, is a sociologist with over 25 years' experience as a consultant. She works mainly with corporate clients helping manage environmental, social and political risks associated with major investment projects worldwide. She has recently completed a two-year assignment with the World Bank's political risk insurance organisation, MIGA, setting up a service to assist clients in Africa manage environmental and social risks, and has in-depth knowledge of the IFC Social and Environmental Performance Standards.

Benjamin K. Sovacool, PhD, was an Assistant Professor at the Lee Kuan Yew School of Public Policy, National University of Singapore; he is now with the Institute of Energy and the Environment, Vermont Law School. There he studies topics related to energy security, renewable energy policy, climate change mitigation, and climate change adaptation. He has served in advisory and research capacities at the U.S. National Science Foundation's Electric Power Networks Efficiency and Security Program, Virginia Tech Consortium on Energy Restructuring, Virginia Center for Coal and Energy Research, New York State Energy Research and Development Authority, Oak Ridge National Laboratory, Semiconductor Materials and Equipment International, U.S. Department of Energy's Climate Change Technology Program, and the International Institute for Applied Systems and Analysis near Vienna, Austria. He is the author or editor of eight books and more than 120 peer reviewed academic articles on various aspects of energy and climate change, and he has presented research at more than 60 international conferences and symposia.

Daniel Spreng was, until 2005, director of the Centre for Energy Policy and Economics (CEPE) of ETH Zurich. Professor at ETH Zurich since 1991. He studied physics at ETH Zurich and did his doctorate in materials science at Northwestern University, USA. He worked for many years at Swiss Aluminium Ltd., ultimately as head of the ecology department. He was a Visiting Fellow at the Institute for Energy Analysis, Oak Ridge TN, a think tank headed by Alvin Weinberg, and a Guest of the Rector at the Institute for Advanced Study in Berlin. Partly retired, he still conducts research on the interrelationships of energy, poverty and development (particularly in southeast Asia).

Michael Stauffacher is a deputy head of the Chair of Environmental Sciences: Natural and Social Science Interface (NSSI), research group leader, lecturer and coordinator of the Major Human-Environment Systems at ETH Zurich. He graduated in sociology at the University of Zurich (Dr. phil.) and since 2009 has served as president of the Swiss Academic Society for Environmental Research (SAGUF). As of 2011, he is an Associate Professor Extraordinary at the School of Management and Planning, Stellenbosch University (South Africa).

Kirsten Trott is currently studying to complete a Master of Laws and Masters in Public Administration at the National University of Singapore. She has an LLB from the University of Western Australia and a post graduate certificate in International Community Development from the University of Queensland, Australia. Admitted to practice as a lawyer in Australia, she has pursued a career including private legal practice and advisory work relating to governance and regulatory policy. Her areas of expertise include whistleblowing, transparency, accountability and consumer banking.

Harold Wilhite is a social anthropologist. He is currently a Research Director at the University of Oslo's Centre for Development and Environment. Wilhite has engaged in research on energy conservation, international development and sustainable consumption in a number of countries. His main research interests have been associated with theorising and promoting sustainable energy use, both in developed and developing countries. He has consulted for a number of international policy efforts on sustainable energy consumption and climate mitigation and was a founding member of the European Council for an Energy Efficient Economy (ECEEE).

Ong Yanchun was a research associate at the Lee Kuan Yew School of Public Policy, National University of Singapore, from 2008 to 2011 where she researched regulatory risks in infrastructure projects in Asia, and business and human rights in Southeast Asia.

Part I

Chapter 1
Introduction

David L. Goldblatt, Jürg Minsch, Thomas Flüeler, and Daniel Spreng

Abstract Hitherto in energy research, social sciences have tended to be used to research and boost public acceptance of technologies and to help facilitate their introduction in the market. There are a variety of reasons for this neglect: Research agendas tend to be framed by technoeconomic perspectives; social sciences are presumed to be irrelevant or useless in the energy field; their approaches and assumptions may collide with the dominant technoeconomic paradigm and as such may be politically unpalatable and intractable; relativism and political instrumentalisation in practice have given the field a reputation for pseudoscience. This publication, especially the in-depth studies showcased in Chapters 4–11, supports the equal collaboration of the social sciences with the physical, natural and engineering sciences in energy research by demonstrating their relevance, compatibility and complementarity with conventional research; their policy usefulness; and their scientific rigour.

Keywords Energy research · Interdisciplinary · Perspective · Research agenda · Social sciences

---

Box 1.1 Gyges and the Ring of Invisibility

Plato's dialogue, *The Republic*, centres on what it is that makes a political system just and good. Glaukon, one of the participants in the discussion, tells the story of Gyges, a poor shepherd in the service of the king of Lydia. One day, as Gyges was tending his flock, a violent thunderstorm was unleashed and an earthquake tore the field open. Gyges descended into the chasm and emerged with a golden ring, which had the ability to render its wearer invisible. Gyges put this power to use, invaded the palace and overthrew the king, usurping the kingship for himself.

D.L. Goldblatt (✉)
Department of Management, Technology, and Economics (D-MTEC), Centre for Energy Policy and Economics (CEPE), ETH Zurich, 8032 Zurich, Switzerland
e-mail: dgoldblatt@mtec.ethz.ch

D. Spreng et al. (eds.), *Tackling Long-Term Global Energy Problems*,
Environment & Policy 52, DOI 10.1007/978-94-007-2333-7_1,
© Springer Science+Business Media B.V. 2012

This tale would not be worth recounting if it merely illustrated the dominance of strong over weak. In this case, the storyteller, Glaukon, is concerned with what lies behind the power of the strong and the impotence of the weak. The secret of Gyges' ring lay in the dichotomy between the visible and the invisible. What happens under the cover of invisibility is removed from moral judgment; it simply happens and is imposed on the people, who are left to suffer the consequences of decisions made in secret. There is nothing to shape in the realm of the invisible. The question of a just government is not raised, and the conscious construction of a just order only becomes possible when the veil of invisibility is lifted.

According to Glaukon, compassion, ethical rules and laws are only effective when a person's conduct is visible – or, more precisely, when these laws, ethics and compassion are visible to the people. In contrast, conduct that remains invisible in this sense destroys societal harmony.

Glaukon explains the ring's design principle in this way: outrages are to be committed in such a way that no one notices them; only a bungler allows himself to be caught (Plato II/4/361 a). Glaukon's brother Adeimantes cites some concrete examples:

> With a view to concealment, we will establish secret brotherhoods and political clubs. And there are professors of rhetoric who teach the art of persuading courts and assemblies; and so, partly by persuasion and partly by force, I shall make unlawful gains and not be punished.
>
> (Plato II/8/365 d)

The discussion probably continued in a lively fashion, with someone likely pointing out that more is involved than just deliberate injustice and defrauding. Inconspicuous daily routines may also play a part, evolving in the shadow of invisibility and unleashing destructive forces. One can exercise power under the cover of invisibility, both consciously and unconsciously.

This holds true today, even though the 'production of invisibility' has become more sophisticated and there are civilised alternatives to conspiracies and violence. Today, the 'art of persuading assemblies' can mean that certain ecological, social and economic aspects of a project or business deal are belittled, dismissed as irrelevant or not discussed at all. This might happen because they are far removed in terms of their time, space or cultural context, because they hide behind high problem complexity, because their low probability of occurrence makes them seem insignificant, because they belong to a different area of technical expertise, because they may threaten economic, political or scientific interests, or because analysing them requires an alternative approach, which for this reason alone can seem ideological and unscientific. Instead of hazarding a public discussion, one settles for promoting acceptance through artful speech.

1 Introduction

A modern-day observer of the dialogue would accuse Gyges of lacking self-reflection and compassion for having used the ring. But it solves nothing to shift responsibility solely to the individual: he or she is morally overtaxed. Glaukon invites his dialogue partners to participate in a Gedanken experiment. Suppose that there were two such rings, and a just man put one of them on, an unjust man the other. What would happen? According to Glaukon's assessment:

> No man can be imagined to be of such an iron nature that he would stand fast in justice and keep his hands off what was not his own when he could safely take what he liked ... and in all respects be like a god among men.
>
> (Plato II/4/360 b+c)

In important societal matters, nobody – and no scientific discipline – should imagine themselves as God among the people. And nowhere should we be more vigilant against the magic of invisibility, the 'art of persuading assemblies', the constriction to narrow technical and economic perspectives, than in the crucial issue of *energy*.

Jürg Minsch

1.1 Point of Departure

The point of departure of this book is the neglect of the full range of social sciences in energy research, a field dominated by engineering and mainstream economics. Where social sciences have been involved, they have tended to play a subordinate, often instrumentalised role in the service of preordained technical fixes. Moreover, they have often been used to research public acceptance of these technologies or to help facilitate their introduction in the market. Limiting research questions primarily to commercial or technological objectives is in marked contrast with the central importance of energy production for economies and for society. The full range of the social sciences should be used to shed light on energy issues that are of vital future importance.

Why is it that most of the social sciences have been neglected in the study of energy and energy-related problems? Lutzenhiser and Shove (1999) described how national funding structures in the US and the UK ensure that energy research agendas are framed by a technoeconomic perspective, which tends to exclude the broader range of social sciences. For their part, social scientists have not been quick to respond to emerging opportunities to collaborate in interdisciplinary energy research.

There are other possible reasons why the social sciences have been neglected:

- *Claim of irrelevance*: Although this book seeks to demonstrate otherwise, a common assumption is that the social sciences lack relevance or useful application in the energy field. This assumption is implicit in the dominant technological approach to energy issues. Energy systems, enabled by technological innovations and developments that began before the Industrial Revolution, rightfully remain the domain of the scientific and engineering disciplines that have matured with them, along with economics. Other social science practitioners bear the burden of proof to show that their disciplines are not outside the sphere of relevance in energy research.
- *Clash with the dominant technoeconomic paradigm*: Inclusion of the full range of other social sciences may expose differences in approaches and assumptions that are irreconcilable with engineering and economic approaches.
- *Politically unpalatable*: Social science sometimes produces results and recommendations that do not fit the existing power structure.
- *Politically non-expedient*: For a variety of reasons, standard engineering and economics lend themselves more readily as tools of the state for traditional development, planning, direction and implementation; or, in more authoritarian polities, imposition of the regime's decisions.
- *Hostile to science*: In some circles, social constructivist theory has evolved to the point of deep scepticism regarding the value, durability and authority of scientific expertise. Instead, the theory has favoured a relativist perspective in which all views are potentially equally plausible and valid as knowledge.
- *Susceptible to pseudoscience*: Compared to natural and physical sciences and engineering, it has been argued that some of the soft sciences are more easily co-opted by radical ideologies and political philosophies. In the worst cases, the social science ends up involving minimal 'science', instead becoming an instrument of politics or political advocacy with a patina of objective scientific enquiry. In the absence of experimental controls or peer review beyond like-minded academics, the resulting pseudoscience runs the risk of discrediting sound social science by association.

The chapters in this book, especially the in-depth studies showcased in Chapters 4–11, should help reduce the resistance to the participation of social science in energy research by demonstrating its relevance, compatibility and complementarity with conventional research; its policy usefulness; and its scientific rigour.

1.2 Energy-Related Challenges

The importance of social science for energy research is clear when one considers contemporary energy-related challenges, which is why this book has taken a problem-oriented approach. *Chapter 2: Energy-Related Challenges* identifies three challenges that stand out in a long-term perspective, taking interrelations between global development and the environment into consideration:

1 Introduction

I. Access and security
II. Climate change and other environmental impacts
III. Economic and social development.[1]

The social, economic and policy dimensions of the challenges define a central role for the social sciences in the academic treatment of these problems.

These challenge clusters emerged from extensive discussions in the interdisciplinary workshops conducted within the framework of ASRELEO (Agenda for Social-Science Research on Long-term Energy Options). This project was carried out between 2005 and 2007 under the auspices of the Energy Science Center of ETH Zurich on behalf of the European Fusion Development Agreement and BP.[2]

1.3 Social Science's Role and Responsibility and Energy Issues in the Public Realm

In a modern economy, the production factor of *energy* relies on well-functioning technical systems. These are developed and operated by people, which therefore makes them sociotechnical systems, and they are also embedded within other economic, political, social and cultural systems. The production, conversion, distribution and use of energy have a number of consequences, both intended and unintended, as analysis of the interrelationships between various parts of these sociotechnical systems has shown. Because decisions concerning the design and development of sociotechnical energy systems are societal matters, the analysis of these systems and the formulation of design recommendations for them fall directly within the purview of the social sciences.

This book makes the case for placing the social sciences on an equal footing with engineering and natural sciences within energy research. In doing so, it helps concretise the specific role and responsibility of social science in energy research. The energy-related challenges point to the centrality of energy for economic and social development as well as for environmental problems. Energy issues are increasingly becoming public issues, and the energy agenda is a societal agenda.

Economic treaties, environmental laws, land use planning and traffic regulations, as well as many social and development programmes often have significant effects on energy production and use. In so doing, they implicitly implement energy policies, although this function is not always recognised or acknowledged. In this wider sense, energy policy reflects a threefold multiplicity:

---

[1] Although the parallelism between our three challenge groups and sustainability's conventional triad of environment, economy and society (as well as equity/access) was not deliberate, it is, in retrospect, not entirely coincidental.

[2] Details of the development and findings of the ASRELEO project are available in the final report (Flüeler, Goldblatt, Minsch, & Spreng, 2007).

1. A multiplicity of diverse energy policies in a narrow, traditional sense, along with broader policies in the economy and civil society. Instead of *one* energy policy within a given political entity, energy policy is a component of diverse policy areas at all levels of public and private exchange.
2. A multiplicity of analytical and problem-solving approaches. The energy-related challenges call upon multiple disciplines; there is no single scientific approach to analysing and working out solutions, nor is there a single social scientific approach. The scope of the challenges calls for openness to a wide repertoire of social science research.
3. A multiplicity of problem perceptions, assessments, visions and interests across society. As societal concerns, energy issues principally involve the entire civil society but are embedded in specific ecological, economic, engineering, political, social and cultural contexts and are informed by the corresponding bodies of practical knowledge.

At first glance, these multiplicities seem to reduce the elegance of simple responses to energy problems and to lower the ease and efficiency with which solutions can be formulated. Nevertheless, they better reflect the real-world conditions that apply to energy issues, and they rule out any reliance on magic bullets. Incorporating variety in policy and applied contexts also means dealing with concrete political-institutional, economic, social, natural and technical factors. Taking particular contexts into consideration is one of the chief areas of competence of the social sciences and, from a purely instrumental perspective, is one of their main sources of 'added value'. Organising diverse strands of research and integrating multiple scientific disciplines and actors are two other competencies, both of which are required in order to meet energy-related challenges.

All three of the posited challenge clusters, as well as many of the difficult problems that fall under their rubrics, fit the description of what Rittel and Webber (1973) termed 'wicked'[3] policy problems. Environmental problems are usually complex and ill-defined (Reitman, 1964; Abelson & Levi, 1985) or ill-structured (Scholz & Tietje, 2002), often long-term and may be associated with equity issues (Flüeler, 2006). Certain global environmental problems such as climate change have even been termed 'super-wicked' (Levin et al., 2009). The epigraphs of several chapters in this book include short excerpts from Rittel and Webber's classic 1973 article entitled *Dilemmas in a General Theory of Planning* in order to ground the discussion with some sobering lessons in policy realism.

The ASRELEO workshops underscored the difficulty of coordinating ideas and building consensus across disciplines and approaches, even relatively close ones.

---

[3] Defining properties of wicked policy problems include the following: (1) Wicked problems have no definitive formulation, (2) they have no stopping rule, (3) solutions are good-bad rather than true-or-false, (4) there are no immediate or ultimate tests of solutions to them, (5) every solution is a one-shot operation (no trial-and-error), (6) potential solutions are not enumerable, (7) each one is essentially unique, (8) each is a symptom of another problem, (9) the analyst's choice of explanation determines the recommended solution and (10) the planner has no right to be wrong (i.e., the political consequences of mistakes tend to be severe) (based on Rittel & Webber, 1973).

1 Introduction

Even in the process of writing this book and coordinating the input of various outside contributors, the authors were made keenly aware of the pitfalls. Still, this is only a pale reflection of the larger problem of knowledge management and integration across disciplines that arises when applying scientific findings to real-world problems.

Chapter 3: The Indispensable Role of Social Sciences in Energy Research discusses the particulars of the social science competencies mentioned above as applied to energy, and it goes into some depth on the conceptual and methodological issue of knowledge management and integration.

1.4 Goals and Contribution to the State of the Art

Chapter 4 of ASRELEO's final report (Flüeler et al., 2007) presented two overviews of the state of the art in energy research in the social sciences, the first of which concentrated on assessments and calls for R&D. It noted that the most commonly explored topics are closely connected to particular technologies and tend to be characterised by a managerial approach, often involving the implementation of technocratic solutions. Detailed explorations of institutional and social contexts for the energy technologies are rare, as are broadly participatory and inclusive decision approaches. Chapters 2 and 3 in Part I of this book make the case for more of this kind of research, with Chapters 4–11 offering several examples.

The longest part of the book is *Part II: Invited Contributions*. This part showcases new contributions from a wide variety of disciplines from the social sciences and even humanities, including anthropology, energy analysis, environmental and resource economics, ethics, history, linguistics, political science, science and technology studies (STS), and sociology, in addition to practitioner knowledge sets. The chapters and case studies in Part II were solicited or selected on the basis of their relevance to the energy challenges as well as their individual multi-, inter- or transdisciplinary breadth. Chapter 12 systematically examines the studies in light of their direct contributions to meeting the energy-related challenges. Chapter 13 summarises and synthesises findings from a research perspective and positions them in a broader taxonomy and transdisciplinary framework.

This book's *goals* are to

- Substantiate social sciences' significance and role in long-term-oriented energy research
- Demonstrate and illustrate this role with salient and convincing examples from research
- Support the equal collaboration of the social sciences with the natural and engineering sciences
- Stimulate broader interest in long-term, interdisciplinary energy research on the part of social scientists
- Highlight potential opportunities for interdisciplinary energy research endeavours for actors in science policy and research.

Part III, the concluding *Chapter 14: Lessons for Problem-Solving Energy Research in the Social Sciences*, provides the broadest synthesis of our findings and highlights the major research streams in problem-oriented energy-related social science. In addition, each thematic section offers a number of research questions that may be useful for designing future research agendas.

References

Abelson, R. P., & Levi A. (1985). Decision making and decision theory. In G. Lindsey & E. Aronson (Eds.), *Handbook of social psychology. 1. Theory and method* (pp. 231–309, 274). New York: Lawrence Erlbaum.

Flüeler, T. (2006). Decision making for complex socio-technical systems. Robustness from lessons learned in long-term radioactive waste governance. *Series Environment & Policy: Vol. 42.* Dordrecht: Springer.

Flüeler, T., Goldblatt, D., Minsch, J., & Spreng, D. (2007). *Meeting global energy challenges: Towards an agenda for social science research.* Final Report for EFDA and BP, Contract EFDA/05-1255. Zürich: ETH. All web links accessed November 16, 2011, http://www.esc.ethz.ch/publications/ASRELEO-Projekt.pdf

Levin, K., Cashore, B., Bernstein, S., & Auld, G. (2009). *Playing it forward: Path dependency, progressive incrementalism, and the "Super Wicked" problem of global climate change.* Paper presented at the 2009 IOP Conference Series. Climate Change: Global Risks, Challenges and Decisions IOP Publishing. IOP Conference Series: *Earth and Environmental Science, 6,* S 50.02 (502002). doi:10.1088/1755-1307/6/0/502002. http://iopscience.iop.org/1755-1315/6/50/502002/pdf/1755-1315_6_50_502002.pdf

Lutzenhiser, L., & Shove, E. (1999). Contracting knowledge: The organizational limits to interdisciplinary energy efficiency research and development in the US and the UK. *Energy Policy, 27,* 217–227.

Plato (trans. into English, 1894). *Plato's Republic: The Greek text* (B. Jowett & L. Campbell, Eds.). Oxford: Clarendon Press.

Reitman, W. R. (1964). Heuristic decision procedures, open constraints, and the structure of ill-defined problems. In W. M. Shelly II & G. L. Bryan (Eds.), *Human judgments and optimality* (pp. 282–315). New York: Wiley.

Rittel, H., & Webber, M. (1973). Dilemmas in a general theory of planning. *Policy Sciences, 4*(2), 155–169.

Scholz, R. W., & Tietje, O. (2002). *Embedded case study methods. Integrating quantitative and qualitative knowledge.* Thousand Oaks, CA: Sage.

Chapter 2
Energy-Related Challenges

Thomas Flüeler, David L. Goldblatt, Jürg Minsch, and Daniel Spreng

Abstract Energy systems and attendant institutions have long-term characteristics basic to the development of economies and societies. Mankind faces a wide range of serious problems connected to the world energy system. All players involved must find a delicate balance between flexibility and stability, between the demand for urgent change and the need for stable, lasting solutions. The following sections provide an introductory overview of the energy-related challenges Access and Security, Climate Change and other Environmental Impacts, and Economic and Social Development. These challenges and their interconnections are explored in depth in later chapters.

Keywords Access · Climate change · Economic and social development · Energy systems · Environmental impacts · Risk · Safety · Security · Social sciences

> *The problems that planners must deal with are wicked and incorrigible ones, for they defy efforts to delineate their boundaries and to identify their causes, and thus to expose their problematic nature. The planner who works with open systems is caught up in the ambiguity of their causal webs.*
>
> *Rittel and Webber (1973, p. 167)*

2.1 Introduction

Physical infrastructure is a dominant feature of energy supply systems and the long lifetime of this infrastructure makes these systems resistant to quick change. In this way, the global energy sector itself resembles a super-tanker, the largest and most iconic of its transport vehicles: It has long lead times, its components have long technical lifetimes and its operation has extremely long-lasting effects on resource

---

T. Flüeler (✉)
Institute for Environmental Decisions (IED), ETH Zurich, 8092 Zurich, Switzerland
e-mail: thomas.flueeler@env.ethz.ch

D. Spreng et al. (eds.), *Tackling Long-Term Global Energy Problems*,
Environment & Policy 52, DOI 10.1007/978-94-007-2333-7_2,
© Springer Science+Business Media B.V. 2012

availability and the environment.[1] Putting a nuclear power plant into operation, including getting it licensed, takes an average of 20 years; the electrotechnical parts of a hydroelectric power plant can be run without being replaced for some 50 years; the residence time of CO_2 in the atmosphere is significantly more than 100 years; effectively, a dam lasts forever.

The institutions that interact with these long-lived physical systems have long lives themselves and require correspondingly long planning horizons. Evidently working from different premises, and despite the partial public good character of the energy supply, the European Union has based its recent energy directives on the notion that energy is a good or service like any other and that the energy market should especially facilitate competition and the easy entrance of new players into the marketplace (EC, 2010). But in markets that firms enter and exit quickly, regulatory and supervisory bodies must provide the long-range perspective required for the public interest. They must actively shape the course of the energy system, with the input and guidance of a broad range of experts. In order to properly fill their role, social scientists must study the long-range development of the relevant institutions and, indirectly, the technical systems involved. For the sustainability of the energy system, the social sciences must consider how the interaction of multiple factors will affect the well-being of present and future generations.

Society faces a wide range of serious problems connected to the energy system. As introduced in Chapter 1, this book groups these problems into three challenge categories:

I. Access and Security
II. Climate Change and other Environmental Impacts
III. Economic and Social Development

Some of the problems and crises of the global energy system, such as accelerating climate change, insufficient access to energy for the 'bottom billion' (Collier, 2007), gross geopolitical disparities in the distribution of energy resources and the resulting large money flows (in some cases to rogue states and groups), seem to demand immediate attention. At the same time, the need for economic efficiency calls for speed and flexibility from market actors.

Consequently, institutions, politics, regulatory bodies, firms and end-users must find a delicate balance between stability and flexibility in their interactions and they must be conducive to technological and social innovation processes that are simultaneously constructive and durable. All energy challenges must be examined against

---

[1] In their latest update to the classic *Limits to Growth* (Meadows, Randers, & Meadows, 2004), the authors also used a ship metaphor to explain how a system's momentum delays the transmission of negative feedback, causing the system to overshoot: 'To steer correctly, a system with inherent momentum needs to be looking ahead at least as far as its momentum can carry it. The longer it takes a boat to turn, the farther ahead its radar must see. The political and market systems of the globe do not look far enough ahead'.

2 Energy-Related Challenges

the backdrop of this tension between the demand for urgent change and the need for stable, lasting solutions.

The following sections provide an introductory overview of the challenges and their interconnections, which are explored in depth in later chapters.

2.2 Challenge I: Access and Security

Although they may evoke completely different issues, energy access and security are actually interconnected concepts. For example, the security of Western Europe's oil supply is entirely different from the security of a power plant or an electricity network, which involves protecting the infrastructure from unwanted intrusion. Local energy security is a cause of concern to consumers when they doubt the reliability of their energy supplier.[2] Nations' supply security is a perennial international issue, particularly for countries that are heavily reliant on a single fuel. The dependence of developed countries on supplies of gas and oil from countries that are openly hostile to their interests and often ruled by dictatorships (and their massive financial transfers to such countries) seems to be an inescapable feature of the world stage. For their part, producer countries concerned about expanding their production capacity without greater certainty of future demand have raised the issue of supply security of demand in recent years.

The term *access* commonly denotes the technical issue of connecting energy demand to supply by means of a power line, for example. In many parts of the world, however, an individual consumer's access to an energy source has as much to do with their economic or social status as their residence's or community's technical configuration. Approximately three billion people lack access to clean cooking fuel (UN AGECC, 2010) and one fifth of the world population (1.4 billion) do not have access to electricity (IEA, 2010).[3] Energy poverty is a cause for misery, ill health and deprivation and is also a hindrance to general development in many countries with high population densities (cf. Chapter 5).[4] Energy security and energy access are both intimately connected with economic and political power (cf. Chapter 7).

Unequal access to resources and energy carriers within countries also presents a daunting challenge. Electricity access requires a reciprocal economic commitment: the provider must serve a specific region and the customer must make use of the

---

[2] The notion of security has to be distinguished from safety and safeguards concepts, which we subsume under 'Environmental Impacts' below. Safety is the protection of human beings and the natural environment from damage caused by hazard potentials. Safeguards are measures against the diversion of nuclear and dual-use material for the construction of atomic weapons.

[3] The World Energy Outlook 2010 estimates 2.7 billion people lack access to modern fuels for cooking and heating (IEA, 2010), while WHO and UNDP (2009) estimate that this number is over 3 billion people (UN AGECC, 2010, p. 14).

[4] Almost 1.2 billion additional people will need access to electricity and 1.9 billion people will need access to modern fuels by 2015 to meet the Millennium Development Goal of halving the proportion of people living in poverty (WHO & UNDP, 2009, p. 32).

service. In developing countries, however, a provider's decision to newly service a region often has a strong political element as well, in that denying or temporarily limiting access to electricity can serve the deliberate purpose of excluding specific parts of the populace from development. Thus, energy access may be used as a tool in political and economic power struggles within nations; the lack of access to energy is an expression of these struggles. Few rural electricity customers in developing countries have access to around-the-clock electricity (cf. Chapter 6).

For potential new customers, accepting access to the grid often involves embracing a new and unfamiliar lifestyle. Kemmler (2007) showed that the electrification of individual Indian households is not so much a question of economics as of consumers' education and access to other 'modern' goods and services.

It is important to consider both energy access and energy consumption when addressing the issue of energy poverty. Energy access can significantly relieve energy poverty, even if the household is not in a position to consume large amounts of energy. Amartya Sen insists that poverty has two dimensions. The poor suffer at least as much from their lack of consumption possibilities, their limited mobility and their inability to change their living conditions (which Sen refers to as *capability*) as they do from their lack of consumption per se (*functioning*) (Sen, 1997).

There is no easy remedy for the increasingly insecure supply of cheap oil and gas to consumers in developed countries. Many high-consuming countries have inadequate domestic fossil fuel reserves or none at all, and as history has shown, unconditional and uninterrupted imports cannot be taken for granted. Some of these countries are realising the need for a dedicated 'energy foreign policy', even a joint multinational energy policy (cf. EC, 2010). This requires diverse knowledge about the geology of energy resources, management of oil and gas fields, and maintenance and development of production facilities. It also requires knowledge of the resource-rich countries' transport potentials, current and future markets and likely future domestic demand, as well as their societies, cultures, and politics.

International and intranational aspects of the relationship between access and supply security on the one hand, and political/economic/military power on the other, are under-researched and poorly understood. Political science analysis is also necessary in order to develop scenarios for international conflicts as fossil energy resources are depleted and the race for access accelerates. The research conducted in these areas indicates that whether the availability of more modern energy carriers promotes peaceful development or is a source of conflict and lasting inequality depends largely on the associated institutional environment.

Diversification of energy sources, in terms of both geographical regions and non-fossil fuel alternatives, is necessary in order to improve developed countries' supply security. However, in order to avoid obstructing the energy access and security of developing nations, developed countries must moderate their energy consumption so that developing countries have the room – in terms of availability and affordability – to increase their consumption to lift their populations above subsistence levels, without producing global levels of consumption and greenhouse gas (GHG) emissions that exceed the earth's ecological carrying capacity. This is a moral issue as well as a matter of technology and economics, which also requires input from

2.3 Challenge II: Climate Change and Other Environmental Impacts

As national energy infrastructures have evolved, their attendant environmental and social problems have tended to be shunted across increasing geographical and temporal scales. These scales range from the village charcoal gatherer and mother cooking over a poorly ventilated open-pit fireplace, to communities – both ecological and human – that are affected by the local extraction or conversion of energy resources, then to entire ecosystems and countries and, finally, humanity at large, which is subjected (albeit unevenly) to global environmental changes (Smith & Ezzati, 2005; Smil, 2008).

Despite and partly as a result of continuing technological and efficiency improvements, the environmental impacts of energy production and use continue to increase in absolute terms around the world. These include acid deposition, fine particulate and ozone pollution, land use changes, depletion of non-renewable resources and degradation of renewable resource stocks. Some energy-related environmental problems are novel, such as the health risks posed by new organic compounds adsorbed on diesel exhaust particulates. Many others are not new but have reached an unprecedented scale or scope. Under 'environmental impacts' we subsume risk and safety issues of energy systems (renewable and non-renewable). The nuclear accidents in Chernobyl (April 1986) and Fukushima (March 2011) demonstrate that risk profiles of 'low probability-high impact' events have to be reconsidered.

Many global environmental problems call for global institutions to address them.[5] Although there is no global juridical system, there are numerous international and multilateral treaties and conventions relating to resource use and pollution that may be useful, if hard to monitor and enforce. On balance, such institutions have proved inadequate to the task of enticing nations, firms and individuals to be mindful of the global commons. How can they be improved and what other kinds of institutions are needed? What prejudices are at work and what interests are at stake? What is the cost of suggested measures to the various parties? Most of the questions that need answering must be addressed by social science research.

Anthropogenic climate change from the emission of CO_2 and other greenhouse gases into the atmosphere has reached the top of many political agendas and has attracted vast media attention. Global CO_2 emissions, still the largest source

---

[5] The examples just mentioned support the proposition that energy systems associated with a high hazard potential require stringent and independent surveillance. By 'surveillance' we mean regulatory oversight as well as integrative review processes (involving pluralistic expertise) and societal embedding, which are needed to overcome the evident information asymmetry among the actors involved. For nuclear energy this encompasses informed and competent *supra*national institutions.

of GHGs, are due largely to the combustion of fossil fuels and the incomplete combustion of biomass. The United Nations Framework Convention on Climate Change (UNFCCC), supported by the Intergovernmental Panel on Climate Change (IPCC), is a political attempt to manage a global common good (the atmosphere) in a responsible manner.[6] This is not a well-trodden path. Developing methods to deal effectively with this kind of very long-term, global environmental threat is a social-learning process that requires support from the social sciences.

Emission permit trading is the main policy tool that the UNFCCC adopted in order to stabilise and then reduce CO_2 emissions. How well this instrument, a product of social science research, would work under varying conditions and with different designs in an international setting is an open question (cf. Chapter 8). The example set by the EU Emissions Trading System (EU ETS) by granting companies free carbon credits rather than auctioning them may not only make the instrument ineffective but may also have other negative consequences. Political deadlock over binding emission reduction commitments from major participants in the UN Climate Change Conference in Copenhagen in December 2009, reinforced by the outcome of the 2010 UN Climate Change Conference in Cancún,[7] may make it necessary to scale back ambitions for a global carbon market in favour of binational agreements. The survival of millions of people may hinge on effectively implementing one of these schemes and ensuring its environmental effectiveness.

Many facets of the UNFCCC are in need of detailed investigation. What is necessary to make the treaty effective and fair? What combination of policy measures and targets for efficiency, emissions and renewable technology penetration, with what degree of international harmonisation, will be most efficacious, cost-effective and politically saleable? There is a risk that compromise will water down the treaty to the point that it will become environmentally ineffective. Without adequate safeguards, increased corruption and other problems may also jeopardise Clean Development Mechanism (CDM[8]) projects.

Until recently, the most commonly proposed technical solution to excess CO_2-emissions has been carbon capture and storage (CCS). Although CCS offers hope for a (limited) solution, implementing and safeguarding it is likely to encounter significant practical and legal problems, many of which are again issues for social science research. High costs, inadequate technical advances and flagging government support, as well as the daunting scope required for successful global implementation, have convinced a number of energy companies to shift resources away from CCS in favour of intensified development of renewable energy.

---

[6] http://www.ipcc.ch. All web links accessed November 16, 2011.

[7] http://unfccc.int

[8] The clean development mechanism (CDM) is one of three 'flexibility mechanisms' of the Kyoto Protocol designed to provide cost-effective emissions reduction opportunities. CDM allows developed countries to purchase 'cheap' CO_2-emissions certificates in developing countries by paying for projects there that, at least on paper, lead to reduced CO_2 emissions. The other two mechanisms are Joint Implementation (JI: receiving credit for implementing emission-reducing projects in other Annex I signatory nations) and emissions trading (ET).

The research community that has been working on another long-term, energy-related environmental problem, namely the disposal of radioactive waste from – mainly – nuclear power plants, seems to have recognised some of its social science research needs (e.g., Solomon, Andrén, & Strandberg, 2010). Results of their analysis of the complex 'NIMBY' (Not In My Back Yard) phenomenon, as well as issues of public trust and social acceptance, regulatory oversight, liability and other issues, may be useful in the case of CCS and other contexts (cf. Chapter 10).

Continued support for renewable energy development is needed in order to reduce CO_2 emissions and the dependence on (imported) fossil energy resources (cf. Chapter 9). The most effective economic measure for furthering renewables to date has been the implementation of feed-in tariffs in energy markets. In Switzerland, for example, feed-in tariffs have helped to forestall the decline of small hydroelectric power projects and have even led to their recent renaissance. Unlike emissions permits, the tariffs are not a product of social science research but both instruments need further development on the basis of solid analysis and recommendations from the social science research community. Feed-in tariffs may be set too low (rendering them ineffective), too high (making them uneconomical) or their phase-out may be too fast or excessively delayed. How can these problems be remedied and in what time frame? How can consensus be reached among the various involved actors whose own interests change over time? What adjustments are needed if energy markets are dramatically affected by the extensive utilisation of feed-in tariffs? How can the tariffs avoid becoming a victim of their own success?

In developing countries with specific combinations of geography, tree-cover, population density, cultural practices and other factors, the extensive use of traditional fuels often causes significant environmental damage. Depending on local conditions, the introduction of appropriate modern commercial fuels may alleviate these impacts, provided the necessary preparatory groundwork is laid. The availability of such fuels depends on many factors, chief among which is price. The adoption of fossil fuels may be inhibited by their high prices, which may spur local environmental degradation in areas whose population relies heavily on a robust local environment to provide biomass fuels. The additional challenge of ensuring that renewables continue to play a role in rural development is, again, only marginally a technical issue and is instead predominantly an issue of social change and, therefore, also of social science research.

The global economy's dependence on fossil fuel is so great that it would not be possible to achieve a switch to carbon-neutral renewables or carbon capture and storage quickly enough to make efforts at curbing energy use redundant. More effective policies and institutions are also needed to both limit energy use and to diversify supply, yet barriers to energy conservation remain controversial and poorly understood. Differing measures of external costs, varying notions of fairness, and perceived advantages for certain individuals and social groups are among the factors that stand in the way of the 'best' solutions. Finding ways of taking the environment into account along with social goals is a huge challenge for all of the social sciences, including economics.

In principle, much is known about what is necessary to improve the performance of energy markets. Internalising external costs of energy and other resources can be achieved by regulatory or market mechanisms such as taxes or GHG emissions permits. Second-best measures include promoting and introducing more environmentally favourable products and services to the market by direct subsidies or via regulation, for example with feed-in tariffs. However, some of these second-best measures work against energy efficiency by reducing energy prices; cost internalisation is preferable on this score.

2.4 Challenge III: Economic and Social Development

Whereas the energy sector resembles a massive lumbering super-tanker in terms of its rate of change, its energy input to economies and societies tends to speed them up. High-powered vehicles are used to get quickly from place to place, requiring orders of magnitude more energy than surface journeys consumed at the start of the last century. Fast mass-production of disposable goods uses vastly more energy than the artisanal production of custom-tailored durables.

Modern civilisation has been built on cheap oil, coal, and gas. Fossil fuels have arguably been the primary sine qua non for the North's industrial and post-industrial development and high energy consumption remains one of modern civilisation's defining characteristics. Technoeconomic systems are precariously dependent on inexpensive, readily available fossil energy carriers to provide the lion's share (currently about 90 percent) of the global commercial total primary energy supply (TPES). However, long-term reliance on fossil fuels is unsustainable, for several reasons:

- As described in the previous section, the global long-term environmental effects of supplying and consuming energy threaten the biosphere and human health, as well as the stability of the energy system and the global economy.
- Most developing countries lack clean and affordable energy sources. High fossil fuel prices also tend to stall their development and can exacerbate inequities and tensions between North and South.
- The geopolitical concentration of fossil fuel resources, combined with huge external demand, generates a huge wealth transfer to resource-rich countries, which may be used for non-competitive, non-productive or outright destructive purposes and/or funnelled to radical non-state groups.

These pressures challenge global society to develop policies and institutions that diversify supply, limit the use of fossil fuels and moderate final energy demand and emissions of greenhouse gases, especially CO_2.

Broadly speaking, gross world economic product maintained a close correlation with TPES throughout the 20th century. Per capita gross domestic product (GDP) and national energy supply also had high correlations (Smil, 2008). Variations in the ratio of GDP to national energy consumption have supported the notion that the two have been or could be decoupled in certain countries. However, any such decoupling, if it has occurred, has been weak, with rebounds taking back a portion of the gains in energy efficiency. Since the 1980s, the growth in energy use (according to metrics that typically exclude aviation and merchant shipping) has been held down in several industrialised countries despite strong economic growth. Without this economic growth, however, energy use could have been reduced in absolute terms (Nørgård, 2009). Technical change has often increased overall energy intensity rather than reduced it. Especially when fuels are appropriately weighted by their relative economic productivity in energy statistics, energy consumption appears rather closely coupled to economic growth (see Sorrell & Herring, 2009).

Nevertheless, the wide range of national energy intensities across developed countries (and for similar income groups across these countries) shows that similar levels of (economic) well-being can be achieved with highly varying energy inputs. The different fuels used for primary energy generation across countries, along with the structure and efficiency of final energy conversions, account for much of this variation (Smil, 2008). A continuous supply of energy to the economy is absolutely necessary and to some extent irreplaceable in the short- to mid-term. However, no natural law dictates that energy use must be maintained indefinitely at the high levels to which industrial society has become accustomed.

Increasing energy efficiency is a necessary but insufficient means of limiting energy use and GHG emissions. Since the beginning of industrial development, improving energy efficiency has been one of engineers' primary goals and one they have met spectacularly. Historically speaking, energy efficiency has experienced a continual increase. In spite of these gains – indeed, partly because of them – overall energy use has also rapidly increased. Technological progress has not only increased energy efficiency but has generated economic growth and has made the use of energy-consuming appliances economically accessible to broader swaths of the population. There are a number of ways to explain the general failure of policies to limit energy consumption (for example, in terms of the political economy, the psychology of end-users or the sociology of population groups (see Goldblatt, 2005). To date, however, a definitive general explanation has continued to elude researchers.

Nevertheless, social science disciplines are integral to the development of effective instruments and institutions for energy conservation. As noted, focusing purely on the technical problem of increasing energy efficiency has not led to lower overall energy use and has even at least indirectly frustrated it. Consequently, research into energy conservation must incorporate both (1) the development and deployment of high-efficiency equipment and processes and (2) the analysis of behaviours, social practices and trends, lifestyles and economic arrangements and institutions with a view to holding down aggregate energy demand (cf. Chapters 6 and 9). Energy conservation and renewables both tend to require high initial investments[9] but have low operating costs. Combining the partially opposing goals of economic efficiency and long-term viability in an organised fashion and systematically devising the methods for economic assessment remain unresolved research tasks.

---

[9] Renewables also have high energy input requirements: Almost all renewable energy conversions have energy returns on investment (EROI) of less than 10 (Smil, 2008).

End-users must also become more cognisant of the implications and impacts of the supply and use of energy. Some amount of energy is irreplaceable for powering economic activities, analogous to an organism's basal metabolic rate; energy is not entirely substitutable. In the course of daily activities, however, much energy can be saved by decelerating activities and doing things smarter (substituting with time, information, or both (see Spreng, 1993)). There is still little consensus among scientists regarding the durability of consumers' aspirations and behaviours and the role that trends and social group dynamics play in end-use energy consumption. Furthermore, the potential of involving consumers in the choice of energy carriers and energy mixes is as yet unknown.

Throughout much of the 20th century, technological innovation and economic competition provided consumers in developed countries with declining or relatively constant energy prices. Price breakouts in the first years of the 21st century seem to have pointed to the limits of these restraining factors in the face of unprecedented increases in global demand without commensurate increases in supply. Until this time, increasing disposable incomes had decreased the relative cost of all forms of energy in developed countries, especially in the US (Smil, 2008). Yet oil price shocks played a part in the American consumer crisis that led to a severe worldwide recession in 2008, and oil imports account for a significant portion of international trade imbalances. The bright side of the combination of recessionary Northern economies and declining middle class incomes, along with increasing world oil prices, may well be the generation of higher long-run price elasticities of demand and greater public demand for substantial energy conservation and alternatives.

In a regulatory environment that is increasingly shaped by open markets, deregulation and competition, and a milieu in which energy models geared towards the public good have fallen into disrepute, energy corporations have been forced to focus on short-term gains in order to survive. However, energy systems are also made up of physical plants and institutions, which means they are about more than just markets. With disappointing results and deregulation debacles (such as the electricity shortage following California's partial deregulation of the state's electricity markets in 2001, see e.g., Bushnell, 2004), the dogmatic faith in unbridled markets has begun to be questioned and the pendulum has started to swing back.

With regard to social development, Section 2.1 above discusses international inequities in access to energy, but other forms of international inequities ripe for research include differences in climate-change-related impacts and mitigation capabilities. For example, there are already indications that the countries likely to be most drastically affected by climate change have the least technical and financial resources required to adapt to it.

Vaclav Smil has noted: 'Every society is moulded by energies it consumes and embodies. A different set of primary energizers must necessarily remould structures and mores in many profound ways' (Smil, 2008, p. 382). There is a long tradition within Science and Technology Studies (STS) of examining the interactions between technological research and society. This and related fields will be increasingly called upon to examine future unintended consequences in the social sphere caused by large- and small-scale technical developments. They also have

2 Energy-Related Challenges

a substantial role to play in identifying and, if possible, aligning divergent modes of knowledge and expectations regarding future energy systems (Bammer, 2005; Wiesmann et al., 2008).

Clearly, it is difficult to overstate the depth and complexity of the energy-related challenges facing society. 'The imperatives of ultimate resource availability, infrastructural inertia, and contrasts between conversion and consumption power densities limit [society's energy] choices ... Escaping the imperatives of scale built into the world's energy system by more than a century of fossil fuel combustion and electricity generation will not be easy' (Smil, 2008, pp. 359–360). In order to have a chance of success, it is essential to use and coordinate existing and emergent knowledge from multiple fields (cf. Chapters 4 and 11). In fact, this can be considered the foremost methodological challenge for the social sciences in energy research. This issue is addressed in depth in the following chapter.

References

Bammer, G. (2005). Integration and implementation sciences: Building a new specialization. *Ecology and Society, 10*(2), 6. All web links accessed November 16, 2011, http://www.ecologyandsociety.org/vol10/iss2/art6

Bushnell, J. (2004). California's electricity crisis: A market apart? *Energy Policy, 32*, 1045–1052.

Collier, P. (2007). *The bottom billion: Why the poorest countries are failing and what can be done about it*. New York: Oxford University Press.

EC, European Commission. (2010). *Energy 2020. A strategy for competitive, sustainable and secure energy*. COM (2010) 639 final. Brussels: European Commission.

Goldblatt, D. L. (2005). *Sustainable energy consumption and society: Personal, technological, or social change?* Dordrecht: Springer.

IEA, International Energy Agency, UNDP & UNIDO. (2010). *Energy poverty. How to make modern energy access universal? Special energy excerpt to the World Energy Outlook 2010 for the UN General Assembly on the Millennium Development Goals*. Paris: OECD/IEA.

Kemmler, A. (2007). Factors influencing household access to electricity in India. *Energy for Sustainable Development, 11*(4), 13–20.

Meadows, D., Randers, J., & Meadows, D. (2004). *The limits to growth: The 30-year update*. White River Junction, VT: Chelsea Green.

Nørgård, J. S. (2009). Avoiding rebound through a steady-state economy. In H. Herring & S. Sorrell (Eds.), *Energy efficiency and sustainable consumption: The rebound effect* (pp. 204–223). Basingstoke: Palgrave Macmillan.

Rittel, H., & Webber, M. (1973). Dilemmas in a general theory of planning. *Policy Sciences, 4*(2), 155–169.

Sen, A. (1997). Human capital and human capability, editorial. *World Development, 25*(12), 1959–1961.

Smil, V. (2008). *Energy in nature and society: General energetics of complex systems*. Cambridge, MA: MIT Press.

Smith, K., & Ezzati, M. (2005). How environmental health risks change with development: The epidemiologic and environmental risk transitions revisited. *Annual Review of Environment and Resources, 30*, 291–333.

Solomon, B. D., Andrén, M., & Strandberg, U. (2010). Three decades of social science research on high-level nuclear waste: Achievements and future challenges. *Risk, Hazards & Crisis in Public Policy, 1*(4), Article 2. doi:10.2202/1944-4079.1036.

Sorrell, S., & Herring, H. (2009). Conclusion. In H. Herring & S. Sorrell (Eds.), *Energy efficiency and sustainable consumption: The rebound effect* (pp. 240–261). Basingstoke: Palgrave Macmillan.

Spreng, D. (1993). Possibility for substitution between energy, time and information. *Energy Policy, 21*(1), 13–23.

UN AGECC, The Secretary-General's Advisory Group on Energy and Climate Change. (2010, April 28). *Energy for a sustainable future*. Summary report and recommendations. New York: United Nations.

Wiesmann, U., Hirsch Hadorn, G., Hoffmann-Riem, H., Biber-Klemm, S., Grossenbacher, W., Joye, D., et al. (2008). Enhancing transdisciplinary research: A synthesis in fifteen propositions. In G. Hirsch Hadorn et al. (Eds.), *Handbook of transdisciplinary research* (pp. 433–441). Dordrecht: Springer.

WHO, World Health Organization, & UNDP, United Nations Development Programme. (2009). *The energy access situation in developing countries. A review focusing on the least developed countries and Sub-Saharan Africa*. New York: UNDP, Environment and Energy Group.

Chapter 3
The Indispensable Role of Social Science in Energy Research

Jürg Minsch, David L. Goldblatt, Thomas Flüeler, and Daniel Spreng

Abstract Limiting the role of social science in energy research to its traditional instrumentalised functions of acceptance research and market introduction is problematic for research and policy and inadequate to the contemporary energy-related – and fundamentally socioscientific – challenges described in Chapter 2. Social science can facilitate societal learning processes involved in the co-evolution of technology and a liberal society. Energy research in the social sciences serves the core functions of generating reflection and target knowledge, analysis, and transformation knowledge. In addition, sustainable technology development calls for equal footing for the social sciences and engineering. Finally, the management and integration of knowledge related to energy requires multi- and transdisciplinary frameworks.

Keywords Energy systems · Co-evolution of technology and society · Knowledge types and integration · Learning society · Policy integration · Reflection · Social sciences · Technical fixes · Transdisciplinarity · Unintended consequences

> *The classical paradigm of science and engineering is not applicable to the problems of open societal systems.*
> *We have been learning to see social processes as the links tying open systems into large and interconnected networks of systems, such that outputs from one become inputs to others. In that structural framework it has become less apparent where problem centers lie, and less apparent where and how we should intervene even if we do happen to know what aims we seek.*
>
> *Rittel and Webber (1973, pp. 159, 160)*

3.1 Introduction

The term *social sciences* generally includes those academic disciplines that theoretically and empirically examine the phenomenon of human and societal behaviour. The social sciences include communication and information sciences, cultural

---

J. Minsch (✉)
minsch sustainability affairs, 8057 Zurich, Switzerland
e-mail: juerg.minsch@bluewin.ch

D. Spreng et al. (eds.), *Tackling Long-Term Global Energy Problems*,
Environment & Policy 52, DOI 10.1007/978-94-007-2333-7_3,
© Springer Science+Business Media B.V. 2012

sciences (including anthropology and ethnology), demography, economics, education, geography, history, law, management sciences, political science, psychology, public administration and sociology, among other disciplines. Social science analyses the structure and function of the interdependence of societal institutions and systems, as well as their interplay with the actions and behaviour of individuals.[1] The general definition of social science in this chapter is deliberate, as it prevents disciplines from being rashly excluded or left unaddressed. All of the social sciences, and in some cases the humanities as well, may be called upon to demonstrate the value of their particular perspectives and, potentially, to participate in energy research.

Traditionally, a primary function of social sciences, especially economics, has been to support the market introduction of new technologies through specific promotion mechanisms such as subsidies and tax breaks as well as through supportive measures in regional planning and development. Such measures have been incorporated in wide-ranging development strategies in the fields of industrial policy, economic development, location marketing and even in public procurement (e.g., defence spending in public administration). Another classical function of social science has been acceptance research, providing scientific support for the creation of public acceptance of new technologies. Heightened acceptance is supposed to facilitate the introduction of new technologies. Social psychology in particular has a long history of involvement in acceptance research for energy technologies. A related role of social science has been barrier analysis, which attempts to account for consumers' failure to behave according to economic rationality and to adopt new technologies (e.g., Stern & Kasperson, 2010).

The scientific support provided by the social sciences for the market introduction and public acceptance of energy technologies has largely met with public approval. Economics and political science have played a particularly prominent part in the phasing in of new technologies. However, it is striking that the social sciences have rarely been called upon to phase out demonstrably non-sustainable technologies or to support the choice not to deploy suspect ones in the first place. In short, where the social sciences have played a role, they have mainly been used as adjutants in the introduction of new energy technologies. Above all, they have been in demand for their role in the process of *addition* to the general technical stock; *substitution* has yet to become part of their repertoire.

3.2 Excursus: The Mercantilism Syndrome

3.2.1 Wealth – From Personal Matter to De Facto Entitlement

Within the development of the political economy, the character of the economic and political participation of the social sciences in the energy field has been shaped by

---

[1] For a definition of terms, see, for instance, Smelser and Baltes (2001).

larger historical trends. For over 200 years, politico-economic policy has centred around the pursuit of economic growth via a cheap energy supply. In modern times, this has made it difficult to advocate a more comprehensive understanding of development that is inclusive of negative ecological, social and political consequences. Scientists voicing such concerns have risked being marginalised.

Most cultures place enormous emphasis on the quest for increasing material welfare; it may even be part of human nature. In any case, it has a remarkable conceptual and historical background. For classical economists (such as Smith, 1776), the pursuit of personal advancement and prosperity was a private matter for the individual. The market economy was regarded as the economic system that channelled this ambition – as if by an 'invisible hand' – into the most productive course for society as a whole.

The quest for wealth was overshadowed by the scientific and political debates surrounding this idealised view of the market economy and quietly transformed from a private affair into a political postulate. The modern state allowed itself to be held increasingly accountable for ensuring rising wealth and personal wishes were turned into de facto rights. Economic growth became the central politico-economic goal (e.g., Erhard, 1957) and the gross national product became the predominant macroeconomic indicator.

3.2.2 The Mercantilist Policy of Cheap Resources

The demand for cheap energy supply and policies that aim to satisfy this demand are long-standing traditions that go back to the age of mercantilism. The overall aim of national mercantilist policies was to eliminate a country's dependence on imports of essential manufactured products by increasing domestic production, especially in industry and manufacturing (cf. Issing, 1984, p. 35 ff.). In addition to the protectionist pursuit of a favourable balance of trade (particularly through import duties), countries pursued an increasingly successful policy of cheap production factors, which was meant to guarantee the competitiveness of exports on international markets. Originally, the central focus was support for the lowest possible wage level; this was pursued through policies favouring work discipline and population growth. This 'low-wage economy' (cf. Heckscher, 1932, p. 130 ff., esp. p. 150) arguably increased the wealth of the state but also led to poverty among large segments of the populace. The policies were supplemented, however, with strategies that aimed to reduce the price of food and other goods, which made it possible to keep wages low and poverty within certain limits.

Over time, a new area for mercantilist price reduction strategies was developed in the form of wood as an energy source and raw material. At the threshold of the industrial revolution, the cheap wood policy was an early model for today's policies of reducing the price of energy resources. Wood was the 18th century's main energy source and, as it was depleted, its rising price reflected its increasing scarcity. In response, regulatory measures increased the supply of wood (and thus made it cheaper) to strategically important fields of economic activity, simultaneously reducing the supply of wood to non-privileged economic sectors and more strictly regulating its use in those sectors. While this strategy did not resolve the problem of scarcity as such, it did transfer it to the underprivileged, where it intensified into a crisis. The continuously worsening wood shortage precipitated further price increases, first in England and then in mainland Europe, and it spurred both conservation efforts and increased use of the alternative energy source of the time: coal, the 'underground forest' (cf. Sieferle, 1982). Today, the pursuit of energy price reductions with accompanying post-hoc interventions that correct for and fine-tune them is reminiscent of policies of the mercantilist era.

3.2.3 Cheap Natural Resource Supply, Asymmetrical Globalisation and Intervention

In the industrialised world, the modern state has adopted, instrumentally refined and universalised the mercantilist recipe of reducing the price of energy and natural resources. In place of the privileged allocation of cheap resources to export-oriented economic sectors, many governments have pursued policies of an unimpeded (that is, cheap) supply of natural resources for the economy and have subtly advanced protectionism. More concretely, these policies involve cheap energy, cheap raw material supply and cheap waste and sewage disposal, unimpeded and cheap mobility, extensive spatial integration (despite absolute land shortages) and a repudiation of major technological risks (cf. Minsch, 2001). The strategies used to implement these policies have ranged from the non-internalisation of negative externalities through various indirect and direct price reductions (tax exemptions or reductions, subsidies, limiting liability, technology promotion, and research policy) to supply-oriented infrastructure policies, up to and including diplomatic and military intervention.

Besides producing the immediate effect of advancing (energy-consuming, material-consuming and natural resource-consuming) economic growth, these practices tend to function in a protectionist way. They reduce the cost of domestic production compared to production in developing countries that do not have such practices at their disposal, a subtle but effective modern form of protectionism. Rather than the products themselves, nature as a production factor is made cheaper; this is a non-transparent factor that is difficult to measure monetarily. This protectionism continues to be supplemented by import restrictions and duties, export subsidies and subsidised export risk and investment guarantees, which can work to the disadvantage of developing countries.

3.2.4 The Social Sciences in the Context of Mercantilism

The policy of cheap resources is not limited to nature as a production factor. Two other important areas in which this policy has been applied include money and knowledge. The policy of cheap money (that is, loose monetary policies by central banks) has been extensively discussed, especially in the wake of financial crises such as the most recent that began in 2008. The tendency to develop knowledge asymmetrically is less obvious and has developed more recently. In the energy field in particular, mercantilism has furthered the technocratic emphasis on knowledge and skills that prove their immediate technical and marketplace usability at the expense of more comprehensive and long-term perspectives (a strength of the social sciences). As noted, the social sciences have tended to be instrumentalised to further the public acceptance of technologies and their introduction into the market. In this way, the traditional role of the social sciences in the energy field is itself an expression of ingrained mercantilism.

The policy of cheap wood ultimately failed because of (absolute) ecological scarcities, which forced the transition to sustainable forestry. Today, the policy of cheap natural resource inputs has also reached its limits, but the challenge is admittedly more difficult. Scarcities and their hazard potential are often hard to discern as they are hidden behind complex dynamic and global interdependencies and tend to be communicable only through probability formulations that are difficult to convey accurately to the public. Moreover, scarcities often loom in the future or arise in developing countries; that is, at a supposedly safe temporal and geographical distance from the instigators of the overuse of the resources and the resulting environmental costs. This usually means it is possible to comfortably ignore these nearly invisible problems (cf. the parable of Gyges, Chapter 1/Box 1.1) or to process them as isolated cases and shift the task of resolving them to specialised policies of ex post correction or simply treating the symptoms. In particular, the fields of environmental policy, regional policy, development cooperation, social policy, education policy and even energy policy are now being given this problematic responsibility.

Today's popular calls for renewable energy should be viewed critically against the background of the dogma of cheap centralised resources. Although there is little doubt that renewable energy will play a significant role in the energy supply in the future, policies to promote renewable energy technologies are often simply political acts of compensation. In the short term, these acts score points that are easy to communicate and attribute to individual politicians, which makes them politically attractive. However, they often distract from the need to initiate more in-depth reform and innovation in the field of energy policy. New techniques, products and business models, even if they are deemed 'ecologically efficient', do not automatically replace old ones. At first, they simply join their ranks. The extent to which they go on to replace non-sustainable products, techniques and business models, and over what period of time, is rarely enquired about. The answer depends not only on the type of technique, product or business model but on the entire system of social innovation, and especially on general economic conditions.

3.3 A Call to Arms for Social Science

While supporting public acceptance and assisting market introduction of new energy-related technologies are legitimate functions, the social sciences should not be reduced to marketing tools for technological developments. The sheer existence of unintended ecological, social and economic consequences counsels against such a misuse of the social sciences. The challenges described in Chapter 2 make it necessary to draw upon the social sciences' wider repertoire.

A simple look at any of these challenges underlines the truth of these assertions. Even describing and elaborating upon the challenges requires the use of socioscientific reasoning. The following will develop several aspects in order to illustrate and reinforce the point that, at their root, the challenges are primarily socioscientific.

(1) *Tensions, conflicts, development prospects and human rights*

There are a range of triggers for changes in the global energy supply, such as the emergence of new providers, the constriction of supply resulting from the exhaustion of resources, and limitations in the range of infrastructure. Such adjustments in supply affect vested interests as well as the potential for regional, national and international development. Consequently, they involve not only economic phenomena in the narrow sense but also political challenges. Finally, deleterious changes in the global energy supply can exacerbate international tensions and conflicts, worsen the human rights situation in energy-exporting countries and poor energy-importing countries and degrade countries' general prospects for development. A central task of social science energy research is to analyse the interaction of these various effects.

(2) *Policy instruments and institutions for energy conservation*

Apart from its technical dimensions, energy conservation is largely a question of underlying political and institutional frameworks, business models, lifestyles and patterns of consumption of energy services. Research must explore why it is so difficult to realise the politico-economic conditions that are conducive to energy conservation. Without being embedded in a larger integrative policy framework, energy-saving technologies may have counterproductive effects (rebound or even backfire) and novel business models may remain confined to market niches. While technological innovation has delivered impressive gains in terms of higher efficiency and reduced energy consumption at the level of individual energy-using devices and systems, it has not led to the stabilisation or reduction of aggregate energy use or greenhouse gas emissions. A central task of long-term-oriented social science energy research is to devise instruments, strategies and institutions that are both environmentally effective and politically and economically realistic, such as a workable and effective emissions trading system. This requires the creation of institutions that foster a long-term perspective by markets and governments.

(3) *Energy access*

It is undeniable that the connection between energy access and development has an important technical dimension. Whether energy access actually contributes to regional development and to a reduction in poverty, however, depends on local institutions, the development model, the political and legal culture (democratic conditions, access to credit, property rights, legal security) and the conditions for international commerce (export possibilities for manufactured products). Two historically recent examples from India provide concrete illustrations of the impact of general economic conditions and development models.

The first example features energy prices and energy subsidies. In India, water scarcity has led to intense exploitation of groundwater resources, a practice that has been encouraged by the policy of not charging farmers or subsidising them for the electricity they use to pump out groundwater (Kumar & Singh, 2001). As a result of this practice, ground water resources have been critically depleted or salinised. Conditions are especially severe in the Punjab, the country's breadbasket and main rice growing area, where the majority of the aquifers are already depleted and both wheat and rice yields have decreased drastically. In other words, faulty energy price signals through outdated electricity subsidies are contributing to the degradation of the long-term ecological basis of the Indian agricultural sector. In other cases, high electricity prices have proved a severe obstacle to access for the lowest economic strata, hindering or even completely preventing economic development. This example, however, shows the devastating consequences of a careless policy of cheap electricity. In this case, the foundations for (agricultural) development are being destroyed.

Another important part of energy access is the feasibility of maintaining a constant residential power supply. Even though households and small businesses in some rural areas of India are physically connected to the power supply system, they do not all enjoy access to electricity around the clock. Some rural villages receive power for only about two hours a day. Although this is enough for irrigation, and thus for the food supply for urban areas, it is not enough for more extensive development of the rural areas themselves.

(4) *Support for renewable energy*

India's subsidies for mechanical groundwater extraction illustrate the problem of innovation policies that target isolated innovations and technical solutions but are not grounded in sage general economic development strategies. The attraction of such isolated measures is that they are politically feasible, can produce fast results and may even be important for assisting the launch of new technologies. However, these specific uses rarely justify their general application to the complexity of economic development. If they are not fitted to a broad development concept, technology-dependent development policies often simply displace problems rather than solve them, or they induce rebound. In some unfortunate cases, instead of promoting a substitution of renewable energy for fossil fuel, a flawed renewable energy policy may contribute to an increase in total energy supply and demand, along with further unintended side effects (such as exacerbated use conflicts with the agriculture and food sectors, as seen recently in the effects on agricultural commodity markets in the 'tortilla crisis', see e.g., Bello, 2008). Since energy systems are large, tightly knit and lethargic structures, energy investment and policy decisions must be made with a long-term view. In their current form, energy markets do not always encourage necessary long-term investments, particularly for renewables with their high up-front costs.

These four examples demonstrate the importance of social science to the effort to meet global energy-related challenges. As Chapter 2 explained, this effort is central to securing the 'good life' on Earth, in the sense of broadly-construed sustainability. Failing to involve the social sciences amounts to a vote to tolerate unintended negative ecological, economic, political and social side effects. It is, in other words, a vote for the tyranny of unintended consequences.

3.4 Equal Footing for Social Science and Engineering in Technology Development

Energy systems can be viewed as a collection of hardware and software elements. Hardware consists of infrastructure such as industrial equipment, machines and facilities, while software includes the know-how and behaviour of actors and institutions, as well as regulations and laws. Hardware and software develop hand in hand and the energy system only functions in its totality (Fig. 3.1).

The interactions between hardware and software are extremely relevant to society and can only be adequately researched by a combination of researchers from 'both cultures' (Snow, 1959) or by researchers who have acquired sufficient competence in both fields of enquiry. Many systemic problems of the energy system can only be addressed at the level of hardware-software interactions.

Nevertheless, the social sciences are still often regarded merely as a substitute for technical solutions ('fixes'). They are called upon after a problem has arisen due to technical failure and, as such, are used to remedy technical difficulties by using 'end-of-pipe' social solutions. This limited role follows from the exclusion of the social sciences from earlier phases of technology research and development. Fixes, which are supposedly more elegant and more quickly realisable, take priority, and as long as things go smoothly, social science is seen as almost optional.

To some extent, a substitutive relationship between technical and social science solutions to 'small questions and small problems' is warranted. Small problems

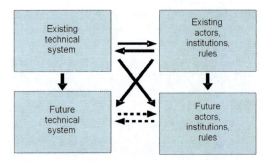

Fig. 3.1 Interactions between hardware and software components form the current and future energy systems. The future energy system, comprising both software and hardware, is a product of both the software and hardware of the current system (*vertical and diagonal arrows*)

can be solved by technical means or through behavioural changes. However, the magnitude of the challenges facing the energy system and society makes it clear that, on a global scale, the logic of such simple either-or choices no longer holds true. In an age of climate change, economic globalisation and increasing scarcities in energy and natural resources, the necessary adjustments in technical energy systems and in society renders such simple choices inadequate. Instead, technical and social science solutions are complementary and should be developed in close cooperation from the start of research and development. The co-evolution of technology and society also implies that engineering and social science are of equal value at all stages of societal development.

3.5 The Leitmotif of a Learning Society

Studies of science's contribution to sustainability have variously recognised social science's broad repertoire and its equal footing with the natural sciences and engineering. For example, in their framework of a broad participatory process supporting sustainable development (CASS/ProClim, 1997), Swiss researchers have described how in order to achieve overall societal goals, science must supply target knowledge[2] and systems knowledge, as well as transformation knowledge (knowledge of how existing practices can be changed to reach the targets). The researchers identified a deficit in the area of social science and recommended placing much greater emphasis on the generation of target knowledge and transformation knowledge, as well as systems knowledge in social sciences, humanities and economics. These recommendations, in line with WBGU, 2011, formulated in the context of general sustainability science, apply in equal measure to long-term-oriented social science research in the energy sector. Box 3.1 offers a separate in-depth example of the function of social systems knowledge in the seemingly technically-dominated field of global climate modelling.

Box 3.1 Social Systems Knowledge in the Climate Knowledge Infrastructure

The Intergovernmental Panel on Climate Change (IPCC) is the main institutional gateway for the climate knowledge infrastructure,[3] which has been built on top of much older weather information systems. Whereas the latter are settled and accepted, the climate knowledge infrastructure is still

---

[2] Target knowledge: knowledge that supports the setting of goals to deal with real-world problems and the monitoring of progress towards these goals.

[3] Paul N. Edwards has written extensively on this topic (for example, in Edwards, 1999 and most recently in Edwards, 2010). The discussion of systems knowledge in this section is heavily indebted to his work.

highly controversial and far from settled, although the fundamental framework of climate knowledge is technically well established.

As Edwards (1999, 2010) identified from a perspective that combines theories of information systems and sociotechnical and large technical systems (LTS), the process of generating technical climate systems knowledge involves a continual re-creation he terms *infrastructural inversion* in which past data and model sets are constantly re-examined and re-interpreted. The metadata generated by examining past data generation methods have led to the revision of existing models, which in turn has suggested new understandings of past climatic conditions.[4]

Since climate change research relies so heavily on modelling ('data-laden models, model-filtered data' (Edwards, 1999)), uncertainties in natural/physical science systems knowledge (including epistemological, irresolvable ones) require social construction at every step of the modelling, empirical data filtering and interpretation process; that is, across the range of even seemingly pure technical systems knowledge. Thus, social/political systems knowledge is involved at every step of the science-generating process, not just in the explicit communication of scientific results or target-setting and policy making. The role of social science in this context is to reveal, explicate and analyse this knowledge (in order to crystallise and characterise social/political systems knowledge), as well as to describe (or even, in a classical transdisciplinary setting, to participate in) its interplay with knowledge of a more purely natural/physical science nature.

Parameterisation, which is a universal and unavoidable feature of all global climate models (GCM) and indeed of all other types of global modelling, aptly illustrates the interplay of technical and social systems knowledge. In parameterisation, semi-empirical parameters are set more or less arbitrarily based on heuristic principles that are derived from observations. In any GCM, a great many parameters must be set for physical processes that are too small (especially sub-grid processes), complex or poorly understood to be modelled directly. For example, the science behind cloud physics is relatively poorly developed and many clouds actually form on a scale that is two orders of magnitude less than the resolution of GCMs. GCM cloud parameters must heuristically represent cloud cover within a grid box as a function of atmospheric conditions there. This is done using a cloud fraction to account for cloud cover in that grid, as well as functions that deal with variable cloud cover at two or more altitudes in a given atmospheric column (Edwards, 2010). The relatively undeveloped state of the physics and the semi-arbitrariness of the parameterisation practice lead to a large degree of uncertainty in current GCM

---

[4] Edwards emphasises that infrastructural inversion is a feature of all knowledge infrastructures. This continual reanalysis of past data and the methods by which they were collected is a hallmark of historical research.

cloud parameterisation; to some extent, this uncertainty is carried over into model results. The situation with aerosol parameters is arguably even worse.

Parameters are routinely tuned, their coefficient values adjusted, or the equations reconstructed to produce a 'better' (for example, more physically plausible) model result. When modellers cannot find links between large- and small-scale processes (that is, when the parameters are not 'physically based'), they may invent ad hoc schemes based on professional judgement and experience in order to make the connections. This practice is highly controversial, even across different schools of modelling communities. An example of this is flux adjustment, which was still used in some models included in the 2007 IPCC assessment.

Therefore, parameterisation and parameter tuning are scientific art forms, and specialised social scientists – ethnographers of climate modelling – have demonstrated the subjectivity of the practices involved (e.g., Lahsen and Shackley in Edwards, 2010).

Table 3.1 uses examples of parameterisation and other features of climate modelling to bring out the involvement of technical and social systems knowledge at every step of modelling and data collection[5]:

Table 3.1 Examples of technical and social systems knowledge involved in global climate modelling (GCM) (based on Edwards, 1999, 2010)

Modelling feature	Technical systems knowledge	Social/political systems knowledge
Parameterisation: radiative transfer	Fast radiative transfer models with nested line-by-line models	Expert judgment, professional and personal experience
Parameterisation: cloud physics	For example, indirect representation within a model's grid box as function of temperature and humidity; e.g., cloud fraction, with parameterisation of cloud overlap between grid boxes	Expert judgment, professional and personal experience
Parameterisation: other difficult, controversial parameters	For example, planetary boundary layer, elevation, land heat retention	Expert judgment, professional and personal experience

---

[5] As the knowledge production process in any global knowledge system evolves into a knowledge infrastructure, it becomes submerged; that is, it becomes more and more difficult to observe the relationships among scientific, technological, social and political elements of this knowledge, to a point at which they are almost invisible. Concurrently, the degree of controversy regarding this knowledge generation tends to decrease (see Edwards, 2010).

The controversy over climate knowledge continues. At this stage of the evolution of the climate infrastructure, separating the ubiquitous social and political systems knowledge from technical systems knowledge still requires a conscious, paedagogic effort.

Modelling feature	Technical systems knowledge	Social/political systems knowledge
Parameterisation and tuning	Aerosol effects: key source of uncertainty in GCM models to date	Expert judgment, professional and personal experience
Model-data symbiosis and model-filtered data	Global data is always filtered by use of models, which introduces artefacts even in empirical observation	Scientific tradition (e.g., laboratory culture); politics of funding for research; power configurations
Frontier (theory-oriented) vs. high-proof (observation-oriented) science	Scientists' basic attitude toward uncertainty and imperfections in climate models	Social negotiation to determine which orientation receives greater emphasis in practice (and policy)

Table 3.1 (continued)

The IPCC has recognised the interplay between uncertainty and social negotiation in the processes that create climate knowledge and has moved away from the notion of model validation or verification, implicitly endorsing Oreskes et al.'s view that models can be, at best, confirmed (Oreskes, Shrader-Frechette, & Belitz, 1994). Nevertheless, the convergence of model results and predictions has grown considerably, as reflected in the 2007 IPCC assessment, and model intercomparisons showing strong agreement across a large number of relatively independent models can increase the scientific credibility of the models' results. However, there are still concerns that subjective tuning and calibration practices, rather than improved model accuracy, may actually lie behind the reduced intermodel spread in climate sensitivity, which may be artificially narrowing prediction ranges of future global surface warming (e.g., Huybers, 2010).

David L. Goldblatt

Within the context of the co-evolution of technology and society, and with liberal society's overarching goals of prosperity, peace, freedom and justice, the special task of the social sciences should be to make creative contributions to shaping a 'learning society'.

What are the design principles behind the development of a society viewed as a learning system? What measures can be conceived that combat the tyranny of small decisions? More fundamentally, are today's democratic-constitutional systems in a position to identify and successfully meet the challenges of the future? Are the institutional prerequisites in place that enable a societal thinking, learning, searching and creating process geared toward sustainability?

The task for institutional innovations in the context of sustainable development is to develop problem-solving rules (institutions) that are suited for specific problems without limiting them in advance to state actors and also without abandoning basic constitutional principles. The starting point for conceptualising such a process of transformation is a polycentric understanding of policy. Instead of leaving the handling of social problems solely up to the state, this concept takes various actors into account at all stages of societal problem solving.

In fact, far-reaching changes are under way in the field of public policy. New institutions are being created and are increasing in importance. In terms of political theory, conventional public policies used to be based on the notion that state action consisted of decision making by the legislative branch and implementation and enforcement by the executive branch. In addition to the element of control, negotiation between state actors and norm addressees is also an integral part of the fulfilment of norms and functions. This is because the state increasingly depends on information from societal actors and needs to motivate them to act and independently change their behaviour. State control mechanisms have been changing accordingly. In this way, the state's task of coordinating and moderating social actors takes its place alongside regulatory, distributive and redistributive policy, a trend that has been described as the transition from central governance to aid for self-governance (cf. Mayntz & Scharpf, 1995; Mayntz, 2006). Since only the problems and instruments of state action have changed, not the state's legal basis or the basis of its legitimacy, there has been a 'change in the form of political control' (Mayntz & Scharpf, 1995). Policy increasingly ranges beyond the actions of state functionaries, and problem solving is increasingly conducted in cooperation with other actors and often completely by self-organised groups, such as networks businesses and industry associations.

To help support more comprehensive sustainable development, the features of a future-oriented institutional landscape and the potential role of individual actors can be derived by analysing the principles of construction and function of modern societies. A distinguishing feature of modern societies is their functional differentiation (cf. Luhmann, 1984), a societal development process that forms subsystems specialised in fulfilling certain functions (such as policy, law, economy and science). The advantages of this differentiation are analogous to the advantages of the division of labour in production, i.e. specialisation, dynamisation and high efficiency and effectiveness (at least at the level of each individual functional system). However, the dynamics of these differentiated subsystems also presents its own set of problems (cf. Minsch, Feindt, Meister, Schneidewind, & Schulz, 1998).

3.6 Central Functions of Socioscientific Energy Research

On a final abstract level, energy research in the social sciences serves three central functions (CASS/ProClim, 1997; Minsch et al., 1998; WBGU, 2011).

(1) *Reflection and target knowledge*

One deficit of today's functionally differentiated societies and knowledge systems is that they do not encourage a holistic perception of and reflection on real-world problems and opportunities for development. Societal subsystems, such as policy, economy, law and science, as well as individual political bodies and scientific disciplines, occupy strictly separated niches. This separation creates far-reaching barriers to communication and shared interests. Knowledge is selectively produced and recorded in each subsystem or body and is usually processed only with regard to the goals of the respective subsystem. As a result, ecological, economic and social side effects of actions in one system, body or discipline that arise in other subsystems are barely or inadequately detected.

There is a certain symmetry between the deficits that arise during perception and reflection and the corresponding deficits in target knowledge. Societal objectives based on a holistic view of society exist: some of them have been set down in national constitutions. However, the challenge for societal actors to remain mindful of 'the big picture' conflicts with the logic of functional differentiation. The problems (and parochial interests) particular to the different bodies or scientific disciplines push the objectives of that subsystem to the fore. Effects that extend beyond the limits of the body or discipline (especially those related to conflicting objectives) are usually disregarded. Any reference to the whole is presented from the perspective of the subsystem.

Institutional innovations must improve institutions' capacity for holistic perception of, reflection on and evaluation of the dangers and opportunities inherent in multifaceted social developments. This calls on the social sciences in two ways. Firstly, in terms of content, the social sciences must bring their perspectives to holistic perception, reflection and evaluation. Secondly, they must assume a leading role in creating the necessary institutional conditions to achieve this aim.

This requirement implies that an important function of social science energy research is perceiving and reflecting on developments that are societal, political, economic (for example, globalisation), ecological (for example, climate change) and technological, all of which are important for the energy sector itself and for society as a whole (both ex ante and ex post). In addition, such research should support society's reflection on its core values and attitudes towards risk, security, trust, rationality, change and tradition. It should also help the sciences reflect on their own societal functions, their potentials and limitations and their responsibilities. This reflection involves the following:

- Identifying the central energy-related challenges to the economy, society and engineering
- Translating these challenges into positive visions and precepts and guiding their implementation
- Scientifically contributing to society's discussion of broader aims
- Supporting, and possibly even initiating, societal, political and scientific discourses for problem solving.

(2) *Analysis*

Analysis is the core function of every science. In the social sciences, analysis explores the trends and challenges identified in the course of social science reflection and probes the basic societal mechanisms related to the energy system. 'Societal mechanisms' is a deliberately broad term, encompassing all the effects and innovation mechanisms of the human/society system that can be described by the social sciences. It also encompasses all societal subsystems (such as economy, politics, law, education, science-technology and culture) and actors at all levels.

The energy policy issues described in Section 3.3 make it clear that in energy-related enquiry, the research questions generally require interdisciplinary and often transdisciplinary work, alongside disciplinary research. Social science energy research is characterised by its contextuality: The perception of a research issue, derivation of concrete research questions and the analysis and elaboration of solutions generally take place in a specific context that bounds and embeds problems and solutions.

(3) *Transformation knowledge*

The third function of social science energy research is to elaborate and sometimes support the implementation of realisable and forward-looking measures and strategies aimed at ecological, economic and social sustainability. This function is intricately linked to reflection and analysis. Similar to analysis, the realisation function applies to the entire human/society system understood in its totality and should be guided by the normative value of sustainable development.

In addition to the perception deficit identified above, four other deficits of modern, functionally differentiated societies, in the context of a 'learning society' for sustainable development, deserve special mention. These deficits are especially relevant to energy issues and call for particular institutional innovations to correct them.

(1) *Self-organisation and participation: partners capable of taking action*

A typical drawback of differentiated societies is the dearth of partners capable of acting in favour of a policy of sustainability. Subsystems often allow for few or no 'positions' that can effectively introduce sustainability concerns into decision-making processes. They would be perceived as a disturbance. Furthermore, because of a lack of integration, the creative potential of available actors often remains unutilised. Institutional innovations should take these circumstances into account and strengthen the capability for self-organisation.

(2) *Conflict resolution: forward-looking handling of conflicts of interest*

Conflicts of interest can lead to blocked action and policy. Environments that are affected by intensifying ecological and social scarcities and conflicts over resource use require novel and much more effective mechanisms of conflict resolution. One example is discursive methods that aim at participatory and enlightened forms of societal problem solving; another is the various forms of so-called mediation models.

(3) *Innovation: a climate of creativity and impulses toward sustainability*

Actors need room to innovate. They need a culture of innovation, participation and self-organisation that encourages experimentation and expressly recognises that mistakes and setbacks are necessary by-products of learning and innovation processes. Modern collaborative research emphasises actor partnerships for the development of innovative milieus (cf. Chapter 11); this is a prerequisite for innovative processes. In addition, reflection and the ability to self-organise and resolve conflicts constructively are always important. However, innovation alone does not guarantee sustainability (cf. rebound effects).

(4) *Enabling societal restraint and foresight: dealing with absolute scarcities constructively*

Increasingly, scarcities are migrating from the private realm (relative scarcities) to affect society as a whole (absolute scarcities). The modern state has been able to honour its promise of prosperity thanks to a mercantilist strategy of making central energy resources cheaper while tacitly accepting the ecological, economic and social side effects; in other words, by shunting these problems. These side effects have slowly but surely mutated into prime shapers of reality. Society needs to be able to operate constructively where scarcity, risk and uncertainty have again become the norm (see, e.g., Princen, 2005; Goldblatt, 2007). In particular, this means being capable of societal self-restraint while preserving the accomplishments of an open society, namely freedom, human dignity, democracy and the rule of law.

3.7 From Knowledge Management to Knowledge Integration Across Boundaries

The discussion in this chapter has outlined what the authors consider to be the indispensable function of the social sciences within the area of energy research. Social science does much more than merely further the design, acceptance and implementation of (apparently) autonomous technological progress and it is not just an optional add-on to autonomous technological development. This is manifestly evident from the discussion of the global energy-related challenges in Chapter 2:

- All of the challenges pertain directly to competence areas within the social sciences.
- The social sciences play a key role in addressing these challenges, which cannot be met without substantial social science competence.

Addressing the challenges calls upon the social sciences in all of their three main functions (reflection and knowledge of objectives, analysis and transformation knowledge) to varying degrees.

In addition to the three content-related energy challenges, and tantamount in importance, knowledge management is a central methodological challenge. The importance of knowledge management and integration for science and policy makes this methodological challenge a linchpin for achieving genuine progress on the three energy-related challenges. For this reason, the remainder of this chapter is devoted to explicating this challenge, specifically as embodied by the framework of transdisciplinarity (TD). Transdisciplinarity, especially Max-Neef's concept of weak transdisciplinarity (introduced below), will be repeatedly invoked in the studies and synthesis in Part II.

Ideally, crafting and implementing coordinated policies and institutional arrangements capable of achieving long-term energy-related goals would be accomplished in open, democratic societies with the use of extensive scientific knowledge and would engage the public and private sectors, citizens and other stakeholders. The sciences and these stakeholders would jointly identify local and global problems, as well as a range of forward-looking solutions.

This would involve the global research community's pursuit of interdisciplinary ventures, expanded cooperation in teaching and research and collaboration among academic niches that are traditionally highly compartmentalised. It would also see a melding of classical education methods with a comprehensive process of inclusive, broadly participatory learning, which would integrate heterogeneous modes of knowledge across traditional divides between science and society.

However, academia has had limited success at applying the vast sum of knowledge and expertise of its members across the globe towards coordinated solutions in society. The lack of cooperation and communication among members of fragmented and highly specialised disciplines is a chief reason for their failure to apply scholarly knowledge to the solution of society's most complex problems (see Nissani, 1997). Despite enormous individual advances, the fragmentation and isolation of the disciplines and their overall ineffectiveness in confronting the scale of the problems argues against leaving academic knowledge entirely to the existing, self-organising, competitive processes that are already at work in the academic marketplace.

Improving this situation calls for both interdisciplinary and transdisciplinary (TD) frameworks. However, the focus here is on transdisciplinarity – describing something of the state of the art and room for improvement – since the frameworks of most concepts of transdisciplinarity subsume interdisciplinarity.

In an attempt to bring results into the 'life world', transdisciplinary research involves exchanges and deliberation about knowledge, practices and values, both between scientific disciplines and between science and society. Systems knowledge, target knowledge and transformation knowledge together comprise transdisciplinary knowledge (Wiesmann et al., 2008). Not all of the knowledge required as an input to a TD project can be determined beforehand, nor should it be; it needs to be developed in the course of the research process. Therefore, TD research design should be recursive; results must be sent back, problems restructured and assumptions changed during the course of the research. Iterative or circular processes are also appropriate for combining TD integration methods (Pohl, van Kerkhoff, Hirsch Hadorn, & Bammer, 2008).[6]

One significant disadvantage of current transdisciplinary research is that in its delicate task of coordinating multiple disciplines, it often has to compromise on disciplinary standards of knowledge production to some extent, even while relying heavily on disciplinary contributions. Standards for TD research itself are best produced by experience but require appropriate incentives from academic and research funding bodies in order to develop further. It is not currently clear if and when a standardised methodology of TD research will emerge or whether TD research will – or should – ever become an established 'discipline' (Wiesmann et al., 2008).

Knowledge integration is crucial to TD projects and such integration cannot be achieved simply by assembling teams of experts in multiple fields. Successful generation, interpretation and acceptance of the results of TD projects require project participants and management to agree on methods and concepts of integration. Systems analysis is the classic modelling approach for integrating technical information but there are many others, especially within systems thinking (see Bammer, 2005) and the various schools of thought are far from internally unified.

One example of a simple conceptual integrative framework is Max-Neef's (2005) 'weak' transdisciplinarity. Knowledge is coordinated across hierarchical levels of disciplines in which each level's purpose is defined by the level immediately above it. Figure 3.2 represents this as a pyramid with an empirical level at the base (for example, physics, chemistry and geology), above which are pragmatic disciplines (agriculture, engineering, architecture, etc.), followed by a normative level (including law, politics and planning) and a fourth level at the apex that deals with values (ethics, theology and philosophy).

For example, contemporary energy analysis[7] can be characterised as a subsection of the full TD pyramid. In practice, its breadth (width) along Level 2 is limited largely to engineering and traditional economic disciplines (see Fig. 3.2). An energy analysis subsection is also bounded in height at Level 3; it is missing Level 4 altogether, as elements of values and ethics are outside its scope of consideration. From a transdisciplinary perspective, then, energy analysis is too narrow and too shallow. However, remedying this is not easy; hasty attempts to extend energy analysis to Level 4 may stretch disciplinary constructs beyond their intended usefulness. Similarly cursory efforts to enlarge the width of Level 2 by bundling various disciplines together bring out the contradictions in the axioms and epistemological assumptions that underlie the suddenly conjoined disciplines, as seen in the clash between assumptions of neoclassical economics and ecology.

---

[6] Integration methods are still nascent and the recent surge in interest and funding has yet to produce real methodological progress. Both integration methodologies and institutional support for TD research are critical: 'One cannot effectively move forward without the other' (Pohl et al., 2008).

[7] Energy analysis is a semi-integrative, multidisciplinary mode of study that emerged in the 1970s as a direct reaction to the societal challenges posed by the energy crises of that decade.

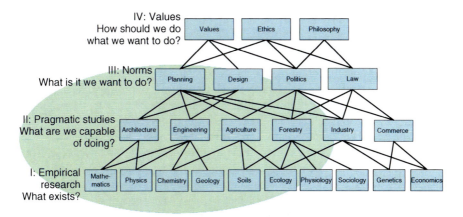

Fig. 3.2 The weak Max-Neef transdisciplinarity pyramid of energy-relevant disciplines, with the field of energy analysis indicated by the shaded ellipse (based on the diagram in Max-Neef, 2005). Graphically, the challenge of 'knowledge management' is to move energy studies from the area highlighted by the ellipse to the full energy-relevant weak-transdisciplinarity pyramid

Max-Neef also posits a 'strong' transdisciplinarity. However, the knowledge integration required to move in the direction of strong transdisciplinarity must occur in the minds of each of the researchers involved, which would require a radical reorientation or great enhancement of current teaching practices in higher education. On the other hand, Brand and Karvonen's (2007) *meta-experts* – who can be viewed as a subset of Bammer's (2005) suggested new type of expert, the integration and implementation scientist – bridge boundaries between disciplines. They translate across areas of expertise that are blurred or clustered by the *interdisciplinary expert*, acting as brokers of expert knowledge and deciding which parts can contribute to sustainability understood as a 'problem of organized complexity'.[8] The meta-experts cannot know the 'whole', but the disciplinary strands they weave together help bring it out of the collective efforts of the members of the TD team.

Incorporating broadly representative concepts and values and employing widely understood communicative terms in TD projects reportedly helps increase the prospects for the political acceptance of the solutions that emerge. In a recent report of a TD climate change project that attempted to straddle the disciplines and world-views of economists and climate scientists, Held and Edenhofer (2008) concluded:

> We develop stylised solutions to the climate problem that have a fair chance of catalysing political decisions in favour of the very decision paths we derive ... We integrate paradigms

---

[8] A third type of expert, the *civic expert*, listens to the public, using his or her experiential knowledge to inform the process of scientific and technological development. This type of expert is important in a classical transdisciplinary setting which, in contrast to the more academic-oriented Max-Neef transdisciplinarity, extensively engages the public in research design and decision making.

of opposing actors in such a way that any of those actors views the boundary conditions of the field in which he or she is a stakeholder (economy or climate) as being represented in a scientifically sound way.

However, Held and Edenhofer also conceded that they needed to advance several disciplinary fronts during the project, which required a 'profound' disciplinary knowledge, or at least a supportive environment. Furthermore, a deep understanding of the conceptual foundations and limitations of both economics and environmental/climate science was crucial to developing a common language necessary to choose appropriate categories and terminology for the joint TD endeavour. This understanding was far deeper than what is typically necessary for strictly mono-disciplinary work within either discipline. This seems to contradict Brand and Karvonen's notion of the generic TD meta-expert, who generally does not and cannot be expected to command a knowledge of all the involved disciplines deep enough to work in this fashion.

References

Bammer, G. (2005). Integration and implementation sciences: Building a new specialization. *Ecology and Society, 10*(2), 6. All web links accessed November 16, 2011, http://www.ecologyandsociety.org/vol10/iss2/art6

Bello, W. (2008). Manufacturing a food crisis. *The Nation*. http://www.thenation.com/article/manufacturing-food-crisis

Brand, R., & Karvonen, A. (2007). The ecosystem of expertise: Complementary knowledges for sustainable development. *Sustainability: Science, Practice & Policy, 3*(1), 21–31.

CASS/ProClim. (1997). *Research on sustainability and global change – Visions in science policy by Swiss researchers*. http://www.proclim.ch/4dcgi/proclim/en/media?1122

Edwards, P. N. (1999). Global climate science, uncertainty and politics: Data-laden models, model filtered data. *Science as Culture, 8*(4), 437–472.

Edwards, P. N. (2010). *A vast machine: Computer models, climate data, and the politics of global warming*. Cambridge, MA: MIT Press.

Erhard, L. (1957, 1990). *Wohlstand für Alle*. Düsseldorf: Econ Verlag.

Goldblatt, D. (2007). Book review perspectives: 'The logic of sufficiency'. *Sustainability: Science, Practice & Policy, 3*(1). http://sspp.proquest.com/archives/vol3iss1/book.princen.html#DG

Heckscher, E. F. (1932). *Der Merkantilismus*. Erster Band. Jena: Gustav Fischer.

Held, H., & Edenhofer, O. (2008). Climate protection vs. economic growth as a false trade off: Restructuring global warming mitigation. In G. Hirsch Hadorn et al. (Eds.), *Handbook of transdisciplinary research* (pp. 191–204). Dordrecht: Springer.

Huybers, P. (2010). Compensation between model feedbacks and curtailment of climate sensitivity. *Journal of Climate, 23*(11), 3009–3018.

Issing, O. (1984). *Geschichte der Nationalökonomie*. München: Vahlen.

Kumar, M. D., & Singh, O. P. (2001). Market instruments for demand management in the face of scarcity and overuse of water in Gujarat, Western India. *Water Policy, 3*, 387–403.

Luhmann, N. (1984, 1995). *Social systems*. Stanford, CA: Stanford University Press.

Max-Neef, M. A. (2005). Foundations of transdisciplinarity. *Ecological Economics, 53*, 5–16.

Mayntz, R. (2006). From government to governance: Political steering in modern societies. In D. Scheer & F. Rubik (Eds.), *Governance of integrated product policy* (pp. 18–25). Aizlewood Mill: Greenleaf.

3 The Indispensable Role of Social Science in Energy Research

Mayntz, R., & Scharpf, F. W. (1995). Steuerung und Selbstorganisation in staatsnahen Sektoren. In R. Mayntz & F. W. Scharpf (Eds.), *Gesellschaftliche Selbstregelung und politische Steuerung* (pp. 9–38). Frankfurt am Main: Campus.

Minsch, J. (2001). Merkantilistische Wirtschaftspolitik und Umweltzerstörung. In R. Costanza, J. Cumberland, H. Daly, R. Goodland, & R. Norgaard (Eds.), *Einführung in die Ökologische Ökonomik, Deutschsprachige Ausgabe* (pp. 296 f.). Stuttgart: Lucius & Lucius.

Minsch, J. (2005a). Das Merkantilismussyndrom. Zu einem Hintergrundgemälde einer Forschung für Natur und Gesellschaft. *GAIA, 14*(3), 273–276.

Minsch, J. (2005b). Nachhaltige Entwicklung: Gedanken zur Weiterentwicklung der offenen Gesellschaft. In F. Radits, F. Rauch, & U. Kattmann (Eds.), *Wissen, Bildung und Nachhaltige Entwicklung*. Innsbruck/Wien/Bozen: Studienverlag.

Minsch, J., Feindt, P.-H., Meister, H.-P., Schneidewind, U., & Schulz, T. (1998). *Institutionelle Reformen für eine Politik der Nachhaltigkeit. Studie im Auftrag der Enquete-Kommission 'Schutz des Menschen und der Umwelt' des 13. Deutschen Bundestages*. Berlin: Springer.

Nissani, M. (1997). Ten cheers for interdisciplinarity: The case for interdisciplinary knowledge and research. *The Social Science Journal, 34*(2), 201–216.

Oreskes, N., Shrader-Frechette, K., & Belitz, K. (1994). Verification, validation, and confirmation of numerical models in the earth sciences. *Science, 263*(5147), 641.

Pohl, C., van Kerkhoff, L., Hirsch Hadorn, G., & Bammer, G. (2008). Integration. In G. Hirsch Hadorn et al. (Eds.), *Handbook of transdisciplinary research* (pp. 191–204). Dordrecht: Springer.

Princen, T. (2005). *The logic of sufficiency*. Cambridge, MA: MIT Press.

Rittel, H., & Webber, M. (1973). Dilemmas in a general theory of planning. *Policy Sciences, 4*(2), 155–169.

Sieferle, R. P. (1982). *Der unterirdische Wald – Energiekrise und Industrielle Revolution*. München: Beck.

Smelser, N. J., & Baltes, P. B. (Eds.) (2001). *International encyclopedia of social and behavioral sciences*. Amsterdam: Elsevier.

Smith, A. (1776). *An inquiry into the nature and causes of the wealth of nations*. London: Methuen.

Snow, C. P. (1959). *The two cultures and the scientific revolution*. New York: Cambridge University Press.

Stern, P., & Kasperson, R. (2010, Eds.). *Facilitating climate change responses: A report of two workshops on insights from the social and behavioral sciences*. Washington, DC: National Research Council.

WBGU, Wissenschaftlicher Beirat der Bundesregierung für Umweltfragen. (2011). *Welt im Wandel. Gesellschaftsvertrag für eine grosse Transformation. Zusammenfassung für Entscheidungsträger*. Berlin: WBGU. http://www.wbgu.de

Wiesmann, U., Hirsch Hadorn, G., Hoffmann-Riem, H., Biber-Klemm, S., Grossenbacher, W., Joye, D., et al. (2008). Enhancing transdisciplinary research: A synthesis in fifteen propositions. In G. Hirsch Hadorn et al. (Eds.), *Handbook of transdisciplinary research* (pp. 433–441). Dordrecht: Springer.

Part II
Invited Contributions

David L. Goldblatt, Thomas Flüeler, Jürg Minsch, and Daniel Spreng

Introductory Note

This part features eight original contributions on a wide variety of energy-related topics, together incorporating a range of disciplines from the social sciences and humanities. Individually, and especially as a whole, the studies provide a compelling illustration of the significance and potential of social-science energy research.

The studies vary in character from essay to research paper and their analyses are conducted at various levels of aggregation. However, with the exception of the first study, each was written to meet both of the following two criteria:

– The study describes a major analytical contribution by the social sciences to one or more of the energy-related challenges laid out in Chapter 2: (i) Energy security and access, (ii) Climate change and other environmental impacts and (iii) Economic and social development.
– The study includes measure(s) or solution(s) the formulation of which was informed by input from two or more social-science disciplines; or, in ex-post analyses, the formulation of which could have been improved through the involvement of two or more social sciences. This implies that each study was required to exercise a degree of self-reflexivity with regard to its disciplinary roots and perspectives.

The first contribution, a content analysis of articles published in *Energy Policy* over the past 10 years, empirically demonstrates and underscores the basic contention that energy research, and even energy policy studies, remain predominantly technically-oriented and fundamentally monodisciplinary. As a meta-study, this chapter is not directly discussed in the subsequent synthesis and reflection sections (Chapters 13 and 14). Chapter 13 characterises, categorises and reflects on the disciplinary perspectives taken in each study.

Chapter 4
What About Social Science and Interdisciplinarity? A 10-Year Content Analysis of *Energy Policy*

Benjamin K. Sovacool, Saleena Saleem, Anthony Louis D'Agostino, Catherine Regalado Ramos, Kirsten Trott, and Yanchun Ong

Abstract This chapter provides a content analysis of 1,453 articles published in *Energy Policy* from 1999 to 2008 involving 3,345 authors and 42,768 references. We selected *Energy Policy* because we believe the journal to be representative of the range of topics on energy policy research available in the field. The typical researcher who publishes in this journal is trained in science or economics, is affiliated with a university or research institute and was working within traditional disciplinary boundaries when their article was published. The most discussed technologies were all fossil-fuel-based and the most popular topic areas were energy markets, followed by climate change and projections of supply and demand. Of the articles that referenced other peer-reviewed articles or books, scientific journals were cited the most, followed by economics journals and *Energy Policy* articles. Social science and arts and humanities journals constituted less than five percent of all peer-reviewed citations. Based on this analysis, the authors of this section recommend that researchers consider using more alternative methods and topics in their contributions to the field.

Keywords Disciplinary boundaries · Energy policy · Energy research · Epistemic community · Interdisciplinarity · Social sciences

4.1 Introduction

Over a decade ago, Lutzenhiser and Shove (1999) called on the community of scholars publishing in *Energy Policy* and the field of energy studies to make a few changes. They observed that energy policy researchers were beginning to

---

B.K. Sovacool (✉)
Formerly with Lee Kuan Yew School of Public Policy, National University of Singapore, Singapore 259772, Singapore

current affiliation: Institute for Energy and the Environment, Vermont Law School, South Royalton, VT 05068-0444, USA
e-mail: sovacool@vt.edu

D. Spreng et al. (eds.), *Tackling Long-Term Global Energy Problems*,
Environment & Policy 52, DOI 10.1007/978-94-007-2333-7_4,
© Springer Science+Business Media B.V. 2012

constitute an international epistemic community, a group of likeminded individuals with relatively uniform perceptions of the energy problems facing society and the solutions to these problems. The authors warned, however, that this emerging community tended to downplay the importance of the social sciences and that many energy dilemmas were understood in ways that reflected old preoccupations with the economic model of rational behaviour (and its close cousin, the theory of market failure).

Lutzenhiser and Shove cautioned that this preoccupation with the 'classic' paradigm of science and economics in energy policy research created a 'blind spot' in 'conventional techno-economic thinking' that masked the 'human elements' of energy technologies and use. As they put it:

> Rather than seeing human choice as critical and controlling in energy use and technology choice, the conventional paradigm focuses on physical-mechanical systems in which human factors are of concern only in terms of possible injury, discomfort or misoperation of equipment. In the 25 years since the first energy crisis, this perspective has changed remarkably little, despite its weakness (p. 217).

Lutzenhiser and Shove accused the research environment (and, ostensibly, articles in *Energy Policy* and similar journals) of disadvantaging many 'alternative' forms of enquiry from the social sciences. They also noted that the shrinking budgets for energy research in the United States had resulted in a regrouping of energy research around the traditional areas of science, technology and economics and that, in the United Kingdom, concern about money and efficient research management favoured tight scripting of well-defined technical programmes rather than broad-based sociological or political enquiry. Indeed, similar calls for more interdisciplinary and diverse research in energy have come from Stern (1986), Lutzenhiser (1992), Stern (1993), Kempton and Schipper (1994), Kempton and Layne (1994), Wilhite, Shove, Lutzenhiser, and Kempton (2000) and others.

While varied, these other critiques of the energy policy literature share some common themes. Firstly, they tend to argue that energy policy research has largely undervalued technology use and choice and the human dimensions of energy use and environmental change, instead treating consumers as rational automatons that respond properly to market signals and changes in price. The fields of history, sociology, philosophy, political science and psychology – with their insights into how consumers and politicians behave and practices of consumption become routine – have been treated as secondary and peripheral to the 'hard' or 'objective' disciplines of economics, statistics, mathematics, physics and engineering.

Secondly, these critiques often assert that existing forms of research discourage interdisciplinary interaction and breadth. Researchers seldom collaborate outside of their discipline or department; they rarely undertake research methods in the 'real world' that take them away from their desk, computer or model. They also tend to come from affluent Western institutions and countries where they study problems facing the developed world, not the developing one. In addition, most researchers are men.

Thirdly, the critiques suggest that a large gap exists between what energy policy researchers think is important and what business persons, utility commissioners and policymakers actually think and do (this could be partly connected to disciplinary isolation and a preference for economics and engineering). Archer, Pettigrew, and Aronson (1992), for example, found that when the California Public Utilities Commission asked a group of academic researchers to evaluate energy efficiency programmes in California, virtually none of the utility executives and policymakers had read any of the literature in the energy policy field on effective programme design or management. It was as if the community of practitioners – those managing energy companies, utility programmes or government agencies and ministries – felt the energy policy field had little to offer. Or, as Archer et al. (1992, pp. 1233–1234) put it, a 'troubling gulf between academic standards for research design and applied practice' was emerging and 'it was as if we [academics] had been invited to a dance only to find ourselves in a mud wrestle'.

In the late 1990s, it was claimed that some of the research being published by *Energy Policy* and its brethren was methodologically suspect, disciplinarily narrow and, worst of all, irrelevant to the real world. Over a decade ago, Lutzenhiser and Shove (1999) hoped that the focus of energy policy research was beginning to shift in favour of more social science and interdisciplinary research in response to wider environmental concerns. By offering a content analysis of a sample of articles in *Energy Policy* from 1999 to 2008, the present chapter asks whether this has really happened.[1]

The chapter begins with an introduction to content analysis, which explains the research method and explicates 12 analytical categories in which 1,453 articles, 3,345 authors and 42,768 references were assessed. The first five categories deal exclusively with the authors of those articles, categorising their affiliation and training, country, institution, gender and interdisciplinarity. The next two categories deal with the funding and the methodological approach used by articles, analysing whether they were externally funded and to what extent they utilised social science, economic, econometric and scientific methods. The next four categories look at the content of those articles, broken down by the countries, comparisons, technologies and topics they address. The final category deals with references or, more specifically, how many times each article cited *Energy Policy*, economics journals, social science journals and 'other' sources.

The next section of the contribution presents the study's results. It finds that the typical author of an *Energy Policy* article is European, male, trained in science or economics, affiliated with a university or research institute and was working within traditional disciplinary boundaries when their article was published. The typical article did not disclose whether funding was provided and, if so, by whom. It was not comparative, it relied on a literature review consisting mostly of non-peer-reviewed sources and it focused on Asia, Europe and/or North America. The most discussed technologies were fossil-fuel-related, with two-thirds of articles discussing electricity in some form and with natural gas and coal being the fuels that were most often commented on. Of the 16 percent of articles that focused on transport and transportation, oil represented more than three-quarters of the discussion. The most popular

---

[1] See also D'Agostino et al. (2011).

topic areas were energy markets, followed by climate change and projections of supply and demand. Of those articles that did reference other peer-reviewed articles or books, scientific journals were cited the most, followed by economics journals and *Energy Policy*. Social science and arts and humanities journals constituted less than five percent of all peer-reviewed citations.

The authors of this study believe the importance of the content analysis is twofold. Firstly, it tests whether the field (or at least the journal) has responded to the call initiated by critics for more balance, breadth and relevance. The very questions researchers pose often frame the solutions that follow. If the chief barrier to a new technology is a budget constraint, engineers can design cheaper technology or regulators can channel money to consumers to purchase it. However, if the barrier is inadequate institutional arrangements, insufficient information, poor communication or a set of values, then improved technical design or financial incentives will only go so far. Further policy interventions will be needed but will become apparent only if the 'right' questions are asked (Stern, 1986, p. 214). An exploration of the shortfalls (if any) in contemporary energy policy research could very well cause a much needed shift in research strategy and the funding of energy programmes.

Secondly, if the community continues to miss important dimensions of its research, especially those relating to the 'social' or 'human' side of the equation, then filling such research gaps may be essential towards addressing common energy challenges. As Kempton and Schipper (1994, p. 186) put it, 'people's behavior – their patterns of heating, cooling and driving – as well as their long-term choices on where and how to live, where to work and where to play, are as important to total energy use as the efficiency of [a given technology.]' If responding to the expected massive global increases in demand for energy will require changes in the level or types of goods and services, then understanding human behaviour is just as important as analysing technology. Buildings will not only need to be heated and cooled more efficiently but smaller residences with more shared or common walls will be needed. It will not only be necessary to increase the fuel efficiency in miles per gallon for passenger vehicles but also to decrease the vehicle miles travelled in smaller automobiles. The entire range of potential solutions to contemporary energy problems becomes larger.

4.2 Research and Theoretical Methods

Content analysis as a research methodology was initially used in the field of communication and communication studies. Early approaches posited the strong influence of the mass media, while later studies suggested that mass media influence was dependent on context and circumstances (McQuail, 1994). The analysed content usually ranged from written text in newspapers and magazines to verbal and non-verbal communication in films and television, later expanding to include other forms of communicated content such as symbols, ideas, themes and messages (Neuman, 1997). Since then, content analysis has been used in a variety of studies in different

4 What About Social Science and Interdisciplinarity? A 10-Year Content ... 51

social science fields, ranging from political science and public policy to consumer research and marketing.

Content analysis has been defined in various ways, most of which have accounted for both quantitative and qualitative forms of analysis. Quantitative content analysis refers to the use of 'objective and systematic counting and recording procedures to produce a quantitative description of the symbolic content in a text' and involves frequency counts of specific words, topics or issues that appear in the content (Neuman, 1997). Qualitative content analysis involves determining and considering the tone, interpretation and context of content in order to 'expose the ideological, latent meaning behind the surface of texts' (Newbold, Boyd-Barrett, & Van Den Bulck, 2002).

Quantitative content analysis is considered 'objective' and 'scientific' because it adopts a deductive approach; that is, the research variables are identified and established before the analysis is conducted. This differs from qualitative content analysis, in which an inductive approach is usually adopted in the analysis to discover research variables (Neuendorf, 2002). One major drawback to qualitative content analysis is that it relies heavily on the researcher's interpretation of the analysed content, which could be subjective and prone to researcher biases (Macnamara, 2006). Inferences drawn from the data that are collected and analysed with researcher biases can affect the validity of the research outcome (Kolbe & Burnett, 1991). Furthermore, the methodology for qualitative content analysis can be complicated and analysing large amounts of data is impractical. As such, qualitative content analysis studies tend to use small samples, which may not always be large enough to draw strong conclusions. There is also the issue of low intercoder reliability for judgement-based qualitative content analysis involving two or more researchers/coders (Kolbe & Burnett, 1991).

Content analyses of academic journal articles have been conducted in prior studies using either quantitative or qualitative techniques or a combination of both, although to the knowledge of this paper's authors, one has yet to be done in the field of energy or energy policy. The present study relied mostly on quantitative forms of content analysis, establishing the categories before coding began and counting and recording when an article met the criteria for those categories. (The two exceptions were the categories of technology focus and topic area, which required coders to make judgements about the subject matter of an article.) This section begins by justifying the selection of *Energy Policy* and explaining the sampling techniques, before elaborating on the four categories (authors, funding and methodology, content and references) and 11 analysed topics (author affiliation, author country, author institution, author gender, author interdisciplinarity, funding, methods, country focus, technology area, topic area and references). The final part of the section presents some of the limitations of this approach. A team of coders was used to keep the work load manageable and an intercoder reliability test, which is explained at the end of the section, was conducted to account for inconsistency.

4.2.1 Sample Selection

The authors selected *Energy Policy* because they felt the journal was representative of the range of topics on energy policy research available in the field. The journal includes a rich variety of topics and methods (not focusing just on electricity or economics like some other journals) and many disciplines and authors, as well as a broader readership that consists of academics and energy analysts along with policymakers, managers, consultants and politicians. Furthermore, while there are many good journals in the field (including *Electricity Journal*, *Energy* and *The Energy Journal*), none of these publish as many articles as *Energy Policy*, suggesting that *Energy Policy* offers a larger (and hopefully more robust) sample to work with. It also has a higher impact factor (1.755) than the other three leading journals mentioned above and is typically ranked as a Tier 1 journal. At the time of writing, no content analysis of the journal had been conducted and the only content analysis ever published *within* the journal was on the use of bioenergy in China (Qu, Tahvanainen, Ahponen, & Palkonen, 2009).

In terms of selecting articles to analyse, only full-length, peer reviewed research articles were included. Shorter and non-peer-reviewed contributions such as editorials, viewpoints, discussions, book reviews, conference reports, errata, addenda, comments, communications and corrigenda were excluded. Because there was a major jump in the number of articles published in 2007 and 2008 (which had as many as eight times the number of articles in earlier years), articles from these 2 years were sampled in order to avoid over-representation, using a random number generator to select one out of every three articles. As Table 4.1 shows, the final sample accounted for 1,453 research articles involving 3,345 authors published from 1999 to 2008, or 70 percent of all research articles.

Table 4.1 Sample of articles included in our content analysis

Year	Number of total articles	Number of articles in sample	% of articles in sample	Number of authors in sample
1999	66	66	100	145
2000	91	91	100	183
2001	105	105	100	248
2002	104	104	100	205
2003	125	125	100	256
2004	146	146	100	334
2005	171	171	100	425
2006	323	323	100	751
2007	539	180	33	454
2008	424	142	33	344
Total	*2,094*	*1,453*	*69*	*3,346*

4.2.2 Author Demographics

The content analysis begins by looking at five categories relating to authors: training, country, institution, gender and interdisciplinarity.

In terms of *training*, the stated institutional affiliation or self-described academic training mentioned by authors at the beginning of each article was taken and coded based on 18 disciplines mentioned by Scopus and presented in Table 4.2, along with two additional categories of 'combination' and 'not available/other'. These 20 fields were inclusive in the sense that an author could have multiple types of training. For instance, an author from the Foundation for Research in Economics and Business Administration received one tick each for Economics, Business and Combination.

For *country*, the coding was based on the number of countries rather than by the number of authors. In other words, if an article had four authors from Greece and one author from Canada, each of the two countries would get one tick. When an author worked in more than one country, their primary affiliation was scored.

For *institution*, affiliations were divided into the following seven categories: law (including attorneys and self-identified legal experts), banking (including financial firms), industry (including manufacturing and construction companies), energy (including energy companies and electric utilities), government (including regulatory agencies), research (including universities, think tanks and national laboratories) and consultants (including interest groups and non-governmental organisations).

Table 4.2 Twenty categories for author training

A	Economics
B	Energy policy analyst/researcher
C	Sciences
D	Area studies/ethnic studies/family studies
F	Communication studies
G	Law/criminology/penology
H	Demography
I	Sociology
J	Psychology/psychiatry
K	Anthropology
L	Politics/political science/international relations/international affairs
M	History
N	Philosophy/ethics
O	Public, environmental & occupational health
P	Public policy & administration
Q	Education
R	Ergonomics
S	Planning & development/urban studies/geography
T	Interdisciplinary/combination
U	Not available/other

For *gender*, authors could be male, female or indeterminable. This last term may strike readers as odd but many names from Asia and other areas can be held by either males or females. In addition, it is common in countries such as India, Turkey and Brazil for authors to provide only a first initial, such as M. Rathmann or B. Agarvvy, which makes gender determination difficult.

For *interdisciplinarity*, an article was considered to be interdisciplinary if it met any of the three following criteria: it involved one author who had training in at least two conventional disciplines from the *training* category above, it involved one author who held a self-identified interdisciplinary position, or the piece involved at least two authors holding positions in at least two disciplines. An author whose discipline could not be determined because of omission did not lead to an article being classified as 'interdisciplinary'. An article with authors from economics and philosophy disciplines would count as interdisciplinary but 'economics' and 'not available' would not. It is possible, therefore, that interdisciplinarity was underestimated given that 4.7 percent of author disciplines were 'not available'.

4.2.3 Funding and Methodology

In order to assess the various funding sources and methodologies employed by articles, two categories were examined: funding and approach.

Funding was coded to depict whether an article was funded by a particular grant or institution, typically stated by the author in a footnote or acknowledgements. An article could receive funding from four sources: government institutions (including a national laboratory), private organisations (including businesses, local and international non-governmental organisations and not-for-profit organisations), universities and 'other'. The remaining articles were classified as non-disclosed/non-funded, although non-monetary forms of support, such as access to data and research assistance, were not considered. Information on funding amounts could not be collected.

Methodological *approach* was classified very roughly into the four following categories: social science methods such as research interviews, surveys, field research, focus groups and questionnaires; economic and scientific methods, including regression analysis, decomposition analysis and forecasting; programme evaluations; and literature review, which encompassed all articles that referenced other books, reports and studies.

4.2.4 Content

In terms of content, the four following categories were examined: country focus, comparativeness, technology area and topic area. Of these, all except comparativeness were inclusive, which means that a single article could have multiple checks for each one. To determine the technology and topic areas, coders read the title, abstract and article keywords (when available) and then scanned the rest of the article, particularly the conclusion. Consequently, these two areas involved a degree of coder subjectivity.

Whenever an article had a case study, mentioned a particular country in its abstract or mentioned a country more than five times in the body of an article, its *country focus* was coded. Intergovernmental or international organisations such as the European Union or Organization for Economic Cooperation and Development (OECD) were included when they were classified as such. Similarly, Hong Kong, Palestine, Taiwan and other political entities were counted as their own country if the authors listed it that way, while states or regions (such as Scotland or Wales in the United Kingdom or California in the USA) were converted to their respective countries.

For those articles that had a country focus, the coders also assessed whether they offered *comparative analysis* that explored or compared at least two case studies or countries. Intra-country comparative case studies were not counted as having comparative analysis.

In terms of *technology area*, articles were classified according to five general categories and more than 40 subcategories, tracking technologies mentioned in article titles, abstracts or at least five times in the body of article. The five general categories were electricity (including the electricity industry generally along with various fuel sources such as wind, nuclear and coal), transport (including the transport sector generally along with specific fuels such as petroleum and ethanol), heating and cooling, energy efficiency and 'other'.

In terms of *topic area*, articles were classified according to 11 categories, much like the technology area. The 11 categories were energy markets (including discussions about restructuring, market failure and prices), climate change, supply and demand (including projections about future supply and demand), investment and trade (including discussions about financing and technology transfer to developing countries), public policy mechanisms (ranging from carbon taxes and credits to feed-in tariffs and subsidies), pollution (including air, water and waste), energy security, industry, research and development (including commercialisation and innovation), land use (including agriculture and forestry) and 'other'.

4.2.5 References

Lastly, the coders looked at what each article referenced in its citations. The number of citations from peer-reviewed academic journals were counted according to three general categories defined by the ISI Web of Science database (home to 7,528 journals): 'Science', 'Social Science' and 'Arts and Humanities'. Two separate categories were then created for *Energy Policy* citations and for those in economics journals. Some journals (such as *Ecological Economics*) were classified in more than one field as science, economics and social science. When this happened, the coders chose the most specific subset (for example, it was classified as economics).

The remaining sources were classified according to books published with a university press and 'other/non-classified/foreign' publications such as materials in any language other than English along with all working papers, reports, dissertations and government documents.

4.2.6 Caveats

While every effort has been made to be methodologically sound, a few caveats must be mentioned. Rather than attempting to interpret what authors meant or implied in their writings, we have instead taken these at face value. Thus, if an author believed their piece addressed market barriers to energy efficiency in China, it was coded as such, even if the author may have only partially addressed these barriers or may have been confused about the meaning of market barriers altogether. To further complicate matters, a number of our categories were 'fluid' and inclusive. For example, an article about Chinese and Japanese efforts to promote energy efficiency in the automotive sector with subsidies would receive checks for China, Japan and comparativeness, along with energy efficiency, public policy mechanisms and manufacturing.

Also, the results will not always correlate with the number of authors, countries or articles. An author could have multiple types of disciplinary training or affiliations, for example, meaning the total number of disciplinary categories reflected in the sample will exceed the total number of authors. This is also true of funding sources; some articles had more than one source, and 'not available' counted as a category, so the total number of entries for the funding column (1,520) is greater than the total number of articles (1,453). Similarly, the number of entries under 'methodology' exceeded the number of articles since most used more than one type of methodology.

Table 4.3 Intercoder reliability results

Category	Number of possible entries	Number of accurate entries	% of intercoder accuracy
Author training	1,100	1,034	94
Author country	60	60	100
Author institution	250	244	98
Author gender	140	131	94
Author interdisciplinarity	25	24	96
Funding	200	185	93
Methodological approach	200	196	98
Content country	55	55	100
Comparativeness	20	18	90
Technology area	1,300	1,240	95
Topic area	1,350	1,265	94
References	1,440	1,396	97
Total	*6,140*	*5,848*	*95*

Finally, because a team of different coders was used for the project, there are some interpersonal coding differences. The lead author sent each of the coders a sample of articles from each of the 10 years and then determined how consistently they coded the articles compared to one another. Table 4.3 presents the results, which show that the coders were consistent with 5,848 of a possible 6,140 entries, suggesting that reliability is approximately 95 percent. This implies that the results presented below should be bounded within an error rate of plus or minus five percent.

4.3 Results

4.3.1 Author Demographics

In terms of their training, authors reported 4,240 institutional affiliations (with some authors reporting as many as four affiliations and others reporting none). As Fig. 4.1 shows, authors were strongly affiliated with the economics, science and energy fields, which constituted 62 percent of the sample. Social science affiliations as a whole, excluding economics, accounted for only 17 percent of affiliations. These were dominated by the fields of public policy (five percent), business (four percent) and planning and geography (four percent). Authors from the energy discipline were often from interdisciplinary research centres and university departments with a focus on science and/or economics that also included public policy. Authors from such interdisciplinary centres would have been coded under the 'combination'

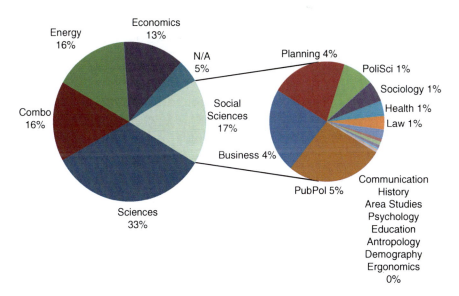

Fig. 4.1 Disciplinary affiliations for energy policy articles. Total 1999–2008 (n = 4,240)

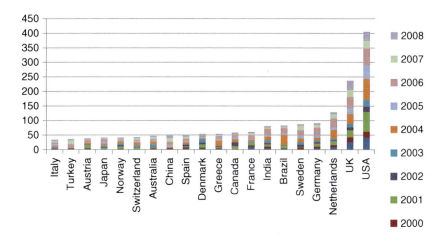

Fig. 4.2 Disciplinary affiliations for *Energy Policy* articles, by year (n = 4,240)

discipline, which could explain why the overall numbers for the two disciplines are similar. Figure 4.2 breaks down affiliations by year and shows that the three primary categories of economics, science and energy accounted for approximately 48 percent of affiliations in 1999 but grew to account for almost 75 percent by 2008. While 16 percent of the authors from the entire sample reported interdisciplinary 'combination' affiliations, which might indicate some diversity in research and perspective, a closer look at Fig. 4.2 reveals a declining trend. The number of authors reporting interdisciplinary affiliations reached a high of 181 in 2004 but dropped to 16 in 2008. Moreover, the number of authors from public policy plummeted from a high of 34 in 2005 to only two in 2008. Interestingly, women's studies and philosophy were coded for initially but these two disciplines had no entries. Other entries that were notably low were ergonomics (one entry), education (six entries), history (four entries), anthropology (two entries), psychology (seven entries), demography (two entries) and communication (six entries).

Authors reported coming from a diverse number of *countries* (84 by our count). However, Table 4.4 shows that these authors tend to be from a consolidated group of countries, with the top 20 countries accounting for 84 percent (1,752) of reported affiliations (2,088). Figure 4.3 breaks down the affiliations by region or continent, revealing that almost half of all authors (49 percent) came from European institutions, followed by about one-quarter (23 percent) for North America and 14 percent for Asia. Also clearly under-represented are South American (five percent), Middle Eastern (four percent) and African (three percent) authors. The five European countries with the most authors were the United Kingdom, Netherlands, Germany, Sweden and France. Authors from the United States made up the majority of all authors, followed by authors from the United Kingdom. The majority of Asian authors came from India (80), almost double those from China (48).

Universities and research institutes clearly dominated the *institutions* that authors reported belonging to. As Table 4.5 illustrates, an overwhelming 82 percent of the

4 What About Social Science and Interdisciplinarity? A 10-Year Content ...

Table 4.4 Country affiliations for *Energy Policy* articles, 1999–2008 (n = 2,088)

USA	405	Hungary	9	Czech Republic	1
UK	237	Iran	9	Ecuador	1
Netherlands	129	Ireland	9	Eritrea	1
Germany	91	Saudi Arabia	9	Ethiopia	1
Sweden	88	Hong Kong	8	Kazakhstan	1
Brazil	83	Jordan	8	Kyrgyz Republic	1
India	80	Lithuania	8	Latvia	1
France	60	Zimbabwe	8	Luxembourg	1
Canada	58	Lebanon	7	Macedonia	1
Denmark	54	Chile	6	Mauritius	1
Greece	54	Columbia	5	Montenegro	1
Spain	50	Croatia	5	Namibia	1
China	48	Portugal	5	Palestine	1
Australia	47	UAE	5	Qatar	1
Switzerland	43	Israel	4	Romania	1
Japan	42	Nigeria	4	Serbia	1
Norway	42	Slovenia	4	Sri Lanka	1
Austria	39	Argentina	3	Syria	1
Turkey	36	Egypt	3	Ukraine	1
Italy	34	New Zealand	3	Uzbekistan	1
Finland	32	Russia	3	Vietnam	1
South Korea	32	Tanzania	3	Yugoslavia	1
Thailand	27	Bangladesh	2	Zambia	1
Mexico	22	Cyprus	2	Brunei	1
Taiwan	22	Estonia	2		
South Africa	20	Indonesia	2		
Singapore	14	Philippines	2		
Malaysia	12	Poland	2		
Belgium	11	Burkina Faso	1		
Kenya	10	Cuba	1		

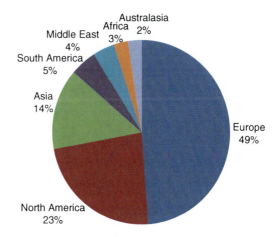

Fig. 4.3 Regional affiliations for *Energy Policy* articles, 1999–2008 (n = 2,088)

Table 4.5 Institutional affiliations for *Energy Policy* articles, 1999–2008 (n = 3,381)

Institutional type	Number	% of total
Research/university	2,778	82
Government	298	9
Consulting/NGO/interest group	193	6
Energy company/utility	55	2
Finance/banking	11	< 1
Industry/manufacturing	8	< 1
Law	1	< 1
Unknown	37	1

3,381 reported affiliations belonged to universities, research centres and think tanks. Only nine percent of authors were from the government or regulatory agencies, slightly more than five percent were from consulting, international organisations and interest groups, and a mere two percent of the authors came from actual energy companies or electric utilities. When bureaucrats, regulators, parliamentarians, policymakers and other government officials did publish articles, they often did so as co-authors and were rarely the first or sole author. This could be explained by the fact that the nature of their jobs may not be conducive to writing academic articles and, consequently, career policymakers may not emphasise publishing as highly as academics and research centres do. Another explanation could be that *Energy Policy* has a low influence on policymakers since so few of them appear to be publishing in it. A third explanation is that the influence *Energy Policy* has on readers is unidirectional, creating inflows of knowledge but not outflows. A fourth explanation is that counting only research articles served to exclude viewpoints and forums, in which a larger number of non-academics may participate.

In terms of *gender*, a staggering majority (69.9 percent) of the 3,346 authors in the sample who published articles were male. As Table 4.6 reveals, only 16 percent

Table 4.6 Gender of *Energy Policy* authors, 1999–2008 (n = 3,346)

Year	Male	Female	Indeterminable	% of female authors
1999	119	19	7	13
2000	146	32	5	17
2001	153	38	57	15
2002	152	41	12	20
2003	192	42	22	16
2004	221	58	55	17
2005	318	66	41	16
2006	518	112	121	15
2007	316	64	74	14
2008	203	56	86	16
Total	*2,338*	*528*	*480*	
%	*69.90*	*15.80*	*14.30*	

4 What About Social Science and Interdisciplinarity? A 10-Year Content ... 61

Table 4.7 Interdisciplinary affiliations for *Energy Policy* articles, 1999–2008 (n = 4,240)

Year	Interdisciplinary affiliations	Total affiliations	% of interdisciplinary affiliations
1999	47	222	21
2000	36	201	18
2001	76	442	17
2002	15	266	6
2003	51	339	7
2004	115	609	19
2005	57	561	10
2006	110	753	15
2007	44	504	9
2008	17	343	5
Total	*568*	*4,240*	*13*

were female and at no point did female authorship ever surpass 20 percent for an individual year. (During some years, such as 1999, the number of female authors dropped to 13 percent, though the high level of indeterminable authors could hedge against this finding.)

Lastly, Table 4.7 documents that only 13 percent of authors reported *interdisciplinary* affiliations, which were used as a proxy for training. This suggests that most authors conducted research within the confines of a single discipline.

4.3.2 Funding and Methodology

Authors reported 1,520 separate sources of *funding* for their articles. While some articles were funded by more than one source, 58 percent (875 entries) of all articles reported either no funding source at all or did not explicitly mention the source of funding. In comparison, 26 percent (296 entries) came from government entities and nine percent (141 entries) from private organisations. Six percent (91 entries) were funded by universities. The remaining one percent (17 entries) were funded by sources other than those listed above.

By far the most favoured methodological *approach* was the simple literature review, with almost 60 percent of articles relying on secondary sources as their main source of data. It is important to emphasise, however, that in most cases the literature review was an ancillary tool (that is, an article that cited studies from the literature) rather than a rigorous, systematic, methodical and exhaustive review of the literature. As Fig. 4.4 shows, around one-third of articles relied on economic tools (such as modelling or regression analysis) and only eight percent used social science methods such as field research, interviews, surveys and focus groups.

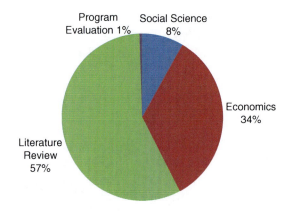

Fig. 4.4 Methodological approaches for *Energy Policy* articles, 1999–2008 (n = 2,529)

4.3.3 Content

Table 4.8 presents every case study from the sample, treating the European Union as a *country* on its own (when classified by authors as such). As Fig. 4.5 highlights, Europe, Asia and North America accounted for 77 percent of case studies and of those articles that had case studies or discussed countries at length, the list was highly concentrated. The top 20 countries (1,072 entries) represented 70 percent of the sample (1,523 entries). Figure 4.6 shows some interesting shifts in country focus, with India and China accounting for nine percent of case studies in 1999 but 16 percent in 2008. Indeed, in 2008, China tied the United States as the most popular country for discussion among authors from the sample. What also stands out is the relatively small percentage of research articles that adopted a regional or global focus rather than a country focus. Only about 14 percent (205 entries) of the sample included articles addressing regional and global organisations and markets such as ASEAN or APEC, as well as geographical 'groupings' of Sub-Saharan Africa, the Persian Gulf, Pacific Rim, Latin America and Eurasia.

Although many articles presented case studies or elaborated on particular countries at length, Table 4.9 reveals that only 11 percent of articles had some type of *comparative* focus (that is, they looked at more than one case study in depth).

In terms of *technology area*, articles discussed 26 subcategories comprising 2,624 entries. Table 4.10 shows that electricity clearly dominated, with almost two-thirds of articles focusing on some aspect of electricity generation, transmission, distribution and use and only 16 percent of articles dealing with transport. Figure 4.7 breaks down the 926 articles that discussed a particular electricity technology, the most popular of which was natural gas (23 percent), coal and clean coal (15 percent), wind energy (13 percent), biomass (12 percent) and solar (11 percent). Figure 4.8 breaks down the 299 articles that discussed a particular mode of transport or transport fuel, with oil discussed 72 percent of the time, hydrogen 13 percent, ethanol eight percent and biodiesel five percent. Readers wishing to assess how these trends have changed across the 10-year period should consult Figs. 4.7 and 4.8.

Table 4.8 Country focus for *Energy Policy* articles, 1999–2008 (n = 1,523)

USA	164	OECD	12	Hungary	3	Central America	1	Senegal	1
UK	119	Malaysia	9	IEA	3	Central Europe	1	Serbia	1
China	99	Zimbabwe	9	Lebanon	3	Chad	1	Slovakia	1
Europe	93	Africa	7	Luxembourg	3	CIS	1	Southern Africa	1
India	92	Indonesia	7	Middle East	3	Cuba	1	South America	1
Sweden	58	Ireland	7	Nepal	3	Cyprus	1	Sub-Saharan Africa	1
Brazil	56	Kenya	7	OPEC	3	Ecuador	1	Syria	1
Global	56	Chile	6	Pakistan	3	Eritrea	1	Turkmenistan	1
Germany	51	Hong Kong	6	Portugal	3	Eurasia	1	UAE	1
Netherlands	48	Saudi Arabia	6	Slovenia	3	G7	1	Uganda	1
Denmark	40	Argentina	5	Ukraine	3	Ghana	1	Uzbekistan	1
Japan	35	Croatia	5	Asia Pacific	2	Guatemala	1	Venezuela	1
Turkey	35	Jordan	5	Bangladesh	2	Gulf states	1	Western Europe	1
Spain	32	Kazakhstan	5	Botswana	2	Honduras	1		
Norway	27	Vietnam	5	Developing Countries	2	Iceland	1		
Australia	25	Asia	4	GCC	2	Iraq	1		
South Korea	22	Austria	4	Israel	2	Kyrgyz Republic	1		
Greece	20	Belgium	4	Latin America	2	Laos	1		
Canada	19	Columbia	4	Latvia	2	Macedonia	1		
France	19	Ethiopia	4	Mauritius	2	Myanmar	1		
Finland	18	Nigeria	4	Morocco	2	Namibia	1		
Russia	18	Poland	4	Philippines	2	New Zealand	1		
Mexico	17	Singapore	4	Scandinavia	2	North America	1		
South Africa	15	Sri Lanka	4	Zambia	2	Northern Europe	1		
Switzerland	15	Tanzania	4	Azerbaijan	1	Pacific Rim	1		
Taiwan	15	APEC	3	Bulgaria	1	Palestine	1		
Italy	14	ASEAN	3	Burkina Faso	1	Persian Gulf	1		
Thailand	14	Egypt	3			Puerto Rico	1		
Iran	12	El Salvador	3	Cambodia	1	Qatar	1		
Lithuania	12	Estonia	3	Cameroon	1	Romania	1		

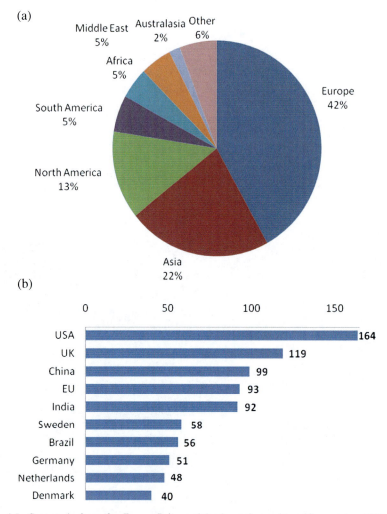

Fig. 4.5 Case study focus for *Energy Policy* articles by region and top 10 countries, 1999–2008 (n = 1,523). **a** Region, **b** Top 10 countries

In terms of *topic area*, articles discussed 28 subcategories comprising 3,899 entries. Table 4.11 documents that the most popular category overall was energy markets (including discussions of electricity restructuring, liberalisation, market barriers and prices), followed by climate change, projections of supply and demand, investment and trade, and public policy mechanisms. Since the Lee Kuan Yew School of Public Policy is expressly interested in public policy, special care was taken to keep track of the specific public policy mechanisms discussed in each article. Figure 4.9 shows that 243 articles discussed a particular public policy mechanism, the most popular of which were subsidies, green power programmes,

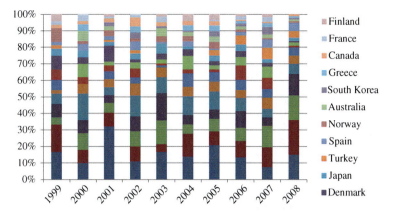

Fig. 4.6 Country case study focus for *Energy Policy* articles, by year (n = 1,523)

Table 4.9 Comparative case study focus of *Energy Policy* articles, 1999–2008 (n = 2,094)

Comparative articles	Total articles	%
13	66	20
17	91	19
20	105	19
20	104	19
17	125	14
30	146	21
8	171	5
60	323	19
29	539	5
26	424	11
240	*2,094*	*11.46*

renewable portfolio standards and feed-in tariffs. Figure 4.10 shows that 252 articles discussed carbon taxes and credits, yet the discussion was dominated by tradable credits, which were mentioned at a ratio of almost 2:1 (64 percent) compared to carbon taxes.

4.3.4 References

Surprisingly, Fig. 4.11 shows that almost two-thirds (64 percent) of more than 42,700 references cited in articles from the sample did not come from peer-reviewed sources and instead came from 'grey' publications such as reports, white papers, newswires and newspaper articles; in some years (such as 2003), almost three-quarters of citations came from non-peer-reviewed sources. As Fig. 4.12 shows, the

Table 4.10 Technology focus for *Energy Policy* articles, 1999–2008 (n = 2,624)

Energy sector	Technology category		Number of articles	% of total
Electricity			1,687	64
	Electricity general	586		
	Natural cas/LNG	217		
	Renewable general	175		
	Coal/clean coal	140		
	Wind	117		
	Biomass	107		
	Solar	100		
	Distributed generation	92		
	Hydro	81		
	Nuclear fission/fusion	53		
	Geothermal	15		
	Ocean	4		
Transport			413	16
	Petroleum	215		
	Transport general	99		
	Hydrogen	40		
	Ethanol	23		
	Alternative fuels general	15		
	Biodiesel	14		
	Oil shale	4		
	Tar sands	2		
	Coal-to-liquids	1		
	Algal fuels	0		
Heating and cooling			128	5
Energy efficiency			325	12
Other			71	3
Total			*2,624*	*100*

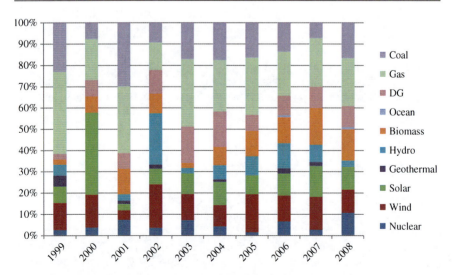

Fig. 4.7 Specific technology focus for electricity articles (n = 926). *Note*: 'Gas' includes natural gas and LNG and 'coal' includes clean coal

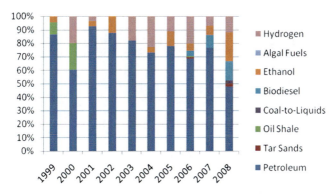

Fig. 4.8 Specific technology focus for transport fuel articles (n = 299)

Table 4.11 Topic focus for *Energy Policy* articles, 1999–2008 (n = 3,899)

Topic category		Number of articles	% of total
Energy markets		852	22
	Electricity markets	505	
	Market barriers	347	
Climate change		546	14
Supply & demand		532	14
Investment and trade		511	13
	Trade	60	
	Investment	223	
	Developing countries	164	
	Financing	64	
Public policy mechanisms		495	13
	Carbon credits	161	
	Carbon taxes	91	
	Net metering	7	
	Systems benefit charges	2	
	Subsidies	73	
	Green power	52	
	Renewable portfolio standards	36	
	Feed-in tariffs	33	
	Tax credits	27	
	Loans	13	
Other		277	7
Pollution		200	5
	Air pollution	144	
	Water pollution	31	
	Waste	25	
Industry		177	5
R&D		135	3
Land use		103	3
	Agriculture	52	
	Forestry	51	
Energy security		71	2

Note: Percentage total is greater than 100 due to independent rounding.
Renewable Portfolio Standard: ensures that a minimum amount of renewable energy is included in the portfolio of the electricity resources serving a state.

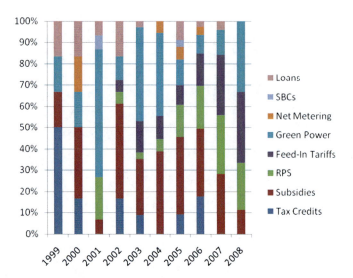

Fig. 4.9 Specific policy mechanism focus for public policy articles (n = 243)

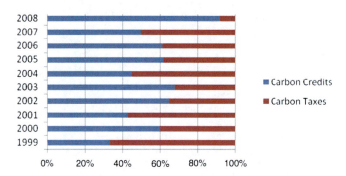

Fig. 4.10 Specific policy focus for carbon tax and cap-and-trade articles (n = 252)

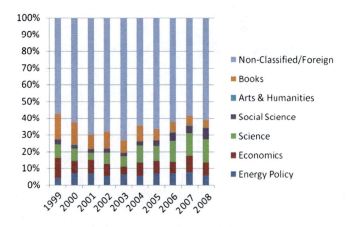

Fig. 4.11 Cited references for *Energy Policy* articles by year, 1999–2008 (n = 42,768)

4 What About Social Science and Interdisciplinarity? A 10-Year Content ...

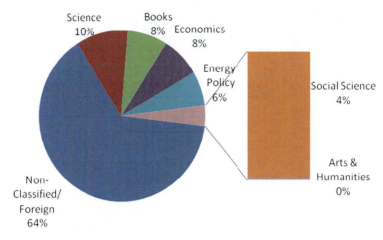

Fig. 4.12 Total cited references for *Energy Policy* articles, 1999 to 2008 (n = 42,768)

largest single class of cited articles (4,380) came from peer-reviewed science journals and 3,281 references were to peer-reviewed books. Economics journals were referenced 3,256 times and *Energy Policy* was referenced 2,778 times. Social science journals (1,643) and arts and humanities (20) represented less than five percent of all citations, implying that research published in such journals is less relevant to the work of *Energy Policy* contributors.

4.4 Conclusion

Energy Policy has undoubtedly produced a voluminous amount of research over the past decade. The present study analysed 1,453 articles published between 1999 and 2008 by 3,345 authors from 84 countries. These articles involved 1,523 case studies covering 136 countries with a variety of topics and methodologies and the sample included about 70 percent of all articles published in *Energy Policy* during this period. The authors hope that this assessment helps shed light on areas for improvement and opportunities for maturation at a time when it is essential for energy to be studied through different perspectives.

The content analysis has uncovered a worryingly low number of women publishing in the journal, a possible under-representation of social science and humanities disciplines and methods, a lack of interdisciplinarity (only 13 percent of articles involved interdisciplinary training or affiliations) and a need for articles to become more relevant to real-world problems (for example, the lack of an external sponsor). For instance, 57.6 percent of articles did not acknowledge funding from an external source and only 18 percent of articles were written by non-academics, although this final figure may be skewed because only research articles were counted. Fewer than 12 percent of articles offered comparative analysis that explored at least two case studies, only eight percent of articles relied on social science methods and tools and fewer than five percent of citations were for social science or arts and humanities journals.

While by no means comprehensive, the content analysis suggests a number of potentially novel methodologies, technologies, disciplines and topic areas that researchers could begin to explore. The point here is not that we must transform all scientists, economists and energy policy analysts into anthropologists and historians, just that researchers may need to do a better job bridging the gap between the 'hard sciences' and social science and humanities approaches. Not a single author reported training or institutional affiliation with philosophy or a philosophy department. Only six authors from the sample reported affiliations with departments of communication. Just 12 articles in the sample dealt with governance and only 73 out of 3,899 entered topics dealt with subsidies. Similarly, water issues represented only 31 out of 3,899 topic entries, yet energy fuels and production cannot be managed without attention to water. Fewer than 16 percent of authors in the sample identified themselves as female and not a single one reported training in women's studies, feminism and gender studies.

Geographically, Africa and the developing regions of Asia and South America had very few authors and case studies. Africa as a whole accounted for only three percent of all case studies and many developing countries or political entities – including Ecuador, Eritrea, Ethiopia, Kazakhstan, Kyrgyz Republic, Namibia, Palestine, Vietnam and Burkina Faso – had only one author in the sample over a 10-year period. Perhaps correspondingly, a number of developing countries, such as Senegal, Uganda, Chad, Ghana, Guatemala, Laos and Azerbaijan, had only one case study, to say nothing of the more than 50 countries that are not represented at all (the case studies in the sample focused on 136 countries and/or organisations, compared to about 190 countries in the world).

To be fair, there may be many articles that have explored these aspects of energy policy that were not published in *Energy Policy* or it may be that few articles on these other topics exist in the first place. However, the results suggest that collaboration with disciplines outside the fields of energy, economics and science and engineering, a focus on technologies beyond electricity and fossil fuels and a consideration of topic areas other than energy markets, trade and investment, climate change and projections of supply and demand may yield fruitful results that enhance our understanding of energy and policy issues. There is potential for the field of energy policy to expand topically, technologically and methodologically. Promising alternative topics of research could include developing countries, water, global energy governance, mundane energy systems, the interaction between technologies and gender, research on sociotechnical systems and the philosophy, rhetoric and historiography of energy technology. Methods that may have been undervalued include research interviews, field research, focus groups, surveys and other 'soft' and 'people-centred' approaches. In addition to the valid and important contribution from scientists and economists to the field of energy policy, the authors call on social scientists, political scientists, philosophers, historians, feminists and others to help broaden scholarship in the journal and the field at large.

Acknowledgments The authors appreciate the helpful suggestions and comments of Richard Hirsh from the Virginia Polytechnic Institute & State University and Barbara Farhar from the University of Colorado. Despite their assistance, all conclusions presented in this study and any errors therein are solely those of the authors.

References

Archer, D., Pettigrew, T. F., & Aronson, E. (1992, October). Making research apply: High stakes public policy in a regulatory environment. *American Psychologist, 47*(10), 1233–1236.

D'Agostino, A. L., Sovacool, B. K., Trott, K., Regalado Ramos, C., Saleem, S., & Ong, Y. (2011). What's the state of energy studies research? A content analysis of three leading journals from 1999 to 2008. *Energy Policy, 36*(1), 508–519.

Kempton, W., & Layne, L. L. (1994). The consumer's energy analysis environment. *Energy Policy, 22*, 857–866.

Kempton, W., & Schipper, L. (1994). Expanding the human dimensions research agenda. *Proceedings of the 1994 ACEEE summer study on energy efficiency in buildings* (pp. 1.85–1.90). Washington, DC: ACEEE.

Kolbe, R. H., & Burnett, M. S. (1991). Content-analysis research: An examination of applications with directives for improving research reliability and objectivity. *The Journal of Consumer Research, 18*(2), 243–250.

Lutzenhiser, L. (1992). A cultural model of household energy consumption. *Energy, 17*(1), 47–60.

Lutzenhiser, L., & Shove, E. (1999). Contracting knowledge: The organizational limits to interdisciplinary energy efficiency research and development in the US and the UK. *Energy Policy, 27*, 217–227.

Macnamara, J. (2006). Media content analysis: Uses, benefits & best practice methodology. Sydney: Media Monitors.

McQuail, D. (1994). *Mass communication theory: An introduction*. London: Sage.

Neuendorf, K. (2002). *The content analysis guidebook*. Thousand Oaks, CA: Sage.

Neuman, W. (1997). *Social research methods: Qualitative and quantitative approaches*. Needham Heights, MA: Allyn & Bacon.

Newbold, C., Boyd-Barrett, O., & Van Den Bulck, H. (2002). *The media book*. London: Arnold (Holder Headline).

Qu, M., Tahvanainen, L., Ahponen, P., & Palkonen, P. (2009). Bioenergy in China: Content analysis of news articles on Chinese professional internet platforms. *Energy Policy, 37*(6), 2300–2309.

Stern, P. C. (1986). Blind spots in policy analysis: What economics doesn't say about energy use. *Journal of Policy Analysis and Management, 5*(2), 200–227.

Stern, P. C. (1993, June 25). A second environmental science: Human-environment interactions. *Science, 260*, 1897–1899.

Wilhite, H., Shove, E., Lutzenhiser, L., & Kempton, W. (2000). The legacy of twenty years of energy demand management: We know more about individual behavior but next to nothing about demand. In E. Jochem (Ed.), *Society, behavior, and climate change mitigation* (pp. 109–126). New York: Kluwer.

Chapter 5
Towards an Integrative Framework for Energy Transitions of Households in Developing Countries

Shonali Pachauri and Daniel Spreng

Abstract Bringing about a transition to more efficient, affordable and sustainable energy sources, carriers and technologies for all of humanity is one of the key challenges of the 21st century. Addressing this challenge effectively requires a complex linkage between multiple perspectives to be assessed within a transdisciplinary framework. Even today, however, there is scant communication across disciplines and among the various actors affecting and affected by such transitions. This section highlights the diverse approaches of past programmes and policies aimed at facilitating energy transitions and discusses some of the key issues associated with their successes and failures. It concludes with some thoughts on how such programmes and policies can be improved in the future by thinking in terms of processes rather than disciplines and taking into account multiple perspectives and the interdependencies between systems, as well as several important ethical dimensions.

Keywords Energy poverty · Household energy · Energy transitions · Poverty · Energy and development

5.1 Applying Max-Neef's Transdisciplinarity Matrix to Energy Transitions

5.1.1 One-Dimensional Perspectives

Transitions from traditional energy sources and techniques to newer ones are often seen purely from a one-dimensional perspective, such as a technical one, considering only direct monetary costs to the user or only environmental or health costs.

Technical/resource perspective: The technical perspective and the resource perspective are two typical examples of such one-dimensional perspectives. According

---

S. Pachauri (✉)
International Institute for Applied Systems Analysis (IIASA), 2361 Laxenburg, Austria
e-mail: pachauri@iiasa.ac.at

D. Spreng et al. (eds.), *Tackling Long-Term Global Energy Problems*,
Environment & Policy 52, DOI 10.1007/978-94-007-2333-7_5,
© Springer Science+Business Media B.V. 2012

to the technical perspective, in light of possible improvements in technical efficiencies, it is unacceptable for a large part of humanity worldwide to still utilise fuels and equipment with extremely low technical efficiency, thereby consuming large amounts of primary energy in order to provide the relatively low levels of energy services they demand. Improving the efficiency of a cooking stove by a factor of three, for example, would reduce wood consumption proportionately. Deployed throughout an entire village, this would reduce destruction of the surrounding forest and, with time, would lead to shorter distances for collecting the wood. According to this one-dimensional efficiency reasoning, there is no good argument why these technical improvements should not be made immediately.

Similarly, resources are often judged in a one-dimensional manner. Solar energy is considered good because it is plentiful; biomass is considered good or bad, depending on whether it is judged on its alleged carbon neutrality or on its potential to cause damage to forests and to compete with food production.

In the past, NGO-donor programmes were often motivated by this type of technical or resource perspective. Even government programmes, especially many rural electrification projects, were frequently treated as solely technical activities, separate from other aspects of rural development.

Even on its own grounds, however, a technical perspective is inadequate by itself. When a household in a remote rural setting is electrified, the electricity may be inefficiently produced and transported and may only be available a few hours per day. The theoretical tenfold gain in efficiency involved in switching from kerosene to electric lighting may be entirely nullified. Of course, it must be said that once the new systems are in place, a high degree of inefficiency implies a large room for improvement.

Direct monetary cost perspective: Another gross simplification[1] considers cost as the driver of all decisions on the ground. A household adopts a new energy carrier and energy appliance if it makes economic sense; that is, if it leads to lower costs. Obviously, development cannot be explained from a situation of zero monetary cost without introducing opportunity cost, such as the time 'lost' in gathering fuels and cooking for hours.

During the 1980s and 1990s, a widely shared view in public administration was that transitions in cooking fuels could easily be influenced by a subsidy policy. Accordingly, a fuel or energy type that would be advantageous to the user and might help households to develop should be subsidised. This was done with little thought as to which subsidies would profit whom. Many recent studies have shown that subsidies are not well targeted (Gangopadhyay, Ramaswami, & Wadhwa, 2005) and are often extremely expensive for the state. In addition, costs are not always a decisive consideration in decisions about energy transitions (see Chapter 6). For example, it is a rather unimportant factor for the household electrification of urban Indian households (Kemmler, 2007). The household's resolve to adopt a modern, technology-friendly way of life is much more important.

---

[1] This alludes to Jacob Burckhardt's prediction in 1889 that the enemies of the (20th) century would be the '*terribles simplificateurs*' (Burckhardt, 1889, ed. 1980).

Towards the close of the 20th century, free market ideology came to dominate, with both fuels and electricity being increasingly traded in free markets and economic efficiency being seen as the measure of success. This ideology was trumpeted as if, for instance, the green revolution, instrumental to saving hundreds of millions of people from starvation, could have been possible without the state having provided electricity to farmers to install millions of irrigation pumps. Today, the introduction of more or less free markets for electricity and fuels has helped the involved industries reach more solid economic footing, even in developing countries. In the absence of appropriate safeguards, however, this has promoted inequities within these countries, as poorer rural customers with lower demand get worse service, if any, while richer urban customers and large industries are favoured. Recent literature on the welfare impacts of energy market reforms have found them to be largely detrimental to the poor unless special legislation and measures are incorporated early in the reform process (Cherni & Preston, 2007).

5.1.2 Multi-Dimensional Perspectives of Actors

User perspective: Users are guided by a multitude of circumstances, opportunities and restrictions. It can be difficult for technocrats, especially those living in geographically and socioeconomically removed circumstances, to appreciate the daily life of poorer householders in developing countries. The technical perspective holds that it would be beneficial for a poor mother in a developing country to switch to a more efficient stove. However, the same woman has many problems to deal with in daily life. Perhaps she has a sick child, an animal that does not produce enough milk or is lame and cannot pull the plough, or maybe the nearby river has flooded her field and her hut needs repair. The woman must assess which one of her problems is most urgent, or even life-threatening, and set her priorities accordingly. There is no time for new things. She is satisfied with coping in the way she has learned to cope in the past. Furthermore, she needs her neighbours and is careful not to alienate them by doing things differently from how they do them. Without sharing the burden with them, she could not survive the hardship. The long daily trek to collect wood is tiring and time-consuming but it has its small rewards. She enjoys the company of her female friends and the short break it affords from other family demands.

The village woman has many tasks. She has a given lifestyle and she is part of a closely-knit community network. She is not waiting for new chores.

Business perspectives: Business perspectives vary greatly depending on the business at hand. Large multi-national companies have many departments, each with their own perspectives on the business, such as business development, marketing, production, etc. A small one- or two-person enterprise also needs a multi-dimensional perspective, although some important dimensions may be missing in both cases.

Take the example of a charburner who transforms wood, a free natural resource, into a transportable and saleable energy carrier. The charburner has to choose a suitable production site, control the air influx to the charcoal kiln, watch the fire and be careful when felling trees and chopping wood. His hard work will not produce the desired results if he has not thought about where and to whom he can sell the energy product. It also helps if he takes into account which season will yield better prices.

The multi-national corporation also has a department of ecology that assesses the company's production processes and products, keeps abreast of and even lobbies to influence the relevant laws to yield economic and ecologically optimal outcomes, to the extent this can be done mutually. The charburner, in the course of trying to make a living, probably has no idea what laws he is infringing or how unecological is the process chain he initiates. In many African countries, charcoal production is rapidly destroying forests and contributing massively to air pollution. As long as charcoal helps the poorest segments of the population survive by providing a relatively cheap fuel far from the forests, it may arguably be close to a socio-ecological optimum in the short-term. However, when the charcoal is exported in massive quantities to rich countries for barbecues, this faulty technology has very little justification. The perspective of the charburner is multi-dimensional but by leaving out ecology and myopically discounting the future, it misses out on a very important dimension and destroys the resource that underlies the charburner's business.

5.1.3 The Max-Neef Matrix

Max-Neef's matrix of transdisciplinarity (Max-Neef, 2005), introduced in Chapter 3, Section 3.7, provides an excellent means of systematising the various ways of looking at, analysing and describing energy transitions in households of developing countries. It illustrates how various scientific approaches can relate to each other along and across four levels (see Fig. 5.1).

Academic research on energy is most often disciplinary research. Such research is vital and important, but it is rarely suited to furnishing sound policy advice.

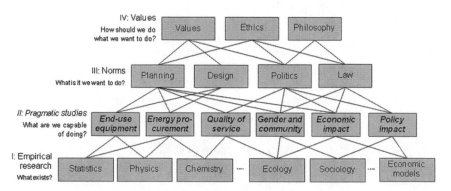

Fig. 5.1 Weak transdisciplinarity concept of Max-Neef applied to the study of energy transitions (adapted from Max-Neef, 2005)

Pragmatic studies of various aspects of energy transitions are more important for this purpose, particularly if they draw on multiple relevant empirical sciences and are then, by definition, interdisciplinary. However, Max-Neef points out that sustainability research must indispensably involve norms and values. As Fig. 5.1 shows, he labels work approaches that draw on all four levels (empirical sciences, pragmatic studies, norms and values) *transdisciplinary*.

As mentioned in Chapter 2, Max-Neef's use of the term *transdisciplinarity* is unusual. To avoid confusing it with the more usual meaning of transdisciplinarity (interdisciplinarity plus stakeholder involvement, to put it simply), the term *Max-Neef transdisciplinarity* will be used here when referring to a study that includes all four levels of the Max-Neef matrix.

Max-Neef applies his concept primarily to research and teaching; i.e., academia (see Section 3.7 in Chapter 3). Here, the concept is adapted to those fields of knowledge that bear some relevance to the study of energy transitions. Obviously, any one discipline on level 1 will not, by itself, be sufficient to study the phenomenon. Even taking them all together and studying energy transitions solely on the basis of first principles – that is, in a multi-disciplinary manner but on the level of research alone – will not work. Experience with various types of end-use equipment, energy procurement, gender and community issues at level 2 will also be necessary in order to understand what improvements are possible. However, even these two levels of knowledge together will not provide an adequate characterisation of the issue or a solid base for formulating recommendations for programmes that might have an impact.

The following section examines various aspects of energy transitions in developing countries, drawing on a rich body of pragmatic and multidisciplinary studies that have been carried out over the past several years. Section 5.3 extends the analysis to the levels of norms and values, showing how the explicit inclusion of these levels adds greatly to the relevance of the multi- and interdisciplinary work.

5.2 Multidisciplinary Studies of Electrification, New Cooking Fuels, New Stoves

Transitions in the amounts, types and ways of using energy have occurred throughout the course of human history. These transitions have contributed to transformations of societies and to a continuous evolution in the scale of human activities and how they are organised and conducted. In the past, energy was procured and consumed locally rather than traded and the pace of the transitions was rather slow. The pace accelerated during the 19th century but only in certain parts of the world. As a result, even today, over one-third of humanity relies primarily on traditional non-commercial sources of energy, such as unprocessed biomass, animal power and human power. This segment of humanity, largely concentrated in rural areas and urban slums in developing nations, engages mostly in subsistence activities in order to meet basic domestic needs. The challenge for these nations is to extend the benefits of modern energy services to these people. This requires huge investments and effective policies to bring about shifts in the energy-related behaviours of people in millions of households. The challenge is made more difficult by the growing environmental constraints that now face these nations. The unsustainability of the existing fossil-based energy system and growing threat of climate change requires them to increase their use of renewable sources and to adopt cleaner ways of producing, delivering and consuming energy. In other words, they need to find ways of developing in an increasingly decarbonising world.

Traditionally, transitions occurred as an outcome of individual innovations and spread slowly via the interactions of social groups. However, it is only over the last two centuries that nationalised systems of energy provision have come into being and governments have assumed a larger role in steering energy transitions through public policy and programmes. In some countries, this has been achieved largely through centralised planning. In other instances, the government has functioned more as a facilitator and/or regulator in providing the conditions in which private or local agencies or cooperatives could operate. In both cases, results were mixed. The following subsections highlight issues that may have been responsible for some of the failures and attempt to shed light on some of the critical success factors. While India is the backdrop for most of the illustrations, other countries are featured in some instances and comparisons are drawn.

5.2.1 Electrification

Electricity was first commercially supplied to the public in the mid-19th century in the United Kingdom, after which it spread quickly throughout Europe and the United States (Smil, 2005). As ex-colonial states attained their independence in the 20th century, providing electricity access to the population was considered a prerequisite to modernisation and progress and was therefore accorded priority by the governments of these countries early on in their development. There was a great deal of political and social pressure to expand electrification in these nations but the financial resources for doing so were limited. Historically, the responsibility for increasing the electricity infrastructure fell to centralised state-owned utilities. The motivation for the electrification programmes was often highly specific to each country but, in many cases, the focus was often one-dimensional. In the case of India, for instance, early electrification efforts were motivated by a desire to provide cheap power to farmers for irrigation pumping. As a consequence, electrification was implemented in the fertile wheat-growing states but the poorest communities were often passed over. The extremely low tariff structures for farmers and households that were put in place then had consequences for the future profitability of the state-owned electricity utilities (Dubash & Rajan, 2001). While the green revolution, which was brought about in part by intensification of irrigation, increased food security and brought benefits to many people in India, there were relatively few benefits of electrification. Even after more than 60 years of Indian independence, rural electrification has not directly benefited the majority of the rural population. Even among those who do have access, the quality of access remains inadequate. Beyond lighting, radios and basic home-appliances, the applications of electricity to activities that might bring economic development to an area are slow to emerge without the presence of institutional mechanisms that are conducive to fostering entrepreneurial activity and the productive use of electricity. In addition, the recent trend towards a more competitive and deregulated electricity sector is forcing utilities to reassess the benefits and costs of many of their internal business practices and to increase their efficiency. This development is, in some ways, at odds with electricity supply to rural and remote areas because of these areas' generally unfavourable contribution to the financial performance of the utilities.

South Africa is often held up as a remarkable success. Prior to 1990, less than one-third of the population had access to electricity. By the end of the decade, that proportion had doubled. However, South Africa holds a unique position in the history of electrification. As summarised by Bekker, Eberhard, Gaunt, and Marquard (2008), it is primarily the nature of the political transition in the country that made it unique. Secondly, the presence of a strong national utility largely spared South Africa from the kinds of problems usually faced by developing country utilities. Thirdly, a dominant industrial load base helped to absorb various forms of cross-subsidies when the programme was initiated. However, the programme was not without its problems and questions have arisen recently as to whether the universal access obligation targets for 2013 will be met (Gaunt, 2005). In addition, the availability of relatively cheap electricity has resulted in an excessively high use of electricity for thermal applications in South African households. This, in turn, has had implications for managing peak power demand, especially after the unprecedented power shortages experienced in parts of the country during late 2005 and early 2006.

China is another major success story, with over a billion Chinese provided with access to electricity over a period of roughly 50 years (Barnes, 2007). In contrast with the case of South Africa, the success of the Chinese programme can be partly attributed to the development of local and regional power companies, which were allowed considerable autonomy but were supported through state investments. The development of decentralised power systems was another key factor.

Although a wide variety of institutional modes were followed in various successful national electrification efforts, they all have certain common factors. In general, one of the important lessons from successful electrification programmes and efforts is the importance of emphasising multiple, especially productive, uses of electricity at an early stage, whether in agriculture for irrigation and mechanisation or in industry (Barnes, 2007). Another common factor is close coordination between different levels of government and private or cooperative implementing agencies, as well as a strong awareness of the development needs and priorities of the benefiting communities.

Problems of Definitions of Electrification

Progress in terms of expanding the extent of electrification depends heavily on the official definitions adopted and the measurement units used. Foley (1990) stated that the definition of rural electrification in particular varies considerably across countries. In one country, *rural* may include provincial towns with a population up to 50,000, while in another it may refer to small farming villages and surrounding areas. The initial focus of Indian electrification was on the loosely defined term, '*village electrification*'. As a consequence, villages were often deemed to have been 'electrified' without a single household having been connected. Changes in the definition of 'electrified village' also impacted the measurement of electrification over time. Until recently, a village was defined as being electrified as long as electricity was used for any purpose within the village boundary (even for powering a single public street light). The official definition was recently made more stringent; now at least 10 percent of households within a village must have access to electricity for it to be classified as an electrified village (Bhattacharyya, 2006). In addition to village electrification, access in India has also been measured, depending on the agency, in terms of the number of electrified households, the electrified population or the number of domestic connections (Fig. 5.2). Of course, even when households gain access to electricity, unreliable power supplies can impede the success of electrification. Ideally, some element of the quality of coverage and supply should be included in any measurement of access but, in practice, there is little data with which to assess this.

In Cambodia, unlike India, the government is not heavily involved in rural electrification and official electrification rates are defined only in terms of connections to a central grid, and these remain very low (15 percent). However, according to Zerriffi (2007), the number of households with access to at least a minimal amount of electricity (for example, enough to power a light bulb and maybe a small television) is

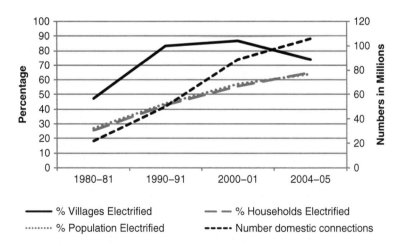

Fig. 5.2 Different measures of electricity access in India

extremely high (Fifty percent of the households have a television and an estimated 85–90 percent have at least one light bulb). Their electricity comes primarily from rural electricity entrepreneurs, who run diesel-based micro-grids, battery charging stations or a combination of the two.

Purposes and Justifications for Electrification

Electricity has the potential to completely revolutionise people's lives and the changes that accompany access to electricity are vast and varied. Most institutions and agencies that support electrification projects have lofty goals that aim to improve lives in the communities they serve, although a full and critical evaluation of these projects is rarely carried out. In many instances, even when such evaluations are undertaken, they often focus on single aspects such as technical efficiency, economic profitability or job creation and on static or immediate benefits. The evaluations rarely take account of how these projects truly affect the lives of citizens in the communities they serve or their full social and societal impact. A fundamental problem with measuring outcomes of electrification programmes is the availability of data. Longitudinal studies with a clear baseline before electrification that also include the full range of energy forms are scant. Consequently, the undertaken assessments often fail to take into account the complexities of the effect that electrification has on low-income households and communities.

In many instances, a complete evaluation of electrification projects is not undertaken because those involved in the process of electrification in developing countries do not clearly distinguish between the economic, social and other objectives of these projects. These objectives often change over time as well. During the 1990s, for instance, with the new drive towards liberalising the energy markets, electricity came to be viewed more as a commodity than a public service. Assessments made solely on commercial or economic bases branded most programmes that were targeted at rural or poor communities as not viable. More recently, the emphasis on achieving the Millennium Development Goals (MDGs) has re-oriented the emphasis towards the social and poverty alleviation benefits of electrification. However, assessments of these benefits remain limited by undeveloped methodologies and scarce data. It is often difficult to isolate the specific impact of electrification within a development process and to certify a causal relationship. Some recent efforts at undertaking such assessments include work by the World Bank for the Philippines (2002) and by Davis, Borchers, and Eberhard (1995) for Botswana.

Immediate applications of electricity in newly electrified households include lighting, appliances, communications and entertainment. Community needs include public/street lighting, refrigeration, health centres and schools, piped water and communication. Electrification also benefits productive enterprises and agricultural activities such as irrigation/water pumping. As mentioned above, the justification for electrification is often rather limited in focus. This has had implications for how the benefits of electrification are distributed and what criteria are used for selecting the areas and the order in which to electrify them.

5.2.2 Fuel Switching and Fuel Stacking

The previous section dealt with some of the challenges of electrification and the varying experience with it in different countries. The focus on electricity so far is not meant to imply that household energy transitions can be equated to electrification. Thermal energy needs are critical to householders' survival and well being and the transitions in fuels are often as, if not more, important than the provision of electricity. While most assessments of the drivers of fuel transitions have focused exclusively on economic factors, a number of geographic, demographic and cultural factors are linked to patterns of energy use and help explain the shift to modern energy sources. In rural areas, in particular, non-monetary factors are often more important in explaining householders' fuel choices. In addition to being viewed as the result of single explanatory factors (income alone or urbanisation), the transitions are often also seen as abrupt changes, rather than gradual shifts brought about by the interplay of multiple drivers.

In the past, the concept of an 'energy ladder' was used to explain how households selected fuels and energy technologies (Leach, 1992; Barnes & Floor, 1996; Smith, Apte, Yuqing, Wongsekiarttirat, & Kulkarni, 1994). In other words, as incomes rise, households metaphorically ascend the ladder as modern energy carriers are preferred for their high levels of efficiency, cleanliness and convenience of storage and use relative to crop residues, dung, firewood and other traditional solid fuels. However, more recently empirical studies have suggested that the energy transition does not take place as a series of simple discrete steps. Multiple fuel use arises for several reasons. Firstly, households often have significant capital invested in 'traditional' technologies (such as wood-burning stoves) and may not have the spare capital to purchase new energy-consuming appliances immediately upon gaining access to new energy sources (Saghir, 2004). Secondly, modern energy sources are usually expensive and are therefore used sparingly and for unique services. Consequently, traditional fuels and technologies tend to exit more slowly than new ones arrive. Finally, multiple fuels can provide a sense of energy security. Complete dependence on commercially-traded fuels leaves households vulnerable to variable prices and often unreliable service. In some instances, the traditional concept of ascending an 'energy ladder' is being turned on its head, with some countries considering the potential for switching back to biomass but with modern, more efficient and less polluting biomass sources (such as biogas, pellets, etc.) and more efficient biomass-burning technologies (improved stoves), rather than fossil-based fuels.

As is the case with electrification, some countries and regions have seen a significant shift in the choice of cooking fuel and/or technology. In Brazil, for example, less than 20 percent of households had access to LPG in 1960 but by 2002 this number had jumped to over 98 percent. Even in rural areas, LPG reached 93 percent of households by 2002 (Jannuzzi & Sanga, 2004). However, the rapid shift in this case took place as a consequence of a sustained and strong government policy support to the LPG sector and high subsidies. The subsidy was removed in 2002, resulting in a 20 percent increase in the price. In reaction, poor households shifted back to using firewood as they could no longer afford LPG. Recent efforts to make LPG subsidies more favourable to the poor have been initiated but the impacts this has had in terms of reversing this trend are not clear yet. In spite of these setbacks, however, the sustained and strong government commitment and support in Brazil has resulted in a significant shift in the fuel use patterns of Brazilian households.

Importance of Cooking Energy in Developing Countries and Main Fuels Used

Hierarchies in household energy services are quite common. Cooking and heating are almost always the first functions fulfilled, followed by lighting and later entertainment. For the poorest people in developing countries, cooking (and space heating in particularly cold climates) can account for upwards of 90 percent of the total volume of energy consumed (WEC & FAO, 1999). Cooking is an energy service that is often associated with strong and highly specific fuel and appliance preferences. In addition, cooking is often only one of a range of services that are delivered from a stove or a fire. For example, coal and wood stoves serve multiple functions, including cooking, space heating, water heating, lighting and social focus (Van Horen, Eberhard, Trollip, & Thorne, 1993). This multi-functionality of some stoves is one important reason why households are sometimes averse to substituting their old stoves for newer more efficient and cleaner technologies.

Households use a variety of fuels for cooking purposes, including wood, dung, crop wastes, charcoal, kerosene, LPG, coal and electricity. Fuels are chosen for their cost effectiveness, ease of access and perceived efficacy in performing specific tasks, and fuel use patterns may differ at different times of the year. Tastes and cultural preferences also play an important role in decisions concerning which fuels to use in the household.

Cultural and Traditional Importance of Cooking Practices and Gender Roles

A considerable amount of research has stressed the importance of cooking practices and gender roles. The cooking of certain dishes, often staple food, is closely linked to traditional stoves in use. Not unlike bread baked on wood stoves in industrialised countries, the Indian 'daily bread' is difficult to make on modern stoves, as open fires give the traditional breads their taste. This may lead a household to keep a traditional stove for these particular foods, resulting in two types of stoves in the kitchen. Although such cultural preferences often have a strong influence on the choice of cooking fuels and technologies, they are often overlooked.

The importance of gender issues in transitions in the use of cooking fuels and/or stoves cannot be overemphasised. Women, who generally do the cooking in most households, often do not have a say in how they are to cook. This is despite the fact that they spend more time collecting biomass than men and, in most countries, cooking activities are considered an all-female domain.

This underscores the role of power relations in energy access, and the household is no exception. Women tend to have limited control over and access to productive assets and income and, therefore, little say in how much can be spent on fuels or new stoves. While not all women would embrace new fuels and stoves if they were given the choice, a proposed change often has to benefit the man in the house for it to have a chance of being adopted (FAO, 2006).

5.2.3 Indoor Air Pollution and Other Hazards of Traditional Fuel Gathering and Cooking

The World Health Organization's 2004 Global and Regional Burden of Disease Report estimated that acute respiratory infections from indoor air pollution (pollution from burning wood, animal dung, other biomass fuels and coal) kills a million children annually in developing countries, inflicting a particularly heavy toll on poor families in South Asia and Africa. Such findings have prompted renewed policies, programmes and projects aimed at facilitating a quicker transition away from such 'dirty' fuels. Even when adopted with the best intentions, the single-minded focus of such policies to 'save' populations from the ill health effects of indoor air pollution have often resulted in programmes being conceived and implemented without regard to their context. This has often set them up for eventual failure (for example, the national improved stoves programme in India, see Barnes & Kumar, 2002). In particular, such a single-minded focus may be misguided if the target population themselves are not aware of the ill effects and do not consider the amelioration of indoor air pollution to be a priority. Educational programmes and campaigns are essential for building such awareness and may be a more effective means for achieving the end in this instance.

A number of health and safety issues for householders relate to cooking, most notably indoor air pollution, the risk of fires and burns, injuries associated with wood collection, neck and back injuries from carrying heavy loads and the risk of attack and rape when collecting fuel wood. Cooking-related health and safety impacts mainly affect women and children since they have the highest exposure to smoke and the other risks connected to cooking with polluting fuels.

Health Implications Associated with Indoor Air Pollution

There is compelling evidence linking indoor smoke to acute respiratory infections in children and chronic obstructive pulmonary disease or chronic bronchitis in women. However, evidence linking indoor smoke to asthma, tuberculosis, lung cancer and adverse pregnancy outcomes is limited and sometimes conflicting. Furthermore, evidence linking indoor smoke to cataract and blindness, otitis media, lung fibrosis and cardiovascular disease is weak or nonexistent (Mishra, 2004). Both the exposure to smoke and the health effects of such exposure depend not only on social, behavioural and environmental factors but also on genetic and other biological factors. In general, though, the evidence is clear and compelling that pollution from burning traditional biomass fuels in inefficient devices causes inflammation and an impaired immune response among women and children who are regularly exposed.[2]

Other Health, Environmental, and Socio-Economic Effects

The time and effort expended in gathering biomass and carrying heavy loads has also been associated with back pain and headaches (Parikh, Sharma, & Singh, 2008). Other risks associated with collecting biomass have also been reported, such as snake bites, violence and other injuries (Wickramasinghe, 2003). The drudgery of the task also leaves little time for women to engage in other productive activities that could earn them an income or secure them an education. Women's involvement in several household, agricultural and other informal sector activities and chores is often overlooked in formal analyses because of the minimal monetary remuneration and the absence of economic valuation of them in the national accounts. Thus, any estimation of the opportunity costs of time spent gathering and using fuels, which are often based on women's wage rates, underestimates these costs.

In addition to the many health and socio-economic impacts of using traditional biomass and other solid fuels for cooking and heating, their use also produces significant quantities of 'products of incomplete combustion' (PIC), especially fine and ultra-fine particles that have a higher global warming potential (GWP) than carbon dioxide (CO_2) (Smith et al., 2000). Bhattacharya, Salam, and Sharma (2000) reported that incomplete combustion of biomass in traditional cooking stoves releases carbon monoxide (CO), nitrous oxide (N_2O), methane (CH_4), polycyclic aromatic hydrocarbons (PAHs), particles composed of elemental carbon or black carbon and other organic compounds. When the biomass burned is not sustainably harvested, the use of these fuels has the added disadvantage of no longer being CO_2-neutral. In addition, recent evidence suggests that the warming effects of black carbon emissions, particularly for arctic and glacial ice, are larger than previous estimates suggest (Ramanathan & Carmichael, 2008).

5.3 Transdisciplinary Aspects: Suitable Statistics and Thoughts on the Nature of Poverty

5.3.1 Including Non-commercial Energy and People Without Monetary Incomes

In past years, India's energy consumption increased more than predicted. While a larger than expected rate of economic growth is certainly one reason for this, there are other reasons that could contribute to this difference. This chapter argues that

---

[2] For evidence linking the use of traditional solid fuels with other child and adult health outcomes, see Smith, Mehta, and Feuz (2004).

two normative effects – that is, effects at level 3 in Max-Neef's matrix of transdisciplinarity (Fig. 5.1) – that are specific to developing countries and have been accorded too little attention, can explain differences in predicted and actual energy consumption. In addition to the effects explained in energy models for industrialised countries, there are two additional factors to consider in developing countries:

- People who are new to the monetary economy
- The substitution of non-commercial with commercial energy.

These two effects cause a higher growth in consumption of commercial energy than would be predicted by models that do not take them into consideration.

Figure 5.3 attempts to illustrate these effects graphically. On the x-axis, household income or expenditure is plotted as increasing towards the right. For simplicity, it is assumed that household income is uniformly distributed from zero to a maximum value. Towards the left, 'distance from the market economy' is plotted. This lends the graph a degree of novelty and recognises the existence of those people who do not participate in the monetary economy. How exactly 'distance from the market economy' should be measured is left open. However, one approach could

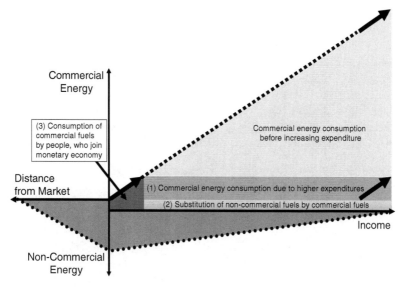

Fig. 5.3 Concept for a model. Household energy consumption is assumed to be linearly dependant on household expenditure. An exogenous rightward shift in all households' expenditure (*three bold arrows*) causes three additional types of commercial energy consumption: (1) The conventional increase of energy consumption due to higher expenditures, (2) a substitution of non-commercial energy by commercial energy and (3) consumption of commercial fuels by people who have newly joined the monetary economy

be to measure it in the number of years that are typically necessary for a household with specific levels of socio-economic indicators (education, health, access to food and infrastructure) to become participants in the market economy.

The y-axis in the positive quadrant plots the commercial energy consumption of income groups or groups with a comparable degree of involvement in the market economy. Consumption of non-commercial energy is plotted below the x-axis. While the dependence of (commercial) energy consumption on income is simply and highly systematically depicted here, countless studies have confirmed this extremely strong correlation. In contrast, the schematic depiction of the dependence of non-commercial energy consumption on household income (x-axis) is a rough approximation based on the results of previous research on India (Pachauri, Mueller, Kemmler, & Spreng, 2004).

Figure 5.3 shows what happens to energy consumption when income increases. For simplicity, a uniform increase in expenditure across all household types is assumed. The large triangle (blank in the figure) depicts the consumption of commercial energy at the initial position; that is, before the exogenous shift. If income increases (assuming everyone's income is uniformly increased by 15 percent to a level 115 percent of the maximum value at the initial position), this can be represented by a shift of the vertically lined triangle to the right and upwards. In industrialised countries, the result would be an increase in energy consumption corresponding to area (1). In developing countries, however, the additional areas 1 and 2 also enter the picture. The use of non-commercial energy declines, both in its horizontal extent towards the left (reduction of the number of people greatly distanced from the market economy – that is, people who are poor), as well as in the vertical extent below (substitution of non-commercial energy by commercial energy). The considerable size of these two effects (areas depicted by (2) and (3)) can be derived from the statistical analysis of extensive 'household consumption surveys'. A study using data for India (Pachauri et al., 2004) showed that only about 20 percent of the population cooked with LPG in 1999, indicating a large need to catch up. This has been partly achieved in the years since. A considerable proportion of the population who only use biomass and kerosene have no or little monetary income. They get by through self-support, bartering and begging. In 1999, 150 million people, or 46 percent of the population who used only biomass and kerosene, lived below the monetarily-defined poverty line. With time, these people will become participants in the market economy and will make additional use of commercial energy carriers. As far as can be ascertained, there are no energy consumption prognoses that challenge these normative determinations regarding energy statistics. It is conceivable that the neglect of these two transitions – from non-commercial to commercial energy use as well as the transformation of people from outside the market economy into active participants in it – may be a reason why recent energy consumption forecasts for India have been lower than the actual consumption.

5.3.2 Transdisciplinary[3] Views on Poverty

On Measuring Poverty

Poverty has traditionally been assessed solely in monetary terms, based on standard income or expenditure/consumption-based indices such as the index of an income of USD1 per day. Over the years, however, poverty has evolved into a multidimensional concept in order to overcome some of the deficiencies of one-dimensional indicators and to enrich the information set and take a plurality of dimensions of well-being into account. The economics literature also generally agrees that poverty is associated with deprivation, which can be seen in terms of constraints on people's choices that limit their ability to access certain material goods, assets, capabilities, freedoms and opportunities.

Amartya Sen devised the concept of 'capabilities' and 'functionings' to reflect two dimensions of the various things a person may value as being necessary for achieving a good life (Sen, 1993). People's capabilities are what they are able to do and become, whereas their functionings are what they actually are and do. Sen emphasised that income is only valuable insofar as it increases the 'capabilities' of individuals and thereby permits 'functioning' in society. Material resources are necessary but not sufficient. Capability poverty focuses on an individual's capacity to live a healthy life, free of avoidable morbidity, having adequate nourishment, being informed and knowledgeable, having a family, enjoying personal security and being able to participate freely and actively in society.

It has not been easy to apply Sen's concept quantitatively to observable situations. Using energy instead of money throughput as a metric may be more amenable to operationalising Sen's approach. A two-dimensional quantitative view of a household's poverty level could also be attained by including not only the question of how much energy it uses ('functionings') but also the energy to which the household has access ('capabilities').

While energy itself is not generally recognised as one of the basic needs, it is clearly vitally necessary for the delivery and provision of basic needs such as food, clean water, shelter and health and educational services (WEC & FAO, 1999). Despite this, there is no consensus on the amount of energy needed to meet basic human needs. The issue is further complicated by the fact that there is no universally accepted set of minimum basic needs. Furthermore, needs vary significantly across countries and regions depending on weather, social customs and a number of other region- and society-specific factors.

The various levels in the energy supply chain[4] are all important in energy analysis but each has a different explanatory power. For poverty assessment, the level of

---

[3] Specifically, in Max-Neef's sense.

[4] Energy consumption can be measured at various levels of the energy supply chain: primary energy, the energy embodied in extracted natural resources (e.g., coal, crude oil, sunlight, uranium); final energy, the energy content of energy supplied to the consumer at the point of end-use (e.g., electricity at the electricity meter); useful energy, the energy that is transformed into the form required for actual use (e.g., the generated heat from a hot plate or the mechanical energy

energy services appears to be the most appropriate. The problem for analysts is that the so-called energy services cannot be measured in energy units since they require many other inputs in addition to energy carriers. Accordingly, there is no way of distinguishing energy services from other services and products. A promising approximation is to measure consumption at the level of useful energy.[5]

The importance of traditional non-commercial sources of energy for meeting household energy needs has been emphasised above. Including non-commercial energy for poverty measurement has the welcome effect of including the consumption of non-monetised goods. A well-off autarchic household might be a large consumer of non-commercial energy. A massive shift at the macro-level away from a barter economy and self sufficiency, with only a small increase in the market economy, will not be mistaken for an increase in well being.

The following energy-access matrix (Fig. 5.4) is a newly updated version of the matrix published in Pachauri et al. (2004).

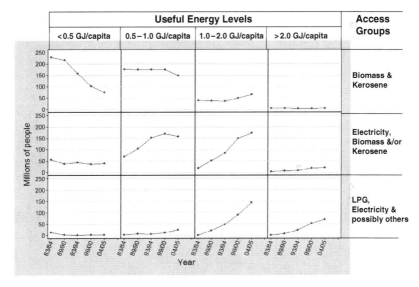

Fig. 5.4 The Consumption-Access matrix furnishes a two-dimensional view of poverty
Source: Pachauri and Spreng (2011)

---

applied to air for air circulation); and energy services, the direct energy-related service demand by households (a cooked meal, a well-lit room, a hot shower, transportation from A to B).

[5] A note of caution must be added here. Analysing useful energy makes a lot of sense when one compares the various ways of providing one kind of useful energy for one specific energy service. However, adding up useful energy employed for different energy services is more problematic. For instance, 'heat supplied to a cooking pot' and 'the energy of the light coming from a light bulb' are two energy flows of very different physical quantity, form and utility, which are produced at different costs (technically, in terms of natural resources and economically). Summing these often produces irrelevant results. However, in the case of measuring energy poverty, cooking is such a dominant energy service that the applicability of useful energy in summing it is not overextended here.

This two-dimensional view of poverty is a significant step toward a somewhat more comprehensive view of poverty. While we cannot be sure that knowledge of a household's energy access and consumption provides a clear, unambiguous indication of the living standard it enjoys, careful statistical analysis (Pachauri et al., 2004; Kemmler & Spreng, 2007) shows that energy consumption also correlates with other indicators (including education or infrastructure variables such as running water or toilet type), as does expenditure, while energy access is an excellent indicator of a household's integration into modern society. Each of the 12 matrix elements contains a wide distribution of households but it is a grouping as good as any other and better than most for dividing households into groups with similar living conditions. In any case, it affords a much more refined view than using a USD1-a-day measure.

A Max-Neef-transdisciplinary understanding of poverty can be pursued much further. For example, additional perspectives can be gained from the two unlikely disciplines of linguistics and history.

Linguistic Examination of Poverty and the Poor

A linguistic examination of the use of the words *poor* and *poverty* reveals many ways of looking at poverty, especially at the divide between poor and rich and the transition from one to the other (Linke, 2007). Linke's study was conducted in German, focusing on the words *arm* and *Armut*, and it does not refer to developing countries. However, its findings warrant a short summary in the present context. The typical economic perspective of a continuum from poor to rich is not common in the thousands of newspaper texts analysed in the course of the study. Much more often, the poor and rich are seen on opposite sides of a deep divide and the poor need to undergo a distinct transition in order to cross the divide.

The following results are of particular interest:

- The word *poverty* is only used by well-off people.
- Poverty, and particularly the word *poor*, are used to express two different things: A low state of well-being and a state that should be pitied.
- The emotional response towards poverty is complex, if not contradictory. Poverty calls for pity, understanding and compassion but it is also something that has to be fought against. The word *poverty* is often used in a string of expressions containing words like *war, epidemic, natural catastrophe*, which have to be eradicated or forestalled.
- Poverty is talked about with reference to the state of a group or a mass of people, whereas richness is usually attributed to (male) individuals.

This linguistic analysis shows that poverty is most often spoken about as a phenomenon separate from the observer, the commentator, and tends to be seen from the outside. Poor people may see themselves as lacking in food or financial wherewithal but they would not see themselves as poor. Their neighbours are unlikely to be much richer. This linguistic observation lends powerful support to the idea that a one-dimensional analysis of 'poverty' may not capture a phenomenon adequately or hold much meaning for the group of people it describes.

History of the Relationship Between Rich and Poor

Historical analysis of the relationship between rich and poor in the West reveals dramatic changes over the centuries (Tanner, 2007). In the Middle Ages in Christian Europe, the rich and poor each had their own specific societal function. Suffering was considered an important aspect of human existence and was relegated to the poor. The rich were happy to give alms to the poor; this relieved them of the religious duty of suffering themselves. In the Late Middle Ages, the orders of the mendicants grew out of this attitude. From the 17th century onwards, a defining difference between rich and poor was that the poor had to work, whereas ownership was concentrated in the hands of the rich. This trend accelerated with the industrial revolution, when the masses of poor greatly enriched the owners of industries. These discontented masses, however, began to threaten social harmony.

This summary of developments in Europe points to the fact that societal relationships in history have changed drastically and may do so again. Masses of poor, suffering without reward, have not been a permanent feature of developing countries in the past and need not be in the future.

The final section of this chapter takes up the ethical issue of whose responsibility it is to right the inequity in humankind's energy access and consumption.

5.3.3 Lower and Upper Limits of CO_2 Emissions

Historically, and even now, national energy strategies and debates have been more efficiency-focused than equity-focused, which has caused financial considerations to overshadow development considerations. The Millennium Development Goals-based analysis and approach has the potential to induce policy dialogues that can (re-)prioritise energy issues at the country level, in line with national development goals and strategies. However, the target and output-based arguments and approaches (for example, how many gigawatts (GW) or how many systems) are already so strongly entrenched that a body of work that provides some guidance on how to be development outcome-oriented (i.e., focused on impacts on income, gender, education, health, macro-economy, environment, etc.) at the operational level has only recently begun to emerge.

The idea that some measure of inequality leads to social conflict is widely accepted, as is the concept that people should have equal opportunities. Slavery, the medieval class system and the Indian castes are systems that have been or are in the process of being abolished. However, the degree of inequality that is beneficial, or at least not harmful, is open to debate. While the 20th century was preoccupied with inequalities within nations, inequalities between nations in this century will likely be as important a topic of political debate.

Inequality is usually discussed in terms of income or some related monetary measure. This chapter argues that per capita energy consumption is a promising measure, particularly if it is agreed that this measure has a natural upper limit imposed by a maximum global aggregate CO_2 emission limit. The existence of a lower poverty limit is accepted both for monetary measures and for per capita energy consumption. However, the upper limit to the average per capita energy consumption does not exist for monetary measures and must be imposed according to alternative, ecological, criteria.

The developments described in some of the stabilisation scenarios of IPCC's Special Report on Energy Scenarios include not only the stabilisation of CO_2 output but also minor economic improvements for developing countries. However, in order for development to be sustainable and free of large social conflicts, the development path must also further a reduced spread (variance) in energy consumption by 2050. The continued large CO_2 emissions by rich countries rob the poorer countries of the possibility of rapid development and the use of the common, cheap CO_2-sink. Figure 5.5 depicts the inequality in today's energy use (an older version of this graph was featured in an undergraduate thesis (Schmieder & Taormina, 2001) and reported in Spreng, 2005).

The graph shows the per capita energy consumption of 73 countries, wherever possible for the year 2001 (IEA, 2007). The novelty here is the inclusion of the estimated energy consumption per capita of the richest 10 percent and the poorest 10 percent of the population (decile)[6] for each country (CIA, 2003; The World

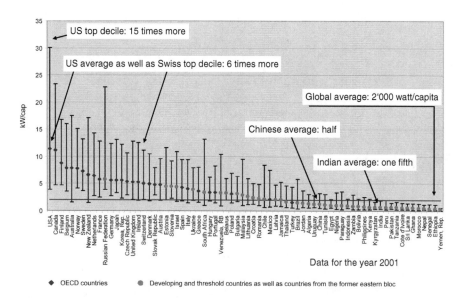

Fig. 5.5 Energy consumption per capita – inequities between and within countries

---

[6] To calculate the energy consumption of the top and bottom decile, the total energy elasticity of income for all countries is assumed to be the same, namely 0.8 (a rounded average number taken from the literature). It has been studied in detail for the Netherlands by Vringer and Blok

Bank Group, 2003). Furthermore, a 2000 W/person line is drawn, as this figure is close to the current world average and can be considered on average compatible with sustainable development in the long-term. With these two additions, the graph presents not only international inequalities but also, at one glance, intra-national inequalities.

The graph shows a troubling situation regarding inequality in present energy consumption, both within and across countries. It is noteworthy that consumption by the richest decile lies under the 2 kW per capita threshold in only 10 countries. Homogenisation of the lifestyles of the global upper classes seems to have already taken place.

International treaties on CO_2 emissions will have to take the lower limit of CO_2 emissions per capita, implied by the Millennium Development Goals, into greater consideration. What should be the guiding principle? Perhaps it would be of some use to refer back to a philosopher who has written about intra- and international inequity. Rawls (1971) was not in favour of uniform social equity across society and did not contest the theory that a rising tide lifts all boats. In *A Theory of Justice*, however, he argued that the way to correctly judge the distribution of goods (material and otherwise) in a society is to look at its most disadvantaged members. If the lot of the poor improves in a nation over time, the distribution of wealth becomes more just.

Can the same criterion be applied in comparing poverty across countries? Rawls (1997) discussed this question in his last book, *The Law of Peoples*, which is of particular interest for the assessment of international inequalities and inequities. Rawls reached the conclusion that there is no similar equity criterion between peoples as within one people. Everyone bears some responsibility for fairness within his or her society and should collectively work to ease the plight of the poor in our society. However, other nations have different traditions and rules. One cannot and may

---

(1995), whose value was 0.83. The corresponding number for Australia is 0.73 (Lenzen, 1998). A value of 0.67 has been reported for urban households in India (Pachauri, 2004) and 1.01 for urban households in Brazil (Lenzen et al., 2006). These numbers refer to the dependence of the total commercial energy (both direct and indirect) on income at a given time within one country.

Figure 5.5 also used 0.8 for the world as a whole, which is as good an assumption as any. Podobnik's (2002) calculation addresses only the inequalities between country averages.

The spread of energy consumption between the top and the bottom decile, $\Delta E = E_{\text{top}} - E_{\text{bottom}}$, is calculated as follows from knowledge of the averages for the per capita income and for the per capita energy consumption, I_{average} and E_{average} and the income of the top and the bottom decile, I_{top} and I_{bottom}:

$$E_{\text{top}} = E_{\text{average}} + \left(\Delta I_{\text{top}} E_{\text{average}} / I_{\text{average}} 0.8 \right),$$
$$E_{\text{bottom}} = E_{\text{average}} - \left(\Delta I_{\text{bottom}} E_{\text{average}} / I_{\text{average}} 0.8 \right),$$

$$\text{where } \Delta I_{\text{top}} = I_{\text{top}} - I_{\text{average}} \text{ and } \Delta I_{\text{bottom}} = I_{\text{average}} - I_{\text{bottom}}.$$

In some cases, the net import of embodied energy in goods and services renders a per capita comparison with other countries less meaningful. For Switzerland, the per capita energy consumption has, for that reason, been augmented by 25 percent to compensate for the large net import of embodied energy. Luxembourg and Trinidad have been eliminated from the list because data on the large exports of embodied energy are unknown to the authors.

not be responsible for how other nations distribute their wealth.[7] However, Rawls asserts that nations do bear responsibility for international treaties, conventions and dealings that impact wealth distribution internationally. If these lead to injustice and contribute to perpetuating poverty in less developed countries, at a minimum we must work to change them.

This excursion into ethics is eminently important for completing a Max-Neef-transdisciplinary view of energy transitions. Whose responsibility is it to promote these transitions and, with them, a more equal distribution of wealth within developing countries? It has been taken for granted that development aid is the responsibility of donor countries and donor organisations. Perhaps there is a basic flaw in this widespread modern view. Perhaps development aid is more a band-aid for the damage of colonialism and post-colonial exploitation. Perhaps developing countries need nothing else but fair conditions for international trade and fair burden-sharing in international agreements on climate change mitigation and adaptation, without interference in the regulation of their internal affairs.

References

Barnes, D. (Ed.). (2007). *The challenge of rural electrification: Strategies for developing countries.* RFF Press. London: Earthscan.

Barnes, D., & Floor, W. (1996). Rural energy in developing countries: A challenge for economic development. *Annual Review of Energy and the Environment, 21,* 497–530.

Barnes, D., & Kumar, P. (2002). Success factors in improved stoves programmes: Lessons from six states in India. *Journal of Environmental Studies and Policy, 5*(2), 99–112.

Bekker, B., Eberhard, A., Gaunt, T., & Marquard, A. (2008). South Africa's rapid electrification programme: Policy, institutional, planning, financing and technical innovations. *Energy Policy, 36*(8), 3125–3137.

Bhattacharyya, S. C. (2006, June). Universal electrification: Will the new electrification programme succeed in India? *OPEC Review, 30*(2), 105–123.

Bhattacharya, S. C., Salam, P. A., & Sharma, M. (2000). Emissions from biomass energy use in some selected Asian countries. *Energy, 25*(2), 169–188.

Burckhardt, J. (1889, ed. 1980). *Letter of 24 July 1889 to Friedrich von Preen.* In M. Burckhardt (Ed., 1980), *Briefe. Vol. 9* (p. 203). Schwabe: Basel/Stuttgart.

Cherni, J. A., & Preston, F. (2007). Rural electrification under liberal reforms: The case of Peru. *Journal of Cleaner Production, 15*(2), 143–152.

CIA, Central Intelligence Agency. (2003). The world fact book. GPO041-015-00231-7. Washington, DC: CIA. Accessed November 16, 2011, https://www.cia.gov/library/publications/the-world-factbook/index.html

Collier, P. (2007). The bottom billion: Why the poorest countries are failing and what can be done about it. New York: Oxford University Press.

Davis, M., Borchers, M., & Eberhard, A. (1995). Rural electrification: A case study in northern Botswana. *Journal of Energy in Southern Africa, 6*(4), 188–194.

Dubash, N., & Rajan, S. (2001, September). Power politics: The process of power sector reform in India. *Economic and Political Weekly.*

---

[7] Rawls was not writing about failed states and post-conflict situations. In these cases and situations, the international community does have obligations directly towards the involved people. Unfortunately (see Collier, 2007), many people face such situations.

FAO, Food and Agriculture Organization of the United Nations. (2006). *Energy and gender in rural sustainable development*. Rome: FAO.

Foley, G. (1990). *Electricity for rural people*. Panos Rural Electrification Programme. London: Panos Institute.

Gangopadhyay, S., Ramaswami, B., & Wadhwa, W. (2005). Reducing subsidies on household fuels in India: How will it affect the poor? *Energy Policy, 33*, 2326–2336.

Gaunt, C. T. (2005). Meeting electrification's social objectives in South Africa, and implications for developing countries. *Energy Policy, 33*(10), 1309–1317.

IEA, International Energy Agency. (2007). *World Energy Outlook 2007*. Paris: IEA.

Jannuzzi, G. M., & Sanga, G. A. (2004). LPG subsidies in Brazil: An estimate. *Energy for Sustainable Development, VIII*(3), 127–129.

Kemmler, A. (2007). Factors influencing household access to electricity in India. *Energy for Sustainable Development, 11*(4), 13-20.

Kemmler, A., & Spreng, D. (2007). Energy indicators for tracking sustainability in developing countries. *Energy Policy, 35*, 2466–2480.

Leach, G. (1992). The energy transition. *Energy Policy, 20*(2), 116–123.

Lenzen, M. (1998). Energy and greenhouse cost of living for Australia during 1993/94. *Energy, 23*(6), 497–516.

Lenzen, M., Wier, M., Cohen, C., Hayami, H., Pachauri, S., & Schaeffer, R. (2006). A comparative multivariate analysis of energy requirements of households in Australia, Brazil, Denmark, India and Japan. *Energy, 31*(2–3), 181–207.

Linke, A. (2007). Wer ist arm? Soziale Kategorisierung im Medium Sprache . In U. Renz & B. Barbara (Eds.), *Zu wenig Dimensionen der Armut*. Zürich: Seismo.

Max-Neef, M. A. (2005). Foundations of transdisciplinarity. *Ecological Economics, 53*, 5–16.

Mishra, V. (2004). *What do we know about health effects of smoke from solid fuel combustion?* East-West Center Working Papers, Population and Health Series, 117. Honolulu: East-West Center.

Pachauri, S. (2004). An analysis of cross-sectional variations in total household energy requirements in India using micro survey data. *Energy Policy, 32*(15), 1723–1735.

Pachauri, S., Mueller, A., Kemmler, A., & Spreng, D. (2004). On measuring energy poverty in Indian households. *World Development, 32*(12), 2083–2104.

Pachauri, S., & Spreng, D. (2011). Measuring and monitoring energy poverty. *Energy Policy, 39*(12), 7497–7504.

Parikh, J., Sharma, S., & Singh, C. (2008, June). *Energy access and its implication for women: A case study of Himachal Pradesh*. (Paper presented at workshop on Clean Cooking Fuels and Technologies, Istanbul, Turkey).

Podobnik, B. (2002). Global energy inequalities: Exploring the long-term implications. *Journal of World-Systems Research, 8*(2), 252–274.

Ramanathan, V., & Carmichael, G. (2008). Global and regional climate changes due to black carbon. *Nature Geoscience, 1*, 221–227.

Rawls, J. (1971). *A theory of justice*. Cambridge, MA: Belknap Press of Harvard University.

Rawls, J. (1997). *The law of peoples*. Cambridge, MA: Belknap Press of Harvard University.

Saghir, J. (2004). *Energy and poverty*. Paper prepared for the International Energy Forum. Washington, DC: The World Bank.

Schmieder, B., & Taormina, N. (2001). *Energieverbrauchsfenster*. Diplomarbeit ETH Zurich, ORL-Institut and CEPE. Zurich: ETH.

Sen, A. (1993). *Capability and well-being*. In M. C. Nussbaum & A. Sen (Eds.), *The quality of life*. Oxford: Clarendon.

Smil, V. (2005). *Creating the twentieth century: Technical innovations of 1867–1914 and their lasting impact*. Oxford: Oxford University Press.

Smith, K. R., Apte, M. G., Yuqing, M., Wongsekiarttirat, W., & Kulkarni, A. (1994). Air pollution and the energy ladder in Asian cities. *Energy, 19*, 587–600.

Smith, K. R., Mehta, S., & Feuz, M. (2004). Chapter 18: Indoor smoke from household use of solid fuels. In M. Ezzati, A. D. Lopez, A. Rodgers, & C. J. L. Murray (Eds.), *Comparative quantification of health risks: The global burden of disease due to selected risk factors, Vol. 2* (pp. 1435–1493). Geneva: World Health Organization.

Smith, K. R., Uma, R., Kishore, V. V. N., Zhang, J., Joshi, V., & Khalil, M. A. K. (2000). Greenhouse implications of household stoves: An analysis for India. *Annual Review of Energy and the Environment, 25*, 741–763.

Spreng, D. (2005) Distribution of energy consumption and the 2000 W/capita target. Viewpoint. *Energy Policy, 33*, 1905–1911.

Tanner, J. (2007). Der Kampf gegen die Armut: Erfahrungen und Deutungen aus historischer Sicht. In U. Renz & B. Barbara (Eds.), *Zu wenig Dimensionen der Armut*. Zürich: Seismo.

Van Horen, C., Eberhard, A., Trollip, H., & Thorne, S. (1993). Energy, environment and urban poverty in South Africa. *Energy Policy, 21*(5), 623–639.

Vringer, K., & Blok, K. (1995). The direct and indirect energy requirements of households in the Netherlands. *Energy Policy, 23*(10), 893–910.

WEC & FAO, World Energy Council & Food and Agriculture Organization (October 1999). *The challenge for rural energy poverty in developing countries*. London: WEC.

Wickramasinghe, A. (2003). Gender and health issues in the biomass energy cycle: Impediments to sustainable development. *Energy for Sustainable Development, 7*(3), 51–61.

World Bank. (2002). *Rural electrification and development in the Philippines: Measuring the social and economic benefits*. Energy Sector Management Assistance Programme (ESMAP) Report. Washington, DC: The World Bank.

World Bank. (2003). *Household energy use in developing countries: A multi-country study*. ESMAP Report. Washington, DC: The World Bank.

Zerriffi, H. (2007). *Making small work: Business models for electrifying the world*. Program on Energy and Sustainable Development, Working Paper #63, Freeman Spogli Institute for International Studies. Stanford: Stanford University.

Chapter 6
A Socio-Cultural Analysis of Changing Household Electricity Consumption in India

Harold Wilhite

Abstract The reasons behind growth in middle class household electricity consumption in India are examined, where household appliances are rapidly taking a place in home cooking, cleaning and cooling consumption. The resulting electricity demand comes with high environmental costs, both locally and globally, since the electricity fuel of choice in India is coal, which is plentiful. The analysis, based on ethnographic research in the Kerala capital city Trivandrum, examines changes in home consumption practices and relates them to political and social changes in India, especially in Kerala, over the past decades. The research demonstrates how an understanding of gender relations; family and household structures; work migration in cross-national ethnoscapes; and changes in India's political relationship to global markets and globalising media are all important to the theorising of changing energy consumption. An important finding is that women are indirectly responsible for increasing consumption as home appliances are purchased to alleviate time pressure. Social performance is also contributing to changing consumption; the purchase of household appliances is not only a sign of 'getting ahead' but also of 'keeping up' with rapidly changing consumption norms.

Keywords Energy consumption · Energy savings · Ethnography · Gender · Globalisation · India · Kinship · Middle class

6.1 Introduction

Household electricity consumption is growing rapidly in many parts of India. In the country with the largest number of people living below the poverty line, this growth is necessary for the development of basic social infrastructure, especially given that only about 50 percent of Indian villages are electrified. Electricity consumption will undoubtedly continue to grow as villages plug into the grid and as hospitals, schools, businesses and local governments electrify their services. Another source of rapidly growing electricity consumption is the Indian middle class. Household

---

H. Wilhite (✉)
Centre for Development and the Environment (SUM), University of Oslo, 0317 Oslo, Norway
e-mail: h.l.wilhite@sum.uio.no

D. Spreng et al. (eds.), *Tackling Long-Term Global Energy Problems*, 97
Environment & Policy 52, DOI 10.1007/978-94-007-2333-7_6,
© Springer Science+Business Media B.V. 2012

electricity appliances such as refrigerators, electric mixmasters, washing machines and air conditioners are rapidly becoming part of the cooking, cleaning and cooling practices in millions of Indian households. This accumulated electricity demand comes with high environmental costs, both locally and globally, since the electricity fuel of choice in India is coal, which is plentiful.

This chapter examines the growth in middle class household electricity consumption. It shows that while rapid economic growth is one factor behind electricity growth, the full range of growth dynamics is only accessible through a lens that accounts for the changing social and political contexts of consumption. The formal economic models for explaining electricity use and predicting changes that have been developed for the rich OECD countries have limited explanatory power for India and many other countries. A theory of changing electricity use is developed that links global processes involving the movement of ideas, goods and money with the changing local contexts around house, home and social practices.

The analysis in this chapter relies heavily on the author's ethnographic research in the Indian state of Kerala from 2001 to 2004, the comprehensive results of which were published in the book *Consumption and the Transformation of Everyday Life: A View from South India* (Wilhite, 2008a). The study is relevant for understanding emerging electricity consumption in many other parts of the emerging South, as well as a source of new thinking on the hows and whys of electricity consumption in the North. The ethnographic approach, with its emphasis on contextualising consumption in social relations and cultural practices, departs from mainstream approaches in consumption studies, which have been dominated by neo-classical assumptions about rational individuals and idealised markets. Neo-classical economic analyses conceptualise consumption as something performed by economically rational individuals operating in a social vacuum. Anthropologists have long critiqued these assumptions. One of the first to consolidate the critique was Karl Polanyi (1957), whose work resulted in the founding of economic anthropology in the 1960s. In the intervening years, however, surprisingly few economic anthropologists have directed their attention to consumption. According to anthropologist Don Slater, who reviewed social science approaches to consumption in 1997, economics still dominates consumption studies and is 'the general model of social order through which the consumer is defined' (1997, p. 41). As Hansen (1997, p. 13, quoted in Wilhite, 2008a, p. 8) put it, 'consumption (in the conventional economic view) is seen as the end point of the economic circuit' and 'focus[es] on commodities rather than on consumers and leaves little scope to explore the workings of consumption and the diverse and changing social relations that individuals and collectivities are constructing through their consumption of objects'. The reasons why Indians are using significant amounts of their income on consumption and why certain goods and practices have been favoured over others are only accessible through a theoretical lens that accounts for the way in which consumption is related to social practices and material infrastructures, as well as to globalising capital, goods and media.

Both the microeconomic and macroeconomic changes in Kerala have defied conventional economic models. Consumption of goods and electricity is growing rapidly in spite of only moderate economic growth compared to that of other Indian states (Surendran, 1999; CI-ROAP, 1998). In terms of overall consumer spending, consumption in Kerala is higher than in any other Indian state (National Sample Survey Organisation, 2001/2002) and growth in household durable goods (electrical appliances and vehicles) is four times the national average. With regard to energy consumption, electricity use doubled in Trivandrum, Kerala's capital city, in the second half of the 1990s, leading to acute electricity shortages and frequent blackouts, both planned and unplanned. The increase in the use of household electrical appliances is an important factor behind the increasing demand for electricity. Household electricity consumption in Kerala grew from 16 to 44 percent of total consumption in the 1990s (Vijayakumar & Chattopadhyay, 1999). This rapidly increasing residential demand for electricity has contributed to severe shortages. Scheduled blackouts are a daily occurrence and unplanned blackouts can occur several times a day. Augmenting the electricity production capacity to meet increased demand is expensive and environmentally problematic. There have been major political confrontations regarding the building of hydropower plants because such construction would necessitate the displacement of people in this densely populated state. It would also affect local ecosystems and contribute to the loss of biodiversity. The least expensive source of electricity, the coal-fired thermal power plant, contributes to several forms of air and water pollution. Kerala's problem is a familiar one worldwide – meeting its growing demand for electricity requires the construction of expensive and environmentally problematic new power sources. These limitations on the production side increase the urgency to understand how and why energy consumption is growing.

In addition to their contribution to increasing electricity use, household electrical appliances also contribute to increased use of water, soaps and other cleaning agents, which in turn contribute to natural resource depletion and the production of waste and pollution. Household waste, agricultural runoff and industrial waste in Kerala is largely untreated. This has contributed to a rapid reduction in access to clean water (Argarwal, Narain, & Khurana, 2001). Today, in many parts of Kerala, drinking water from all sources must be boiled or filtered, adding to energy demand and to household chores, most of which are performed by women. Boiling water requires the use of costly bottled gas or kerosene, or biomass.

Kerala's socialist political history makes its place at the top of the post-1990 wave of Indian consumption interesting and, on the surface, enigmatic. Kerala's first state government in 1959 was Marxist-led and subsequent governments have been dominated by socialist-inspired political programmes. The 'Kerala Model' of development has led to significant achievements in universal literacy, public education, land reform and health. According to the Indian census of 1991, Kerala spent 50 percent more on education than the average Indian state. More than 95 percent of children in Kerala are enrolled in primary education and the state has the lowest dropout rate in India. Its overall literacy rate is over 90 percent, compared to 52 percent nationally, and its literacy rate among females is 84 percent compared to 39 percent nationally. Kerala's per capita health care expenditure in 1991 was twice the national average. Kerala ranks first among Indian states in the provision of hospital beds and in the number of doctors per capita. Kerala's infant mortality rate is the lowest in India, at 17 deaths per 1,000 births, compared to 91 deaths per 1,000 births nationally. These achievements of Kerala's development model, with its priority on social and healthcare goals, inspired UNDP to create the human development index, an alternative measure of development to gross domestic product (GDP), the standard measure of economic growth. In short, the Kerala Model has been based on a vision of public-driven increases in social capital (education and health). The Kerala model has assumed that social and human development are prerequisites for sound economic development. Public planning, regulation and spending have been favoured over the market as a driver of social and economic progress.

6.2 The Elusive Middle Class

The focus of this study of changing consumption in Kerala has been the middle class. However, it is difficult to come up with a precise definition of the Indian and Keralan middle class. Depending on which criteria are used, estimates of the size of the Indian middle class range from 100 to 300 million people (out of India's approximate population of one billion). 'Middle class' has been defined, for example, by income, job, social status and, in India, by caste. Middle class has also been defined by possessions, as is typical in market and consumer research. The Economist (1994 online, cited in CI-ROAP, 1998, p. 83) defined 'middle class' as consisting of those who could 'afford durable goods such as fans, refrigerators and motorcycles.' Sociological and anthropological definitions tend to draw on the Weberian tradition of using social or cultural traits. Liechty (2003) used this approach to study middle class identity in Kathmandu, Nepal. He found that middle class Nepali work to identify themselves as middle class through cultural projects like education and consumption. The present study did not identify such a degree of striving for a middle class identity in Kerala. Four decades of redistributive social and economic policies have resulted in a dynamic middle-class, consisting of families who have seen their fortunes improve as a result of educational opportunities or land redistribution programmes, or fall due to a breakup of family landholdings. In view of this mix, this study used a local (emic) identification of middle class that is strongly linked to neighbourhood. Certain neighbourhoods of Trivandrum are clearly identified as middle class. The research for the study focused on four middle class Trivandrum neighbourhoods, a survey of which confirmed that income, possessions and education levels were consistent with predominant middle-class indicators in Kerala.

6.3 Changing Household and Family

The analysis begins with the factor that is most often ignored or omitted from studies of changing electricity consumption, namely the social and cultural changes that induce, accompany and result from the consumption of energy-using products. In

India and other countries of the South, social life and cultural practice is centred, to a large degree, on the family. In order to understand almost anything about social life in Kerala, including consumption, it is important to account for the role and power of the extended family. The importance of family cuts across all of the three major religions of Kerala: Hinduism (about 60 percent of the population), Christianity (30 percent) and Islam (10 percent). Family were implicated in and mediators of consumption in a multitude of ways, in spite of significant changes in how families are organised into households. One of the important changes in family began a century ago and is attributable to the gradual break up of joint family households into nuclear households. Today, fewer than 10 percent of Trivandrum households reside in a joint family household (though 30 percent of nuclear households have at least one of the head-of-household's parents living in the home). This change is due in part to land reforms that broke down large joint family holdings but also to Kerala's various social and religious reforms over the past century that have promoted small, patrilineal families (Devika, 2002).

Another of Kerala's unique legacies is the practice of matriliny *(marumakkathayam)*, which was practiced by about 60 percent of Kerala's population in the late 19th century and involved matrifocal residence and matrilineal inheritance (Gough, 1962).[1] Extended families lived in joint family households consisting of matrilineal kin (called *taravad*). Husbands did not cohabitate with their wives but rather lived in the *taravad* of their matrilineal related kin. Discussions with elderly people and the work of earlier ethnographers revealed that the matrilineal principle of household organisation did not mean that women were in positions of power in the household. *Taravad* were generally ruled by the senior male in the household, referred to as the *karnavar*, who controlled all household affairs, including the consumption of the household as a whole, as well as its individual members. In a practice that was to continue well into the 20th century, the *karnavar* decided how to allocate both food and clothing among members of the household. He also selected the marriage partners for female members of the *taravad*, thus exercising the role normally assigned to the senior male in patrilineal families (Gough, 1962, p. 352).

The practices of *marumakkathayam* were the targets of a sustained assault by social and religious reform movements of the 19th and early 20th centuries. Quoting practices in the 'civilised' world (by which they meant Europe), reformers insisted that it was a man's right to bequeath his self-acquired wealth to his wife and children instead of to his *taravad*. Because of the loose marriage relationship *(sambandham)*, which involved neither a church nor a civil ceremony and could be dissolved by the mutual consent of the partners, its practitioners were characterised as 'promiscuous'

---

[1] The Nair, some members of the Ezhava castes, smaller castes such as the Samantans, Vilakkitallas (barbers), Veluthedatus (washer men) and, according to Fuller (1976), even some Muslims, practiced matrilineality and matrilocality well into the 20th century. The Hindu Nambuthiri Brahmin caste, which comprised less than two percent of the population, practiced patriliny, but the Nair matrilineal practice of *sambandham* allowed Brahmin men to have concubinal relationships with matrilineal Nair women. Christians and the vast majority of Muslims, who together constitute about 30 percent of the population of South Kerala, have always practiced patrilineality and patrilocality.

and the children of *sambandham* relationships were characterised as 'bastards'. Saradamoni (1999) writes that Europe-educated Kerala men were some of the most ardent reformers of *marumakkathayam*, especially junior male members of *taravad*, who had much to gain from an acknowledgement of their rights as husbands. The reformist thrust contributed to a series of laws in the early 20th century that changed the rules regarding family, property and inheritance. A law was passed in 1920 making *sambandham* illegal (Puthenkalam, 1977), while a series of laws gave inheritance rights to the husband's children rather than to his sister's children. Nephews were deprived of all claims to the property of their uncles and husbands gained legal guardianship of their wives and children. Divorce, which was settled informally under *sambandham,* was made a subject for the courts.

The legacy of *marumakkathayam* has left a widespread impression that women had more rights and advantages than under the patrilineal system that replaced it. Saradamoni (1999) suggested that there was a spirit of independence and security among women because the woman's right to residence in the *taravad* was 'permanent and uncontested,' even when liaisons with husbands were terminated. Men and women both had the freedom to break the *sambandham* bond without stigma and widows were allowed to re-enter *sambandham* with another man. Rights to property were legally traced through women and a *taravad* estate was held in trust for the women and their descendants. It was common practice for girls to be allowed to attend school, at least through the elementary grades.

The legacy is quite different, however, with regard to household chores. Under *marumakkathayam*, women had full responsibility for housework and family care. From pre-teen years, young women were placed under the supervision of senior women and assigned household tasks. According to Gough, 'Women cooked communally in the kitchen, bathed and washed their own and any children's clothes in company in the bathing pool, and at the harvests moved in household groups to the fields to help the serf women of their families in cutting paddy' (Gough, 1962, p. 352). Furthermore, women were expected to perform all domestic services and to give *karnavar* ('obedience and devotion'). In other words, while the nature of inheritance, marriage and residence changed in the early 20th transition from matriliny to patriliny, the responsibility for housework remained constantly defined as women's work.

In the nuclear family households which are now predominate in Kerala, chores that were previously shared by the women of the joint household fall on the shoulders of the wife alone. Increased educational opportunities and greater openness to women working outside the home have led to a steady increase in women in the workforce, yet women still have full responsibility for housework. This change has resulted in tremendous time pressure on wives, who must compress all of the household chores into early mornings and late evenings. These time pressures have contributed to increased interest in convenience appliances such as washing machines, refrigerators and cars. As a result, families in which wives are working are much more likely to have virtually all of the household appliances listed in Fig. 6.1. This figure illustrates the ownership of certain convenience appliances, with a distinction based between families according to whether the wife has a full-time

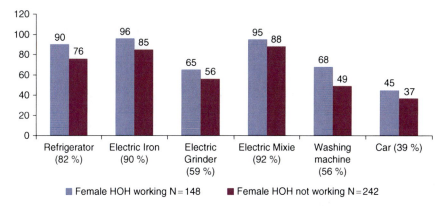

Fig. 6.1 Ownership of household appliances in households with female heads of household working outside the home and households with female heads of household not working outside the home. Based on the author's survey of 408 households in four neighbourhoods in Trivandrum, Kerala
Source: Wilhite (2008a)

job outside the house. Families with wives working full time outside the home are more likely to have all of the convenience appliances listed.

Women who do not have a washing machine wash clothes by hand, a physically demanding and time-consuming practice. Clothing must be soaped, beaten to remove dirt particles, rinsed, wrung out and then laid out (or hung up) to dry. A washing machine relieves women of hard manual labour and frees up approximately one hour or so a day (diaries kept by 20 women showed that they spent an average of an hour per day washing clothes). The time-saving potential is an important motivator, which is reflected in the differences in ownership of washing machines in families with wives working outside the home. Sixty-eight percent of households in which women work full time have a washing machine, compared to 49 percent of households in which the wife is not working. The greater purchasing power of families with two incomes could partly explain the difference in ownership (the mean income in households with working wives was about 23,000 rupees per month, compared to 13,000 per month for households with non-working wives). However, entertainment appliances (such as televisions) and comfort appliances (such as air conditioners), show no significant differences in ownership in families with and without wives in salaried jobs. This supports the interpretation that time considerations are responsible for the rapidly increasing demand for washing machines. Based on sales data provided by Trivandrum retailers, sales of washing machines increased by 500 percent from 1991 to 1999 and by 30 percent each year from 1999 to 2001.

Food preparation is another time-consuming activity. Convenience appliances associated with cooking, including refrigerators, mixmasters ('mixies') and microwave ovens, offer the attraction of saving time and work. Women's diaries revealed that they spent over three hours a day preparing food, on average, much of which involved grinding the spices and condiments typically used in Kerala. Grinding, cutting, scraping and crushing spices and condiments take time, but electric mixies and grinders provide a way to significantly reduce the time devoted to these tasks. Figure 6.1 shows that 92 percent of the households surveyed owned mixies. In the rural areas around Trivandrum where houses had recently connected to the electricity grid, interviews with housewives revealed that the mixie was among the first appliances purchased (along with lamps, electric irons and televisions).

The messages used in television advertising are similar to those used when appliances penetrated North American, European and Japanese markets at various points in the mid-20th century. A generic theme in appliance advertising characterises the modern housewife as modern and efficient. Electrical household appliances are portrayed as gender-liberating because they free up women's time. This message characterised marketing in North America in the 1930s and 1940s (Cowan, 1989), in Japan in the 1960s and 1970s (Wilhite, Nakagami, & Murakoshi, 1997) and elsewhere in Asia, notably China, from the 1990s onwards (Hooper, 1998). However, North American feminists began to question the liberating promises of household appliances in the 1960s. Cowan's (1989) study compared housework in the United States in 1940 and 1965 and found that, despite dramatic increases in the ownership of washing machines and dish washers, women spent approximately the same amount of time on housework in 1965 as women in 1940. Women in 1965 spent less time on some individual household tasks but because they retained full responsibility for housework, new tasks filled the freed-up time. In India and other parts of Asia and the South, where gender roles in housework remain firm, the same potential exists for women's liberated time to be filled with new tasks, new appliances and increased electricity use. A review of recent research in other Asian countries (see Hooper, 1998; Sen & Stivens, 1998), Africa (Hansen, 1997) and from the Caribbean (Freeman, 2000) revealed similar discourses about the modern housewife. For example, Freeman writes that the myth of the Caribbean 'strong matriarch' is still used to justify assigning women hard work in salaried jobs outside the home in addition to full responsibility for housework and children. Writing about China, Hooper (1998, p. 179) cites MacKinnon's (1989) finding that 'Women become as free as men to work outside the home while men remain free from work within it.' Hooper calls this a 'double burden', which she sees as relevant not only for China but for many of the Asian countries with emerging economies. The consumption of electricity is deeply implicated in women's efforts to negotiate this double burden.

6.4 The Globalising Indian Economy

These changes in household, family and gender are examples of social changes that have contributed to an increased interest in new forms of convenience and time-saving appliances. Since the early 1990s, this consumer interest has been complemented by a huge increase in the availability of new products at reduced prices. In India, the availability of new energy-using products has increased dramatically over the past two decades, due to three distinct but related forms of globalisation. This section discusses how changes in consumption are related to the globalisation of the political economy, work and transnational media.

6.4.1 The Politics of Consumption

The political economy of consumption started to change dramatically in India in the early 1990s, when India was opened up to global markets and transnational agents. The 'opening' in 1991 changed the amounts and kinds of products that were available and led to a sharp reduction in the price of many of things that had previously been classified as luxury goods (such as televisions and air conditioners). It allowed foreign investment, permitted foreign media outlets to establish themselves in India and created opportunities for joint ventures between Indian and foreign-based companies.[2] Furthermore, India negotiated an International Monetary Fund (IMF) loan of USD 2.8 billion, bringing with it demands for deeper 'structural reforms' to the Indian economy, including downsizing the public sector, privatising public enterprises and establishing conditions for freer markets (Corbridge & Harriss, 2000). Multinational corporations, which had been virtually locked out of India (or thrown out as in the case of the expulsion of Coca-Cola in 1977), were invited in. Multinational corporations from North America, Europe, Japan and South Korea used direct investment and subsidiary creation to access new markets for their extensive range of products for cleaning, cooking, food storage and thermal comfort, accompanying them with new financing schemes and heavy marketing.

These changes in the political economy of consumption marked a sharp turn from Gandhian and Nehruvian post-independence policies regarding economy, consumption and Indian identity. Gandhi and Nehru were not always in agreement but they agreed on the importance of local development, led by Indian visions and skills. Both leaders were sceptical of Western development and its agents. Mohandas Gandhi was particularly wary of Western consumerism and countered it with his ideologies of thrift, frugality and indigenous development. These ideas affected Indian international relations and its politics of consumption for over 30 years. Rajiv Gandhi was the first Indian leader to regard consumption as a potential contributor to Indian economic development and to encourage middle class consumption. As Fernandes wrote, 'If the tenets of Nehruvian development could be captured by symbols of dams and mass based factories, the markers of [Rajiv] Gandhi's India shifted to the possibility of commodities that would tap into the tastes and consumption

---

[2] This shift towards consumption as an integral part of Indian development was made during a period of declining economic growth, growing national debt and a growing trade deficit (Khilnani, 1998; Corbridge & Harriss, 2000). The Gulf War of 1990 exacerbated these problems because it led to declining revenues from work migrants (Non-Resident Indians). It also contributed to higher oil prices, which put stress on many parts of the Indian economy. The collapse of the Soviet Union in the late 1980s made a further contribution to the deterioration in trade and 'removed the only alternative ideological model to the capitalist market' (Khilnani, 1998, p. 96).

practices of the urban middle classes' (Fernandes, 2000, p. 613). The 'opening' of India legitimised and encouraged consumption as part of both national and personal development.

The research revealed that most South Indians view the Indian opening to global markets and global products as positive. Many see it as evidence that Indians are no longer regarded as 'second-class world citizens'. As one respondent related:

> Each and every day there are new developments. We also have to participate in that. In this period we can't say that it is a foreign idea and avoid those. Before independence, English people did not give us anything. They were only taking from us. They took from here and they produced the output there and sold to us. To get knowledge about the world and to go along with the movements of the world this is necessary. This was a part of the freedom movement.

Despite this interest in things foreign, it would be wrong to say that Indians are obsessed with anything that is produced abroad. William Mazarella (2003) coined the idea of 'close distance' to capture what middle-class Indians want in their products. They want foreign goods but they also want them to be adapted to Indian culture and tastes. Close distance can be detected in the marketing strategies of everything from beauty products to household electricity appliances. One example is a popular television air conditioning advertisement, which opens with a sadhu (holy man) lying on a bed of nails in a hot, dusty street. The scene changes and the sadhu is transported to a modern bedroom with an air conditioner (Carrier, a US brand) humming in the window. His meditational position, dress and pleased expression fit comfortably into an air-conditioned environment. As Fernandes (2000) wrote, 'The core of the Indian tradition, the image suggests, can be retained even as the material context of that tradition is modernized and improved'.

This air conditioning advertisement is one of a flood of sophisticated images promoting consumption from the early 1990s. Part of the reason for the expenditure on advertising is the explosive growth in television ownership. Today, approximately 105 million homes in India have televisions. Ninety-eight percent of middle-class Keralan households have at least one television and 60 percent of television owners subscribe to cable or satellite services. Through cable, families have access to up to 100 channels, about half of which are foreign networks, and most of the remainder of which are Indian national, regional channels and local channels. The most popular programmes are locally produced, especially the Malayalee (Kerala) serials, which combine a generic, global format, the soap opera, with locally grounded stories. While watching these shows with local people, it was interesting to find that the plots and characters did not in any obvious way promote consumption (as many US-based soaps do that have been recycled in India, such as *Dallas*, *The Bold and the Beautiful*, and *Falcon Crest*). Instead, they took as their principal subjects the dilemmas facing families and women in this period of rapid social change in the country. Plotlines involve endless struggles and lots of tears but, in the end, the moral of the story is usually that those who adhere to local ideas and local values fare best.

Even if the stories themselves do not promote consumption, they nonetheless create a perfect format for advertisers because of their huge and loyal audiences.

The growth in television advertising is reflected in the data on spending, which grew from an average of 21 percent a year from 1995 to approximately USD 1.6 billion in 2005. Despite the high frequency of television advertisements, which appear every three minutes during prime time, on average, most Keralan viewers seem interested in advertisements. The author encountered many situations in which someone went out and bought precisely the same product they had seen advertised, down to the model and colour. In this early phase of exposure to television advertising, many people equate commercials with information; as exposure continues over time and consumption matures, consumers may become more critical. For now, while not all commercial advertisements are equally effective, their elaborate productions and the sophistication of many of their themes make advertising an important change agent in India.

6.4.2 Transnational Workscapes

Another contributing factor to changing electricity consumption is the globalisation of workscapes (a term introduced by Appadurai (1996) to capture the expanding geographical space that encompasses home and work). In Kerala, work migration is extensive (Zachariah, Mathew, & Irudaya Rajan, 2002). Keralan workscapes encompass the countries in the Gulf of Oman (Kuwait and Saudi Arabia) but also extend further to Singapore, Europe, Australia and North America. Working migrants are mainly from lower middle-class or poor families, who pool their resources and send one member abroad. In Trivandrum, 40 percent of all families have at least one family member working abroad. Moreover, work abroad is not viewed as temporary or short-term but as a semi-permanent lifestyle that involves dual residence. Migrants maintain strong ties with their family in Kerala, travel back to Kerala often and, in many cases, split the nuclear family, with the male head of household working abroad while his wife and children remain in Kerala.

In their workplaces abroad, migrants encounter new household routines and new goods that are not available in India or are used in very different ways. Goods that are viewed as luxuries in India, such as cars, washing machines and air conditioners, are routine aspects of everyday life in many of the places of work abroad. This encounter is expressed by Gopal (cited in Wilhite, 2008a), a lower-middle-class Indian who spent parts of his childhood in Kuwait and parts of it in Trivandrum.

> Most of the people who have gone from here to the Gulf are not very affluent actually. They go there, make an earning. You have all these facilities. When somebody left from here, he never had a refrigerator at home, he never had a TV at home, he never had an A/C at home, he never had a car. These are all luxuries. But when you go you cannot live in that hot condition without an A/C, and you cannot live without a refrigerator, that is a necessity. You have to have something cool [to drink] after a hard day's work. And you need transport. Over there petrol was not a concern. Petrol was cheaper than drinking water. So having a car was not a luxury but was more an easy conveyance, so most families used the car. And the houses over there are bigger houses. When they return they wish to have those things over here.

Practice theory (Bourdieu, 1977; Reckwitz, 2002; Warde, 2005) provides a useful theoretical approach to understanding what happens to consumption when it confronts new ways of doing things. Bourdieu articulated his theory of practice in *An Outline of a Theory of Practice*, published in 1977. He later built and modified his ideas in his book, *Practical Reason: On the Theory of Action* (1998). More recent theorists, such as Reckwitz, have developed and applied practice theory to the understanding of behaviour. Reckwitz (2002, p. 249, cited in Warde, 2005) define practice as follows: 'A "practice" (Praktik) is a reutilized type of behavior which consists of several elements, interconnected to one another: forms of bodily activities, forms of mental activities, "things" and their use, a background knowledge in the form of understanding, know-how, states of emotion and motivational knowledge.' Practices involve the interaction of people, their social contexts, things and routines. They encompass reflexivity and intentionality but also tacit knowledge, which is usually absent from the theorising of energy consumption. Much of the theorising about energy consumption focuses virtually entirely on reflexive knowledge, thereby largely ignoring the agency in tacit knowledge. New routines challenge tacit knowledge and stimulate reflection about action. A work migrant's life is characterised by confrontation with and habituation to new routines. In their initial meeting with their new homes abroad, migrants are confronted with new practices and new things, which moves the tacit into the realm of the reflexive. In this way, the dual residence of Kerala work migrants can be disruptive and sometimes lead to a reorganisation of daily practice. Establishing a residence in the Gulf, in a home equipped with washing machines, air conditioners and cars, disrupts classifications that are taken for granted. New practices become normal, while luxury goods and the related consumption practices eventually become tacit and taken for granted. The migrant's dual residence moves these routinised practices through the transnational workscape and back to his or her residence in Kerala, displacing old routines with new, more energy-intensive ones.

The author's survey of the ownership of various household appliances in 2002 revealed that families with a close member working outside India possessed more of every major appliance (Wilhite, 2008a) (Fig. 6.2). There are two possible explanations for this. Either the commodity itself had been physically transported back into Kerala or there was a transfer of money through family networks that was then used to purchase new goods.

Work migration is common around the world, including many parts of Asia (see Gamburd, 2000; Gardner, 1995; Johnson, 1998). Insights into how workscapes, dual residence and transfers of practice contribute to consumption change are important for understanding this important facet of globalisation and how it is contributing to growth in household electricity use.

6.4.3 New Technologies and Their Consumption Scripts

When new energy-using technologies move into homes at the rapid rate that they have in India, they contribute to increasing energy consumption in several ways.

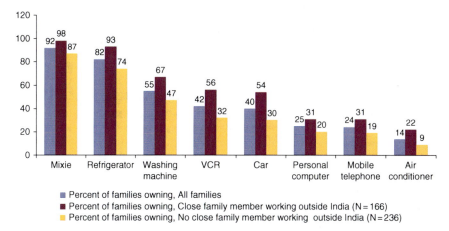

Fig. 6.2 Ownership of selected appliances, families with a member outside India (NRI) versus families with no family member working outside India. Based on a survey of 408 families in four middle class neighbourhoods
Source: Wilhite (2008a)

The most obvious change is that home practices become more energy-intensive as appliances take their place in practices (that is, the electricity input used to prepare food, clean and cool the house increases). Longer term changes in practices, stimulated by the latent potential for change that is inherent in many of the more sophisticated household technologies, are more subtle.

The theoretical aspects of this technology-consumption relationship were explored in Wilhite (2008a, 2008b). Household technologies such as refrigerators, cooking appliances, washing machines and air conditioners involve what Akrich (2000) called 'scripts' for practice. This view contends that rather than being passive objects, technologies are agentive in the sense that they have the potential to influence practices. For example, refrigerators have the potential to influence food preparation in a number of ways. In South India, where the storage of cooked foods is traditionally seen as unhealthy, the refrigerator's potential to store foods prepared in bulk was not exploited to any great extent by the first generation of users (1960s through 1980s). However, the time and work pressures on women, discussed above, have made this storage potential attractive. This has activated one of the refrigerator's time-saving scripts: Food can be prepared in bulk, stored and reheated over the course of several meals. In this way, a changing social context works together with a new technology to change practices. This two-way agency in the human-technology relationship has been characterised as 'distributed agency' (Wilhite, 2008a).

Energy policy has largely ignored this multi-faceted potential of energy-using technologies to affect energy use in ways that are sometimes unanticipated. Energy debates have focused almost exclusively on device efficiency. The thinking has been that promoting efficient technologies will reduce the growth in electricity use that would have taken place had the technologies been less energy efficient. As the examples above show, there is no clean equation that links the acquisition of a (more) efficient device and the consequent reduction of electricity used for a given practice. It is conceivable that technology scripts could lead to greater reductions in electricity use than a pure efficiency assessment would predict. For example, electricity and fuel use for cooking could be reduced by cooking food in bulk rather than several times a day, to compensate for the added electricity used to run the refrigerator. However, as shown above, refrigerator-based food practices have other scripts that have the potential to intensify energy use. Refrigerators also increase interest in the acquisition of complementary appliances such as the fast-heating microwave oven and the electric mixmaster (used to grind and mix spices in bulk). In India, the refrigerator and freezer have paved the way for the microwave. Since the late 1990s, sales of microwave ovens in Kerala have increased by 15 percent per year, the fastest growth of all household appliances. Ninety-two percent of middle class households in Trivandrum now own mixmasters (see Fig. 6.2). This refrigeration regime (refrigerator, freezer, mixmaster, microwave) involves a supply chain of energy-intensive technologies for refrigerated delivery and retail (Garnett, 2007). Similar technology 'regimes', with the potential to intensify electricity use, are part of virtually all home practices, including entertainment (television, DVD player, electronic games) and home cooling (see Rip & Kemp, 1998 for a discussion of technology regimes).

Another refrigeration technology, air conditioning, has the potential for enormous changes in the energy intensity of home cooling. A new regime of building practices accommodated for air conditioning swept the southern United States, Japan and Australia in the early and mid-20th century. Air conditioning was the future; building contractors, banks and energy companies cooperated to provide incentives for air-conditioned homes (see Cooper, 1998). In a 30-year period from 1960 to 1990, Japan moved from having almost no air conditioning to having it in virtually 100 percent of commercial buildings and homes (Wilhite et al., 1997). As the use of porous building materials, screen porches, canted roofs and designs for capturing drafts began to disappear, in Japan and elsewhere, mechanical cooling became essential. This same pattern is now occurring with alarming globalising homogeneity in India, China and other emerging economies.

The story of the change from passive to active cooling began in Kerala in the early 20th century. Prior to this time, construction was done mainly by caste-based craftsmen taking Kerala's hot and humid climate into account. They used wood, mud, unburnt bricks, bamboo, straw and leaves as building materials; all porous materials that allow natural ventilation. The building industry in India changed in the mid-20th century from the use of local artisans and local climate-adapted designs to capitalist principles of least costs, cheap materials and homogeneous designs. Cement became the building material of choice and designs began to incorporate elements such as large windows, south-facing views and flat concrete roofs, which do not perform well in hot climates. The stage was set for air conditioning, which has surged in India since the 1990s. The result is that, today, 'even the most tradition-minded, nature-oriented or environmentally-concerned family would have difficulty living in a modern Trivandrum (capital of Kerala, India) home without air conditioning' (Wilhite, 2008a, p. 193). The advent of air conditioning involved changes in a broad range of building technologies. As experiences have shown in warm climates around the world, once this air conditioning regime has been established, it is difficult to change (Wilhite, 2009). The increase in electricity consumption and in peak loads due to air conditioning regimes is one of the most worrisome developments from the perspective of local economies (meeting peak power load), the local environment and climate emissions.

Air conditioning is rapidly becoming a normal element in home cooling in India. Having established itself as a normal practice, a new consumption dynamic has begun in which people consume a commodity so as not to be considered different or 'behind the times'. Television viewing is a good example: 15 years ago it might have raised eyebrows if a Kerala household did not have a black-and-white television. Only a few years later, the same could be said of a colour television. Today, 96 percent of the families in middle class Trivandrum neighbourhoods have a television and half of those subscribe to cable. In some cases, families that did not subscribe to cable television were stigmatised as being stingy or poor. The agentive power of the normal or conventional is highly relevant to understanding consumption change in India. In this context, the middle class virtue of frugality or thriftiness, so important for the Indian middle classes over a long period of time, is losing its importance. New ways of consuming, involving things like air conditioners and automobiles, are no longer seen as stigmatising and this creates a significant social dynamic behind increasing energy use.

6.5 Conclusion

Energy policy suffers from chronic superficiality in relation to both the urgency and complexities of climate change but also in its portrayal of the social world and its conceptualisation of social change and of how to make change happen. Understanding changing electricity consumption in India demands a more active role from the social sciences and an acknowledgement that energy is bound up in complex social and material interactions that have, to date, been under-theorised in energy research and policy. The practice theory outlined in this chapter has viewed both the technologies and the practices into which they fit as agentive, creating predispositions for action that become activated in the interaction between people, their socio-cultural contexts and changes in their life patterns. One of the important elements in this new approach is an acknowledgement of the grounding of energy consumption in household practices. Virtually all energy use in homes is bound up in practices (food, commuting, keeping clean, staying comfortable, having fun) that consist of clusters of routines and things.

This practice-cluster approach is intended to provide a theoretical framework for addressing socio-material complexities associated with work, gender, media and globalising capitalism. The conceptual manoeuvre that this paper proposes – from viewing energy use as something performed by individuals and individual devices to something that is created in the interaction between things, people, knowledge and social contexts – has subtle but important implications for strategic policies aimed at energy sustainability. As part of this manoeuvre, policy aims in India and for other countries, both developed and emerging, would move from individually-directed economic instruments and efficiency-directed technology incentives to policies that aim to reduce the energy intensity of energy practice clusters. Taking the home cooling cluster as an example, this would mean acknowledging and strengthening local knowledge about buildings and how to make them comfortable without significant additions of mechanical cooling. These passive designs should be promoted as modern (or beyond modern) and subsidised in public energy programmes in order to make them competitive with cheap, homogeneous designs. For automobility, the energy sustainable cluster should involve convenient, cheap and comfortable public transport alternatives, which are sorely needed in all of India's cities.

References

Akrich, M. (2000). The de-scription of technical objects. In W. Bijker & J. Law (Eds.), *Shaping technology/building society* (pp. 205–224). Cambridge, MA: The MIT Press.

Appadurai, A. (1996). *Modernity at large: Cultural dimensions of globalization.* Minneapolis/London: University of Minnesota Press.

Argarwal, A., Narain, S., & Khurana, I. (2001). *Making water everybody's business.* New Delhi: Centre for Science and Environment.

Bourdieu, P. (1977). *Outline of a theory of practice.* Cambridge: Cambridge University Press.

CI-ROAP. (1998). *A discerning middle class? A preliminary enquiry of sustainable consumption trends in selected countries in the Asia Pacific Region.* Penang, Malaysia: Consumers International Regional Office for Asia and the Pacific (CI-ROAP).

Cooper, G. (1998). *Air conditioning America: Engineers and the controlled environment, 1900–1960.* Baltimore: The John Hopkins University Press.

Corbridge, S., & Harriss, J. (2000). *Reinventing India: Liberalization, Hindu nationalism and popular democracy.* Cambridge: Polity Press.

Cowan, R. (1989). *The ironies of household technology from the open hearth to the microwave.* London: Free Association Books.

Devika, J. (2002). *Domesticating Malayalees: Family planning, the nation and home-centered anxieties in mid-20th century Keralam.* Working Paper Series 340. Trivandrum: Centre for Development Studies.

Fernandes, L. (2000). Nationalizing 'the global': Media images, cultural politics and the middle class in India. *Media, Culture and Society, 22*(5), 611–628.

Freeman, C. (2000). *High tech and high heels in the global economy: Women, work, and pink collar identities in the Caribbean.* Durham: Duke University Press.

Fuller, C. J. (1976). *The Nayars today.* Cambridge: Cambridge University Press.

Gamburd, M. R. (2000). *The kitchen spoon's handle: Transnationalism and Sri Lanka's migrant housemaids.* Ithaca/London: Cornell University Press.

Gardner, K. (1995). *Global migrants, local lives: Travel and transformation in rural Bangladesh.* Oxford: Clarendon Press.

Garnett, T. (2007). *Food refrigeration: What is the contribution to greenhouse gas emissions and how might emissions be reduced?* Food Climate Research Network, Centre for Environmental Strategy. Surrey: University of Surrey.

Gough, K. (1962). Nayar: Central Kerala. In D. Schneider & K. Gough (Eds.), *Matrilineal kinship.* Berkeley/Los Angeles: University of California Press.

Hansen, K. T. (1997). *Keeping house in Lusaka.* New York: Colombia University Press.

Hooper, B. (1998). 'Flower Vase and Housewife': Women and consumerism in post-Mao China. In K. Sen & M. Stivens (Eds.), *Gender and power in affluent Asia.* London/New York: Routledge.

Johnson, M. (1998). At home and abroad: Inalienable wealth, personal consumption and formulations of femininity in the Southern Philippines. In D. Miller (Ed.), *Material culture: Why some things matter* (pp. 215–238). London: UCL Press.

Khilnani, S. (1998). *The idea of India*. London: Penguin Books.

Liechty, M. (2003). *Suitably modern: Making middle-class culture in a new consumer society*. Princeton/Oxford: Princeton University Press.

MacKinnon, C. (1989). *Towards a feminist theory of the state*. Harvard: Harvard University Press.

Mazzarella, W. (2003). *Shoveling smoke: Advertising and globalization in contemporary India*. Durham/London: Duke University Press.

National Sample Survey Organisation. (2001/2002). *Household consumer expenditure and employment situation in India, 2001–2002*. Delhi: Ministry of Statistics and Programme Implementation, Government of India.

Polanyi, K. (1957). *The great transformation: The political and economic origens of our time*. Boston: Beacon Press.

Puthenkalam, S. J. (1977). Marriage and the family in Kerala. *Journal of Comparative Family Studies Monograph Series* (pp. 127–166). Calgary: University of Calgary.

Reckwitz, A. (2002). Toward a theory of social practices: A development of culturist theorizing. *European Journal of Social Theory, 5*, 243–263.

Rip, A., & Kemp, R. (1998). Technological change. In S. Rayner & E. Malone (Eds.), *Human choice and climate change* (pp. 327–400). Columbus, OH: Battelle.

Saradamoni, K. (1999). *Matriliny transformed: Family, law and ideology in twentieth century Travancore*. New Delhi: Walnut Creek and London: Sage, Altamira.

Sen, K., & Stivens, M. (Eds.). 1998. *Gender and power in affluent Asia*. London/New York: Routledge.

Slater, D. (1997). *Consumer culture and modernity*. Cambridge: Polity Press.

Surendran, P. (1999). *The Kerala economy: Development, problems and prospects*. Delhi: Vrinda Publications.

Vijayakumar, K., & Chattopadhyay, S. (1999). *Energy consumption pattern: A comparative study*. Report by the Centre for Earth Science Studies. Trivananthapuram.

Warde, A. (2005). Consumption and theories of practice. *Journal of Consumer Culture, 5*, 131–153.

Wilhite, H. (2009). The conditioning of comfort. *Building Research & Information, 37*(1), 84–88.

Wilhite, H. (2008a). *Consumption and the transformation of everyday life: A view from South India*. Basingstoke/New York: Palgrave Macmillan.

Wilhite, H. (2008b). New thinking on the agentive relationship between end-use technologies and energy using practices. *Energy Efficiency, 1*(2), 121–130.

Wilhite, H., Nakagami, H., & Murakoshi, C. (1997). Changing patterns of air conditioning consumption in Japan. In P. Bertholdi, A. Ricci, & B. Wajer (Eds.), *Energy efficiency in household appliances* (pp. 149–158). Berlin: Springer.

Zachariah, K. C., Mathew, E. T., & Irudaya Rajan, S. (2002). Migrant patterns and their socio-economics. In K. C. Zachariah, K. P. Kannan, & S. Irudaya Rajan (Eds.), *Kerala's Gulf connection: CDS studies on international labour migration from Kerala State in India* (pp. 13–47). Thiruvananthapurum: Centre for Development Studies.

Chapter 7
The Changing Context for Efforts to Avoid the 'Curse of Oil'

Jill Shankleman

Abstract The concept of the resource curse provides a useful framework for understanding the unintended consequences of oil production in developing countries. A consensus has emerged over the past decade regarding the key strategies needed to avoid this 'curse'. The core elements are economic policies to spread the benefits of oil wealth over time and limit the de-stabilising effects of revenue volatility, transparency about oil revenues and their use in order to increase political accountability, and environmental and social standards for exploration and production projects to reduce the risk of damage, particularly from land use or pollution. However, since the mid-2000s, the context of the global oil economy has changed drastically. Increasing demand and higher prices have brought new companies into the international oil business and new countries into the ranks of producers. The balance of power between governments of oil-rich countries and oil companies, donors and NGOs has shifted in favour of governments. In particular, Chinese oil companies, supported by the Chinese government, have grown very quickly to match the size and international reach of the largest Western oil companies. These companies are investing in the oil and minerals sector and providing loans to governments to finance large-scale development of national infrastructure. This offers oil-rich countries an alternative approach to revenue management, complicating but potentially enriching the efforts to avoid the curse of oil.

Keywords China · Government · Oil · Resource curse · Transparency

Preface

The political instability of oil- and gas-exporting countries has been a scourge of Western governments for decades and has also inflicted a great deal of misery on indigenous and regional populations. The chronic political and economic difficulties that many such countries experience seem to suggest a paradoxically negative

---

J. Shankleman (✉)
Woodrow Wilson International Center for Scholars, Washington, DC 20004-3027, USA
e-mail: js@shankleman.com

D. Spreng et al. (eds.), *Tackling Long-Term Global Energy Problems*,
Environment & Policy 52, DOI 10.1007/978-94-007-2333-7_7,
© Springer Science+Business Media B.V. 2012

relationship between large fossil-fuel reserves and progress in development. In economics, the potential negative effects that a country's large resource endowment and exports have on its economic and political development are known as the 'resource curse'. The record shows that many oil- and gas-rich countries have predatory and entrenched governments and weak democratic institutions. Their record on poverty alleviation is relatively lacking, they often witness internal conflicts, sometimes exporting violence and revolution as well, and their public finances, especially rents from fossil fuels and minerals, tend to be centralised, non-transparent and open to corruption (Stevens, 2004). This is without even considering the direct effects that resource extraction and export have on the local environment and labour force.

The last several years have seen increasing dissent among economists regarding the established Sachs-Warner (Sachs & Warner, 1995, 2001) findings of the generally negative effect that countries' resources have on their gross domestic product (GDP). For example, Brunnschweiler (2008) used a measure of per capita resource abundance rather than exports and actually found a positive correlation with real GDP growth between 1970 and 2000. In addition, she found no evidence of resources, especially minerals, having a negative effect on institutional quality. The implication is not only that resources may more often be a blessing than a curse for countries[1] but that past analytical errors may have led to explanations and policy advice to remedy generally non-existent phenomena, which in the worst cases may have generated more problems than they solved.

There is, therefore, a lack of consensus on the curse of natural resources and how to overcome it, if and when it occurs.[2] To some degree, this reflects differences in opinions and approaches among economists, between pure economists and political economists and between academics and practitioners like the author of this chapter. However, the universality and strength (and even necessarily the direction of causality) of the relationship between revenues from energy resources and general economic development are not at issue here. As a former consultant with the World Bank and currently a senior scientist at the Woodrow Wilson International Center for Scholars, Jill Shankleman naturally combines several disciplines without being beholden to the doctrines of any single one.[3] As is clear from the start of the paper, Shankleman views the resource curse or 'Dutch disease' as shorthand for a variety of unintended consequences of energy resource development, not just the putative and increasingly challenged negative effects on macroeconomic performance. It is sufficient for her to know that such problems have arisen in the countries where she has worked – without their nature or mechanism necessarily being the same in all countries with large hydrocarbon reserves. She is perhaps glad to leave the academics to argue over the universality of the resource curse or the breadth of its application.

---

[1] This seems to better fit the historical record of the rise of great powers partly as a result of their large natural resource endowments and those of their colonies.

[2] Such heated debate on one issue within a single discipline suggests just how much more difficult it can prove to achieve consensus in integrative multi- or transdisciplinary work.

[3] At the same time, it also potentially leaves her analysis more vulnerable to the precision and critical rigour of highly developed fields such as economics.

Instead, this chapter focuses on the growth and development problems that certain countries have experienced in connection with their endowment and exploitation of fossil resources. It also discusses how shifting international market power configurations may challenge or complement traditionally prescribed Western remedies for them.

David L. Goldblatt

7.1 Introduction

This chapter looks at the interaction of oil and money through the lens of the concept of the resource curse. It starts by outlining the concept and its component economic, political and local dimensions. It then presents the 'Western' consensus set of approaches to avoid the resource curse and looks at how these have been applied. The chapter goes on to discuss the changes to the global oil economy wrought by rapidly growing demand from China and outlines the approaches of Chinese companies and the Chinese government to economic and political development in oil-producing states. The argument is illustrated with the example of Angola. (Countries such as Nigeria, Sudan, Gabon or Equatorial Guinea could have been selected. The basic structure of the argument would be similar, although the specifics would, of course, have been different.)

7.2 What Is the Resource Curse and Why Does It Matter?

The term *resource curse* refers to the mix of macroeconomic, political and local effects of oil production in a country that can de-couple growth in GDP from improvements in the standard of living and quality of life of the majority of its population and, in some circumstances, can exacerbate conflicts to such a degree that conditions deteriorate. The ideas underpinning the 'resource curse' were first expounded in the 1970s. The notion of 'Dutch Disease'[4] was developed to explain the unexpected negative effects of the production of North Sea gas on the rest of the

---

[4] An economic condition that, in its broadest sense, refers to negative consequences arising from large increases to a country's income. Dutch disease is primarily associated with a natural resource discovery but it can result from any large increase in foreign currency, including foreign direct investment, foreign aid or a substantial increase in natural resource prices. This condition arises when foreign currency inflows cause an increase in the affected country's currency. This has two main effects for the country with Dutch disease: (1) A decrease in the price competitiveness, and thus the exports, of its manufactured goods, (2) An increase in imports. In the long run, both these factors can contribute to manufacturing jobs being moved to lower-cost countries. The end result is that non-resource industries are hurt by the increase in wealth generated by the resource-based industries. The term 'Dutch disease' originates from a crisis in the Netherlands in the 1960s that resulted from discoveries of vast natural gas deposits in the North Sea. The newfound wealth caused the Dutch gilder to rise, making exports of all non-oil products less competitive on the

economy of the Netherlands. In 1975, OPEC co-founder Juan Pablo Pérez Alfonso of Venezuela described oil as 'the devil's excrement': 'It brings trouble, waste, corruption, consumption, our public services are falling apart. And debt.' (Pérez Alfonso, 1976). The resource curse explains the phenomenon by which countries that are rich in oil have, with few exceptions (of which Norway is the best example), failed to use this wealth to secure sustained economic development, significant improvements in the quality of their citizens' lives or democratically accountable governments. Increases in the national wealth of oil states, as indicated by GNP per capita, are typically not matched by parallel improvements in development (as indicated in the UN Human Development Index), better governance (as indicated in the composite measures reported by the US Millennium Challenge Corporation's country assessments) or in the business climate for the non-oil economy. Oil states are characterised by high levels of inequality, as measured by the Gini coefficient. For example, despite an average annual rate of economic growth of over 14 percent between 2002 and 2008, Angola scored below the average for sub-Saharan African countries as a whole on all of the World Bank indicators of the business climate.[5] Although it is now classified in economic terms as a 'lower middle income country', Angola languishes in the bottom 10 percent of countries ranked according to the United Nations' Human Development Index.[6] Thus, while the lack of indigenous oil and the need to import it using hard currency can be an obstacle to a country's economic development, so too can be the possession of oil resources in exportable quantities.[7]

The key factors that create the 'resource curse' are government ownership of subsurface resources and the earnings that governments receive when oil is produced from these resources.[8] Except where oil concessions have been poorly negotiated, governments secure the majority of their profits from oil production – the oil 'rent' that comprises the difference between the costs of production and the sale price. Shell, for example, published data for its operations in Nigeria showing that, based on 2006 oil prices, the government's tax take was 95 percent of the profit (Shell Nigeria, 2006, 2011). This rent is divided between the companies producing the oil and the state, through a variety of mechanisms including taxes, royalties and production-sharing agreements. When oil prices are high, so too are rents. Despite the scarcity of good data on profit shares and production costs, it is evident that oil production costs vary widely, depending on the location, scale and type of reserves.

---

world market. *Source*: Investopedia, http://www.investopedia.com/terms/d/dutchdisease.asp. All web links accessed November 16, 2011.

[5] Dramatically below the average measures of the efficiency of the insolvency process, time to enforce a contract, time to register property and costs and time to start a business (World Bank, 2007).

[6] See Human Development Index (UNDP, 2010) and Collier (2007).

[7] For a fascinating account of the response to the absence (until recently) of oil, see McCashie (2008, pp. 313f).

[8] Oil companies clearly profit too, as evidenced by record-breaking profits during 2008 for companies such as ExxonMobil, Shell, Chevron and BP.

7 The Changing Context for Efforts to Avoid the 'Curse of Oil'

For example, oil produced from oil sands through a quasi-mining process is likely to cost over USD 30/barrel,[9] while production costs in the Middle East are perhaps a quarter of this.[10] This is in marked contrast to the cost of crude oil prices in mid-2008, which was well over USD 100/barrel. In 2008 prices, the oil rent from many producing fields was at least USD 70/barrel and, in many cases, even more.

7.3 Economic and Political Problems Associated with Government Revenues from Oil Rents

The key characteristics of oil rents that cause economic problems are their size and yearly variability (depending on production levels and prices). Economists argue that, for most oil-exporting states, the magnitude of government oil revenues inflates the value of the domestic currency, resulting in a less competitive non-oil sector. In countries that have not historically had much traded production, such as new oil-producing countries like Chad, it can be difficult to establish non-oil industries. In other countries, such as Nigeria and Angola, production and export of agricultural commodities has plummeted alongside the growth of the oil industry. (Other factors have also been important, notably Angola's prolonged state of war between 1975 and 2002, which was also linked to the country's oil wealth, cf. Shankleman, 2007.) The results of 'Dutch Disease' include unemployment, which in turn aggravates poverty and, potentially, political instability. Variability in government oil revenues, especially where these form the backbone of government income, makes effective government spending difficult. The effects of oil price volatility were clearly evident in 2009. For example, in January of that year, Angola sharply reduced spending limits for line ministries in order to reflect the sudden reduction in oil revenues. Such fluctuations in government spending in any state contribute to inefficient spending.

Within producing countries, the political risks associated with oil revenue dependence are threefold. The first is corruption, the second is the lack of incentives for governments to be accountable to their citizens and the third is conflict among factions of the elite or between producing regions and the rest of the country over control of oil revenues. Corruption risks are centred primarily around the process of selling concessions to explore for and produce oil and gas. Unless these are auctioned in a transparent way, through a public process and with full disclosure of the terms of winning bids, the incentives for corruption and side deals are strong. While the terms of oil concessions are in the public domain in developed oil-producing states like Canada, the US and Norway, it is rare for this to be the case in non-OECD countries.

---

[9] Suncor Corporation reported that cash operating costs of oil sands production in Alberta in 2007 averaged just under USD 28/barrel (Suncor Energy, 2007).

[10] A. Kaletsky argued that, 'Even including exploration expenses, the total cost of production of OPEC oil is well below $10 a barrel' (Kaletsky, 2008).

Where governments are insulated from the need to raise domestic tax revenue because of the regular injection of oil money into the treasury, elites become less accountable. The active support of a country's citizens is less important to governments with the resources to buy off or repress opponents than it is for governments that need a level of public support to be able to raise taxes. Furthermore, external pressures for better governance are weak when governments do not have to rely on the assistance of foreign countries or international lending bodies. This was brought into sharp focus in Chad, when a hard-negotiated deal between the World Bank and the government of Chad over spending oil revenues on development was largely overturned by the government once the oil revenues started flowing (Gould & Winters, 2007).

Daniel Yergin rightly titled his compelling history of the oil industry 'The Prize' (Yergin, 2003), in which he documented the struggles within and between states and companies for control of oil and the associated wealth. Such struggles have continued in recent decades. For example, control of areas known to have oil resources has been contested by Nigeria and Cameroon, by Australia and East Timor and by China and Japan, although in each case these disputes have been resolved without recourse to violence. Efforts are being made to resolve the dispute between Uganda and the Democratic Republic of the Congo over the demarcation of their border in the oil-rich Lake Albert region (Augé, 2009).[11] Some boundaries within the Caspian remain unresolved because of the different allocation of underwater oil potential that each littoral state would get depending on whether the rules on boundaries for seas or for lakes are applied.

Within states, there has been a trend towards secession movements in oil-rich areas, especially where these have shown significant differences from the rest of the country, such as by being populated by a religious or ethnic minority. As Amartya Sen has shown, the power of different aspects of identity can shift over time (Sen, 2006). What was a weak aspect of community identity, such as ethnicity, can become a dominant one when there is wealth to be fought over. This has been evidenced in Scotland, where a once-marginal Scottish nationalist movement has, in the 40 years since North Sea oil was discovered, attained unprecedented political autonomy and is now credibly pressing for full independence. As another example, the prospect of a viable independent state financed by oil money certainly strengthened South Sudan in its long civil war against the North, which is one of the factors contributing to the insurgency that is reducing oil output from the Niger Delta.

These economic and political factors can become even more damaging through their interaction. In a recent paper, Paltseva (2010) examined relationships between political autocracy, devolution and growth using the game theory-based analytical tools increasingly favoured in the study of this area of political economy. She concluded that autocracies with natural resources such as oil tend to be economically stagnant and resistant to political change: 'If a country is sufficiently rich in natural resources, an autocratic ruler will always resist political change because of the

---

[11] See http://www.ifri.org

lost stream of revenues' (Paltseva, 2010). She argued further that where the ruler's benefits from control are high, the local private sector never starts to invest because they realise that capital will eventually be expropriated. This, in turn, inhibits the devolution of power to other groups; the autocrat sacrifices capital accumulation in order to keep the benefits of the oil revenue stream for himself. This analysis matches the observed weakness of the business climate in oil states (World Bank, 2007; Transparency International, 2010, annual reports).

There is also a local dimension to the resource curse. Unless resources are developed by paying extreme care to avoiding environmental damage, with full and appropriate compensation for people who lose land, homes, fishing grounds and the like to the industry and with deliberate and effective measures to make sure that local people and businesses are employed, then the local impacts of oil development can be highly damaging.

The clearest example of the effects of resource curse can be seen in the Niger Delta, where almost 50 years of large-scale oil production has not improved living conditions, opportunities or the environment. A toxic mix of popular anger, crime and political violence has been created through a combination of failed oil revenue management and poor practices by the industry, particularly in its early years. As a result, the population live in insecurity and the world's oil-consuming countries suffer the impact of price spikes every time oil facilities are shut down through sabotage or for security reasons, as happened frequently in 2008.[12] To date, extensive social spending programmes by oil companies, military crack-downs, formal negotiations and locally-based conflict resolution processes have all failed to improve the situation or even to halt deterioration.

7.4 Avoiding the Resource Curse: The 'London Washington Consensus'

In the 1990s, new academic research on civil war drew attention to greed for resource wealth as a driver of conflict. Pressure by environmental and human rights NGOs pushed companies, governments and the Bretton Woods institutions to consider the negative impacts of oil (and mineral) wealth in many African countries. The result was a revival and extension of the observations from the 1970s into a more highly organised concept of the resource curse. Importantly, in the early years of this century, this was followed by a range of policy prescriptions and operational frameworks intended to reduce the risk of the resource curse. These can be accurately described as the 'London-Washington consensus' because the key players responsible for developing these ideas were experts at the International

---

[12] See, for example, Nigeria oil cut by rebel attack in 2008 (BBC News, July 28, http://news.bbc.co.uk/1/hi/world/africa/7528550.stm).

Monetary Fund (IMF) and World Bank, executives from the largest oil companies based in the USA and the UK[13] and London- and Washington-based activists from international non-governmental organisations. Unusually, many of the prescriptions developed in London-Washington were developed through innovative (though sometimes fraught) collaboration between these groups, hence the term 'consensus'.[14]

Three broad ideas frame this consensus. Firstly, at the macroeconomic level, governments must insulate the rest of the economy from inflation and volatility introduced by oil revenues. The IMF and other organisations have consistently recommended techniques such as limiting year-on-year spending of revenues, saving earnings for the future in oil funds, building capacity to forecast revenues and reforming tax and commercial laws in order to encourage the non-oil economy. The IMF Guide on Resource Revenue Transparency, first issued in 2005 and then revised in 2007, provides detailed guidance for governments that links together many of these aspects (IMF, 2007).

Secondly, there has been a major effort to introduce transparency into the oil economy – making information public about oil concessions, oil revenues and the way these revenues are spent. The key initiative for promoting transparency is the Extractive Industries Transparency Initiative (EITI),[15] which sets standards for transparency by companies and governments. In 2008, the World Bank launched a programme called EITI Plus Plus (World Bank, 2008b). Through this programme, technical assistance supporting transparency and revenue management is provided, on request, to governments of oil states along the entire chain of management for extractive industry resources. This assistance ranges from the process of granting access to those resources to monitoring operations, collecting taxes, sound macroeconomic management and distribution of revenues and spending resources effectively for sustainable growth and poverty reduction. To date, two countries are participating in EITI Plus Plus – Mauritania and Guinea.[16] The emphasis on transparency is also being taken up by national policymakers. Most importantly, in July 2010 the US included requirements in the Dodd Frank Financial Reform Bill requirements that the United States-listed oil, gas and mining companies disclose to the Securities and Exchange Commission (SEC) details of taxes, royalties and other

---

[13] Shell, BP, ExxonMobil and Chevron were more or less active participants in these developments alongside major European and US-based mining houses.

[14] Collaboration between some of the major extractive industry companies, governments – particularly those of the UK and the US, and international NGOs – in efforts to understand and mitigate resource curse issues was pronounced and lead to the development of innovative instruments such as the Kimberley process, EITI and the Voluntary Principles on Security and Human Rights. Many of these organisations remain involved in these processes. In general, however, the NGO participants remain more openly sceptical than governments or companies about the practical possibilities of turning oil wealth from a curse into a blessing.

[15] http://eiti.org/node/22

[16] See footnote 15.

payments in the countries where they operate.[17] Such transparency is expected, over time, to create effective public pressure for accountable government and wise use of revenues and to crowd out oil-related corruption.

Thirdly, sets of standards have been created for managing oil projects with a view to protecting the environment and people in their vicinity. The principal standards are the Social and Environmental Performance Standards[18] developed by the International Finance Corporation (IFC) (the private sector banking organisation within the World Bank Group) and now adopted by most of the commercial banks involved in project finance through the Equator Principles.[19] These standards set out how the direct and indirect impacts of projects should be assessed and avoided, minimised or compensated for. They address important issues such as the compulsory acquisition of land for oil fields and pipelines and they require that where land is acquired under any kind of involuntary system (that is, other than through a consensual buyer-seller negotiation), people should be fully compensated and projects executed that ensure that the standard of living of displaced people is at least as high after land acquisition as before. Following a series of incidents in which security forces overreacted to demonstrations and strikes, a consortium of extractive industry companies, governments and NGOs also developed the complementary Voluntary Principles on Security and Human Rights (VPSHR),[20] which guide how oil and mining companies should ensure that security for their operations does not come at the expense of the security of members of neighbouring communities.

The Natural Resources Charter, launched in 2009, brings together these ideas on macroeconomic management, transparency and environmental standards and establishes a set of 12 principles specifically designed to help governments of resource-rich countries better manage this wealth.[21]

This London-Washington consensus has already had some positive impacts, most notably with respect to macroeconomic management. During the recent oil price boom, several governments introduced systems to try to prevent 'Dutch Disease' by controlling inflation and saving some of their oil revenues. With revenues exceptionally high, many oil states established sovereign wealth funds. For example, Angola established a Reserve Fund for Oil to act as a cushion for Angola's future fiscal

---

[17] Brief Summary of the Dodd-Frank Wall Street Reform and Consumer Protection Act.

[18] IFC (2007) and US Senate Committee (2010) on *Performance Standards on Social and Environmental Sustainability*.

[19] In October 2002, a small number of banks decided jointly to try to develop a banking industry framework for addressing environmental and social risks in project financing. This led to the banks drafting the first set of Equator Principles, which were then launched in Washington, DC on June 4, 2003. A subsequent updating process took place in 2006 leading to a newly revised set of Equator Principles that were released in July 2006. The Principles commit signatory banks to apply the IFC Social and Environmental Performance finance criteria to projects outside the OECD. See http://www.equator-principles.com

[20] See *The Voluntary Principles on Security and Human Rights* (2010). Foley Hoag LLP. http://www.voluntaryprinciples.org

[21] See *Natural Resources Charter* (2009). http://www.scribd.com/doc/31116467/Natural-Resource-Charter

deficits. Excess reserves, designated as profit generated from oil prices above a reference value of USD 45/barrel, are managed in the fund by the National Bank of Angola. In terms of transparency, more information is becoming available on the revenues that governments receive from oil (and minerals). There is growing media coverage about concession contracts and revenues – especially where governments are seeking to change contract terms in order to secure a larger proportion of the boom-time profits.[22] As of mid-2010, the governments of 30 oil- or mineral-producing countries were candidate participants in the EITI and, as such, have made commitments to comply with detailed rules concerning the transparency of the oil (or mining) revenues they earn.[23] The structure of EITI, which mandates that civil society organisations take part in national programmes, ensures that there is discussion of the risks presented by oil wealth and of risk-reduction strategies. At a country level, information on federal government revenues from oil and mechanisms for distributing these to state-level governments has been published in Nigeria for several years. Countries such as Angola, which once had a reputation for secretive selling of oil concessions, have adopted more transparent auction processes for recent bid rounds, despite remaining outside the EITI. Worldwide, some oil production projects have been developed with much greater attention than was typically paid previously to understanding and avoiding negative impacts on local people. Examples of this include the Chad Cameroon production and pipeline projects and the Baku Tbilisi Ceyhan production fields and pipelines. For oil projects that receive financial support from the World Bank or commercial banks that are signatories to the Equator Principles, efforts are made to apply high environmental and social standards. Finally, some countries are taking heed of the risks of secessionist conflict over the distribution of oil revenues and seeking to negotiate agreements with sub-national governments before production starts.[24]

Resource curse issues have become widely discussed in the media in current and putative oil states, for example, in Ghana following the discovery of large off-shore oil reserves. The widely-reported Chad experiment has played an important role in stimulating debate, while NGOs and academics are beginning to enhance their own capacity to develop locally relevant resource curse strategies.

However, it is far from clear that the London-Washington consensus initiatives are being implemented on a scale that enables their full potential benefits to be realised or that they are sufficient to overcome the resource curse, even if they are more widely implemented. There are three key concerns. Firstly, it is doubtful that the macroeconomic prescriptions account sufficiently for the near total lack of functioning transport, power and social infrastructure in many oil-producing countries or for the political and developmental costs of doing nothing to remedy this issue.

---

[22] The re-negotiation of oil and mining contracts is too large a subject to cover in this chapter. Note, however, that in jurisdictions as diverse as Alberta (Canada), Venezuela, Russia, Chad and Nigeria, the terms of contracts have been changed by governments, or such change is under consideration.

[23] See footnote 15.

[24] See *Zanzibar Says No Oil Sharing* (2008) for an example of the ongoing and difficult negotiation between the Tanzanian mainland and Zanzibar about sharing potential revenues from the islands.

Saving oil revenues to avoid the structural problems of using them is a politically difficult message to communicate and can appear to condemn poor societies to remain without the keys to progress, such as access to electricity, clean water, schooling and medical services. Secondly, it remains difficult to see how the political elites who benefit from controlling oil wealth will permit sufficient transparency and debate about revenues, budgets, spending and decision-making to trigger greater accountability[25]; such resistance is clear in the case of Chad. The government of Chad was required to establish a system of revenue management that incorporated transparency, public oversight and a fixed allocation of revenues to social services as a condition of World Bank loans associated with development of the country's first oilfield. Once the oil started flowing, the government first sought to change the agreed revenue allocation in order to give itself a greater share over which it had discretion and then chose to repay its loans as quickly as possible. From September 2008 onwards, having done this, Chad was no longer bound by the World Bank's revenue management conditions (World Bank, 2009). Thirdly, the new social and environmental standards for production operations are difficult to implement in practice (how does a company ensure that subsistence farmers' livelihoods are maintained after their loss of land?) and may not secure the 'local licence to operate' that companies expect in the absence of effective local and national government. In a March 2008 working paper examining the mining sector in Peru, Arellano-Yanguas argued that there is a growing incidence of conflicts related to mining operations, even though the government has complied 'faithfully' with the 'new natural resources policy agenda' and the mining industry has taken steps to be environmentally responsible and development-centred (Arellano-Yanguas, 2009; EIA, 2008).

7.5 Enter the 'New Kids on the Block' and the 'Beijing Consensus'[26]

The configuration of the global oil economy has changed significantly since the turn of the century, when innovative collaborations between the Western majors in the oil industry, governments, international organisations and NGOs were developed to address the resource curse. The key change is that China, and to a lesser extent India, have become significant importers of oil as their economic growth exceeds the oil that is available to them through domestic production. This is the structural factor that has triggered the step change in oil prices through the first decade of this century. Higher prices have, in turn, brought a raft of new companies into the international oil sector. These include the Chinese oil companies, partly government-owned, that have become significant international players over the past

---

[25] Consider, for example, Angola's 2008 decision to cease permitting the United Nations Office of Human Rights to have an office in the country.

[26] Ramo (2004) and Dirlik (n.d.).

few years, as well as state oil companies from India, Brazil, Malaysia and Russia and a raft of small, entrepreneurial 'independents'.[27] Higher prices and competition for oil concessions generate higher oil revenues for producer-country governments and reduce the leverage of donors seeking, for example, to exert influence on revenue management through loan conditionality.

The China National Petroleum Corporation (CNPC) and, on a smaller scale, Sinopec (China Petroleum & Chemical Corporation) and the China National Offshore Oil Corporation (CNOOC), have invested widely in oil concessions world-wide over the past few years. CNPC, which has the largest reach internationally, owns oil and gas assets and interests in 27 countries and provides oilfield services, engineering and construction in 49 countries worldwide. These three companies have sprung up over a short time from being large domestic companies to rank among the world's largest listed global companies. The Financial Times' ranking of the world's largest companies by market capitalisation in 2008 placed Petro-China/CNPC at number two, behind Exxon Mobil.[28] Thus, in Angola, for example, all three companies were reported as being among the likely bidders for the shares on one large off-shore block being sold by the current owner, the US oil company Marathon, alongside state oil companies from India and Brazil (Miles, 2008). (As it transpired, the share was ultimately purchased by Angola's state oil company, Sonangol, exercising their pre-emptive right to block sales to third parties.) However, it is important to note that all of the countries in which Chinese companies have invested have added to the set of companies with exploration or production operations, do not wholly displace Western companies and are rarely the operator that controls activities. For example, a joint venture between the Angolan state oil company (Sonangol) and Sinopec called SSI (Sonangol Sinopec International) holds non-operator shares in four of the 29 Angolan blocks and is a 50 percent joint venture partner in one other.

The relationship between the rise of the Chinese oil companies and approaches to avoiding the resource curse is complex and changing. It can be expected to change further in the future as the complexities of oil-related commercial and political relationships unfold. It may not be as negative as is often assumed. Thus far, two principal areas of impact have some discernable trends and a third can be hypothesised.

One impact is on the approaches proposed in the 'London-Washington consensus' – macro-management of oil revenues, transparency and high environmental and social standards for oil production and pipelines. Initially, China's emergence on the global oil economy clearly undermined the traction of this consensus, as was evidenced in Angola when the government broke off negotiations with the IMF in

---

[27] As well as the Chinese state oil companies, other state oil companies have started to invest internationally, providing increasingly strong competition for the Western oil majors. Prominent among these companies are Malaysia's Petronas, Brazil's Petrobras, Russia's Rosneft, and Norway's Statoil.

[28] See http://www.ft.com/reports/ft5002008. Note that the Forbes Magazine ranking, for which the criteria are unspecified, had PetroChina in 30th place in 2008 (with ExxonMobil at number five).

2004 following its agreement to borrow several billion dollars from China instead.[29] Since then, however, there have been signs of Chinese engagement in elements of this consensus, including attention to mitigating environmental and social impacts and, possibly, steps towards supporting revenue transparency.

The Chinese (or other Asian and Latin American) state oil companies are not currently participants in the Extractive Industries Transparency Initiative or in the Voluntary Principles on Security and Human Rights. However, the three large Chinese oil companies are all active participants in the United Nations' Global Compact, an 'entry level' organisation for companies making their first attempts to grapple with issues of corporate social responsibility for issues that have conventionally been understood as pertaining to government. These issues include human rights, labour standards, environmental protection and combating corruption (United Nations, 2010). Chinese banks, as well as direct investors internationally, are becoming aware on several levels of the need to consider the local environmental and social impacts of their overseas investments. For example, the October 2007 Congress of the Chinese Communist party mandated that state-owned companies must take ecological and social factors into account as well as profit and economic growth.[30] Moreover, in countries such as Nigeria, Ethiopia and Zambia, Chinese companies have started to experience the local hostility that induced Western companies to pay greater attention to their impacts.[31] As a result of these pressures, some Chinese banks and oil companies are developing their own environmental standards and edging towards acceptance of the Equator Principles (Xiaohua, 2008). The State-Owned Assets Administration (SASAC) has made it mandatory for each state-owned enterprise to have an enterprise risk management system.

However, alongside the June 2008 meeting of the G8, the energy minister of China, as well as those of Korea and India, joined the G8 in a public statement of support for the EITI: 'We welcome the efforts of countries exporting oil and gas as well as minerals that are implementing the Extractive Industries Transparency Initiative (EITI) on a voluntary basis to strengthen governance by improving transparency and accountability in the extractives sector' (G8, 2008). This signals potential future involvement by the government, Chinese companies or both. It is also noteworthy that when countries become full participants in EITI, all the oil companies with operations in that country are required to disclose information on

---

[29] 'But at the same time, Chinese support has placed Angola in a position where it could break ties with the International Monetary Fund (IMF) over economic support programmes that require, among other things, governance and transparency' (EIA, 2010).

[30] The October 2007 Party Congress concluded that 'sound and rapid economic development' should be accompanied by actions to 'better safeguard the people's rights and interests as well as social equity and justice, markedly increase the cultural and ethical quality of the whole nation, improve every aspect of the people's well-being, and promote a conservation culture.' See the full text of the resolution on CPC Central Committee report, October 22, 2007, http://english.cpcnews.cn/92245/6287753.html

[31] This Western corporate response is explored in greater detail in Shankleman (2007, see chapter 4 on Corporate Social Responsibility).

128 J. Shankleman

payments to the government. CNPC, for example, has investments in several countries that are candidate members of EITI such as Nigeria, Peru, Kazakhstan and Azerbaijan. To the extent that the governments of these countries take the requirements of EITI seriously, Chinese companies with shares in oil concessions will then become involved in disclosing payments to these governments. In 2010, Iraq joined EITI and CNPC became one of the four oil company representatives on Iraq's EITI Stakeholder Council.[32]

The entry of Chinese interests into the global oil economy is also changing the approach to avoiding the curse of oil via the very large-scale programme of making infrastructure loans to developing countries through the China Exim Bank. A 2008 World Bank study documented Chinese financial commitments to African infrastructure projects starting from less than USD 1 billion per year in 2001 – 'rising to at least $7 billion in 2006, then trailing back to $4.5 billion in 2007 . . . large enough to make a material contribution toward meeting Africa's infrastructure financing needs of $22 billion per year' (Butterfield, Chen, Foster, & Pushak, 2008). This investment involves soft loans,[33] largely – but not wholly – for power and transport projects, particularly hydropower and railways. Similar programmes are in place in Central Asia and Latin America. According to the World Bank, more than 35 countries in Sub-Saharan Africa have engaged with China on infrastructure deals, with the biggest recipients being Nigeria, Angola, Sudan and Ethiopia; that is, three oil-rich countries and one that might prove to have some commercially viable resources (Butterfield et al., 2008).

Since 2002, the government of Angola has had access to lines of credit from China, mostly through EXIM Bank, totalling at least USD 4 billion. The loans are tied to an agreement that initially included supply of 10,000 barrels of Angolan crude per day, approximately 10 percent of the then daily output of almost one million bpd[34]. The volume of oil has increased in subsequent loans, as has Angola's level of production (Allves, 2010; BP, 2008). They are also obligated to use Chinese contractors, selected through a tendering process, to execute much of the work. In the case of Angola's 2004 loan of USD 2 billion, the agreement was that 70 percent of the work would be awarded to Chinese state-owned enterprises selected through a tendering process (Butterfield et al., 2008, p. 27). As of mid-2008, five electricity projects supported by Chinese finance had been completed in Angola, as had an initial project to rehabilitate the Luanda railway, one water project and two telecom projects. Two road projects were under construction and a further tranche of financing had not yet been allocated to specific projects.

This infrastructure financing has a potential impact on 'resource curse' issues in several ways. One, the potency of which will only become clear in the longer-term, is that improvements in infrastructure should, in and of themselves, assist the

---

[32] See http://eiti.org/Iraq

[33] 'Chinese loans compare favourably with private sector lending to Africa but are not as attractive as ODA'(Butterfield et al., 2008, p. 7).

[34] *(Oil) barrels (159 liters) per day.*

development of the non-oil economy and contribute to poverty reduction by facilitating transport, communications and economic activity. They should turn oil wealth into productive infrastructure. Recent empirical work by Calderón and Servén on Sub-Saharan Africa showed that 'infrastructure development – as measured by an increased volume of infrastructure stocks and an improved quality of infrastructure services – has a positive impact on long-run growth and a negative impact on income inequality' (Calderón & Servén, 2010, p. i53). Campos and Vines, in a working paper published in 2008, argued that 'Although not easily measured, Chinese investment has contributed to poverty reduction in Angola. The construction and rehabilitation of electrical and hydroelectric infrastructure by the Chinese has expanded electricity access to over 60,000 new clients in Luanda. The rehabilitation of water supply systems across the country has granted thousands of people access to clean water' (Campos & Vines, 2008, p. 12). Furthermore, borrowing to invest in the construction of physical infrastructure represents a way of converting oil revenues into potentially productive in-country assets, as opposed to saving in a fund that invests offshore and, therefore, does not directly trigger increased capacity in the domestic economy. Although there were initial questions about the quality of construction work, attributed by some analysts to limited inspection and supervision by government authorities (Centre for Chinese Studies, 2005), Chinese contractors are proving themselves to be internationally competitive (as measured by their success in winning open tenders issued by the World Bank and the African Development Bank).[35] The presence of Chinese contractors introduces much-needed competition into Angola's engineering and construction sectors, which had previously been dominated by a small number of firms. The challenge for the incoming construction firms will increasingly be to meet the local expectation and local economic need for a larger proportion of the workforce to be Angolans rather than Chinese expatriates.

Competitive forces are also at work. In mid-2008, the World Bank announced that its financing and technical assistance on infrastructure could rise to USD 72 billion from 2008 to 2011 through the new Sustainable Infrastructure Development Plan (World Bank, 2008a). Although it is too early to tell how far the promise of this expansion of infrastructure can and will be delivered in practice, there is no doubt that the arrival of Chinese oil companies and Chinese finance in oil-producing states has opened up the debate on the resource curse beyond macroeconomics and governance and forced a new perspective onto conventional thinking that redirects the focus away from governance and corporate responsibility issues and towards roads, railways, light and power.

A third way in which Chinese involvement might have a beneficial effect on resource wealth management is through the demonstration effect on elite behaviour. The government of China professes a philosophy of engagement that is different

---

[35] 'In recent years, they (Chinese contractors) have accounted for more than 30 percent by value of civil works contracts tendered by these two multilateral agencies, which makes them substantially more successful than contractors of any other nationality' (Butterfield et al., 2008, p. 15).

in tone from the language of conditionality that dominates the London-Washington consensus. Speaking in London in February 2008, Liu Guijin defined the two principles underpinning relationships between China and Africa as 'treating others as equals and pursuing win-win cooperation' (Guijin, 2008). The history of China-Africa links allows for a shared vocabulary of 'anti-imperialism', even by politicians on both continents whose radical days are behind them. One aspect of the resource curse is elite reliance on oil revenues and a lack of focus on broader economic and social development. Yet, is it possible that engagement with a Chinese political elite that has managed to maintain political power, acquire wealth and create a policy environment enabling rapid and diversified economic growth could stimulate African oil elites to be more ambitious for their countries?

7.6 Looking Ahead

The trends of expanding oil production by a growing number of countries and the involvement of Chinese oil companies abroad can be expected to continue over the long term as China develops economically. Oil consumption per capita in China is less than one-tenth of that of the United States. Even with the predictable efforts to develop a less oil-intensive economy, substantial increases in demand from China will continue, albeit at a rate influenced by the state of the global economy. Without an effective model to overcome the resource curse, the exploitation of non-renewable resources will not secure a better future for the citizens of oil states and will present energy security and conflict risks to both the West and China.

The challenge for oil companies, governments, donors and NGOs is how to adapt to this swift change. The 'London-Washington consensus' on avoiding the resource curse has, arguably, neglected the importance of using oil wealth for investment in infrastructure and training and has underestimated the extent of anger in oil-producing countries about exporting energy from countries where the majority of the population does not yet have electricity. The Chinese approach risks underestimating the capacity of government elites who are wealth-empowered but corrupt and self-serving to undermine effective implementation of infrastructure programmes. It also risks potential anger over using Chinese labour for such programmes in regions where unemployment is already very high and developing oilfields and mines without attention to avoiding environmental impacts. However, the combination of the two approaches might just have the potential to trigger sufficient change in oil-rich states to overcome resource curse effects.

That said, the changing context provided by the opening up of the global oil industry to a far larger set of players, combined with the deep and sudden recession that hit the industry from late-2008, could create the political space for change. Alongside the economic development focus and record that China brings, these developments in the global oil economy could provide the opportunity to develop the new model that is required. A new model would seek to build on the unavoidable interdependencies between governments of oil-producing and oil-importing states and oil companies, wherever they come from, to create a development-centred

7 The Changing Context for Efforts to Avoid the 'Curse of Oil'

approach. Such an approach would apply a degree of rigour in managing the macroeconomic impacts of oil revenues, investing revenues in productive assets, building the local business and skills base to service the industry and avoiding environmental damage equal to the rigour now applied to developing the engineering and technology to be able to extract oil from fields deep below the ocean. Feasibility and impact studies should explicitly address revenue and development issues and adapt the design and pace of oil field developments accordingly. The emergence of the Chinese 'new kids on the block' and their commitment to promote economic development provides an opportunity to expand this dialogue.

References

Allves, A. C. (2010). *The oil factor in Sino-Angolan relations at the start of the 21st century.* South African Institute of International Affairs (SAIIA), Occasional Paper, No. 55. Johannesburg, South Africa: SAIIA. University of the Witwatersrand.

Arellano-Yanguas, J. (2009). *A thoroughly modern resource curse? The new natural resource policy agenda and the mining revival in Peru.* Institute of Development Studies. Brighton: University of Sussex. All web links accessed November 16, 2011. http://www.ids.ac.uk/files/dmfile/Rs300.pdf

Augé, B. (2009). Border conflicts tied to hydrocarbons in the great lakes region of Africa. In J. Lesourne (Ed.), *Governance of oil in Africa: Unfinished business.* Paris: Institut français des relations internationales.

BP. (2008). *BP Statistical review of world energy 2008.* http://www.bp.com (Reports and publications: Statistical review of world energy).

Brunnschweiler, C. N. (2008). Cursing the blessings? Natural resource abundance, institutions, and economic growth. *World Development, 36*(3), 399–419.

Butterfield, W., Chen, C., Foster, V., & Pushak, N. (2008). *Building bridges: China's growing role as infrastructure financier for Sub-Saharan Africa.* Washington, DC: World Bank.

Calderón, C., & Servén, L. (2010). Infrastructure and economic development in Sub-Saharan Africa. *Journal of African Economies, 19*(suppl 1):i13–i87.

Campos, I., & Vines, A. (2008, June 4). Angola and China: A pragmatic partnership. *Center for Strategic & International Studies.* http://csis.org/publication/angola-and-china

Centre for Chinese Studies. (2005). *China's interest and activity in Africa's construction and infrastructure sectors.* http://www.dfid.gov.uk/pubs/files/chinese-investment-africa-summary.pdf

Collier, P. (2007). *The bottom billion: Why the poorest countries are failing and what can be done about it.* Oxford: Oxford University Press.

Dirlik, A. (n.d.). Beijing Consensus: Beijing 'Gongshi.' Who recognizes whom and to what end? In *Globalization and autonomy online compendium.* http://www.globalautonomy.ca/global1/servlet/Position2pdf?fn=PP_Dirlik_BeijingConsensus

EIA, Energy Information Administration. (2008, March). *Angola energy data, statistics, and analysis – oil, gas, electricity, and coal.* Washington, DC: US Department of Energy.

EIA. (2010, January). *Angola.* http://www.eia.doe.gov/emeu/cabs/Angola/Full.html

G8. (2008, June 8). *Joint statement by energy ministers of G8, The People's Republic of China, India and The Republic of Korea.* Aomori, Japan.

Gould, J. A., & Winters, M. S. (2007). An obsolescing bargain in chad: Shifts in leverage between the government and the World Bank. *Business and Politics, 9*(2). http://www.bepress.com/bap/vol9/iss2/art4

Guijin, L. (2008). *Darfur and Sino-African relations.* London: Chatham House.

IFC, International Finance Corporation. (2007, 31 July). *Performance standards on social and environmental sustainability* http://www.ifc.org/ifcext/sustainability.nsf/AttachmentsByTitle/pol_PerformanceStandards2006_full/$FILE/IFC+Performance+Standards.pdf

IMF, International Monetary Fund. (2007). *Guide on resource revenue transparency*. http://www.imf.org/external/np/pp/2007/eng/051507g.pdf

Kaletsky, A. (2008, July 3). The world must kick its addiction to oil. *The Times*. http://www.thetimes.co.uk

McCashie, T. C. (2008). The United States, Ghana and oil: Global and local perspectives. *African Affairs, 107*(428), 313–332.

Miles, T. (2008, August 11). Chinese big three, ONGC seek marathon Angola stake. *Reuters*. http://www.reuters.com/article/2008/08/11/marathon-angola-idUSN1135685920080811

Paltseva, E. (2010). *Autocracy, devolution and growth*. http://www.kellogg.northwestern.edu/research/fordcenter/documents/100524%20Paltseva.pdf

Pérez Alfonso, J. P. (1976). *Hundiéndonos en el excremento del diablo*. Caracas: Editorial Lisbona.

Ramo, J. C. (2004). *The Beijing consensus: Notes on the new physics of Chinese power*. London: Foreign Affairs Policy Centre.

Sachs, J. D., & Warner, A. M. (1995). *Natural resource abundance and economic growth*. NBER working paper, No. 5398. Cambridge, MA: Harvard University, Center for International Development and Harvard Institute for International Development.

Sachs, J. D., & Warner, A. M. (2001). Natural resources and economic development: The curse of natural resources. *European Economic Review, 45*(4–6), 827–838.

Sen, A. (2006). *Identity and violence: The illusion of destiny*. New York: W. W. Norton.

Shankleman, J. (2007). *Oil, profits and peace: Does business have a role in peacemaking?* Washington, DC: US Institute of Peace Press.

Shell Nigeria. (2006). *People and the environment*. Shell Nigeria annual report. http://narcosphere.narconews.com/userfiles/70/2006_shell_nigeria_report.pdf

Shell Nigeria. (2011, April). *Our economic contribution*. Shell in Nigeria. http://www.shell.com/static/nga/downloads/pdfs/briefing_notes/economic_contribution.pdf

Stevens, P. (2004). Resource curse and investment in energy industries. In C. Cutler (Ed.), *Encyclopedia of energy* (Vol. 5, pp. 451–459). London: Elsevier.

Suncor Energy. (2007). *Annual report*. http://www.suncor.com/pdf/ic-annualreport2007-e.pdf

Transparency International. (2010). *Corruption perceptions index*. http://www.transparency.org/policy_research/surveys_indices/cpi/2010

United Nations. (2010, November 22). *Global compact*. http://www.unglobalcompact.org

UNDP, United Nations Development Programme. (2010). *Human development index*. http://hdr.undp.org/en/statistics

United States Senate Committee on Banking, Housing, & Urban Affairs. (2010). *Brief summary of the Dodd-Frank Wall Street Reform and Consumer Protection Act*. http://banking.senate.gov/public/_files/070110_Dodd_Frank_Wall_Street_Reform_comprehensive_summary_Final.pdf

World Bank. (2007). *Doing business*. Washington, DC: The World Bank.

World Bank. (2008a, July). *Sustainable infrastructure action plan*. Washington, DC: The World Bank Group.

World Bank. (2008b). Extractive Industries Transparency Initiative Plus Plus (EITI++, 2008). *The World Bank Group*. http://web.worldbank.org/WBSITE/EXTERNAL/EXTSITETOOLS/0,,contentMDK:21727814~pagePK:283622~piPK:3544780~theSitePK:95474,00.html

World Bank. (2009). *Press statement*. Press release no: 2009/073/AFR, September 9, 2008.

Xiaohua, S. (2008, Jan 25). *China to bring in green loan benchmark*. The Equator Principles. http://www.equator-principles.com/index.php/media-and-news

Yergin, D. (2003). *The prize: The epic quest for oil, money, and power*. New York: Free Press.

Zanzibar says no oil sharing (2008, July 17). http://allafrica.com/stories/200807180026.html

Chapter 8
Contributions of Economics and Ethics to an Assessment of Emissions Trading

Adrian Muller

Abstract Emissions trading (ET) is an economic policy instrument designed to manage environmental pollutants such as carbon dioxide. Economic analysis provides many valuable insights into the design and implementation of ET systems, especially on efficiency-related issues. However, with regard to topics outside its core competences, such as justice, economics needs to be complemented by contributions from other disciplines. After an overview of the economics of emissions trading, this chapter assesses ethical aspects of emissions trading as a showcase for how other disciplines can complement the economic approach. The chapter finds that the large majority of ethics-based objections to emissions trading are based on perceived problems of injustice. The only other such argument that can be consistently adduced is indulgence trade in carbon reduction; that is, that it matters, from an ethical perspective, who does the reducing.

Keywords Economics · Efficiency · Emissions trading · Ethics · Greenhouse gases · Justice · Policy instruments · Regulation

8.1 Introduction

Emissions trading (ET) is the trade of a special, newly created 'good', namely 'the right to emit a certain amount of a pollutant' (for example, one ton of CO_2) in a special, newly created market. Since there are hardly ever circumstances in which an energy user's ability to emit a gaseous pollutant is locally constrained by physical or chemical factors, this constraint must be politically generated; actors must be coerced into perceiving the right to emit as a scarce good.

A. Muller (✉)
University Research Priority Programme in Ethics (UFSPE), University of Zurich, 8008 Zurich, Switzerland

Institute for Environmental Decisions (IED), Chair of Environmental Policy and Economics (PEPE), ETH Zurich, 8092 Zurich, Switzerland
e-mail: amueller@env.ethz.ch

D. Spreng et al. (eds.), *Tackling Long-Term Global Energy Problems*,
Environment & Policy 52, DOI 10.1007/978-94-007-2333-7_8,
© Springer Science+Business Media B.V. 2012

Once a government has made the policy decision to reduce emissions of a pollutant in its jurisdiction by means of an emissions trading system, it must decide on the scope of participation (coverage). This may be a majority of emitting industries only, all industries, industry and households or another combination. Next, it sets a cap on the level of aggregate emissions in order to generate a scarcity of emission rights. For example, it may adopt its national reduction goal under the Kyoto Protocol, which differs by country, but lays at 5 percent less than 1990 baseline emissions by 2012 on average. Thirdly, it establishes a mechanism to generate demand and supply at positive prices. It does this by allocating to participants a quantity of emissions permits that is equivalent in aggregate to the cap. This distribution determines the level of participants' individual endowments and thus – in combination with their abatement activities – their demand for additional permits or the supply of excess permits. Consequently, a price for emission rights emerges. Participants engage in emissions abatement up to the level at which the marginal cost of additional abatement equals the price for an emissions permit. Beyond this level, they will choose to purchase emissions permits at the market price. ET is therefore a market-based policy instrument that harnesses market forces to achieve its goal.

Since emissions trading is primarily a product of economics, economic analysis naturally goes the farthest of any social science towards understanding ET's intricacies. However, other disciplines also offer crucial insights. Law contributes to answering the myriad legal questions that arise when a new policy instrument is introduced and natural sciences are needed to understand the dynamics of the pollutant to be regulated. If the underlying natural science is known and accounted for and if adequate rules, monitoring and enforcement mechanisms are established, under ideal conditions, ET will guide society towards a welfare-maximising level of emissions.

Economic analysis has much to offer with regard to challenges related to efficiency, both on a descriptive and a prescriptive level.[1] However, important aspects besides efficiency, such as justice, are beyond its grasp, as are philosophical, psychological, cultural and sociological questions such as whether it is 'right' to make pollution a good that can be traded as any other and the extent to which such commercialisation is accepted in society. Beyond efficiency, economics may provide a good description of past, current and hypothetical situations – an important contribution in itself – but here it has nothing prescriptive to offer. Furthermore, depending on the importance of these additional issues in a specific case, economics' blindness to them may actually invalidate economic solutions. Given that other social sciences are necessary to complement economic analysis, this chapter explores how the collaboration of different disciplines may function for ET.

This chapter is organised as follows: the next section states the paper's main goals. Section 8.3 introduces the European Emission Trading System (EU-ETS) as an example of an emissions trading system and Section 8.4 discusses how economic

---

[1] Increasing economic efficiency refers to increasing (aggregate) welfare, for example by achieving an equivalent aggregate consumption level at lower cost. In the energy debate, increased efficiency often reflects the internalisation of external costs.

8 Contributions of Economics and Ethics to an Assessment of Emissions Trading 135

analysis can deal with several of its important aspects. Section 8.5 discusses four key topics that can only be assessed by going beyond economic analysis and Section 8.6 concludes the discussion.

8.2 Goals

This chapter has two goals. Firstly, it seeks to illustrate how a disciplinary approach can be useful for examining certain aspects of sustainable development, especially when analysis focuses on the discipline's core competences. Secondly, it also aims to show that a disciplinary approach has fundamental topical and methodological limitations and that, for areas that lie beyond its core competences, it needs to be complemented with insights from other disciplines.

This is illustrated using the example of emissions trading and ethics. Because ET is primarily an economic contribution to the problem of pollution externalities, this study discusses the wide range of issues economics can fruitfully deal with and the considerable challenges it faces when doing so. It then illustrates how ethical analysis complements economics. Clearly, the humanities and many other social sciences also complement economics and would need to be considered in any all-encompassing assessment of ET. Ethics plays a special role, however, since recurrent core topics in the humanities and the social sciences are ultimately linked to ethical aspects (such as equity and justice). In addition, broad coverage of other disciplines is beyond the scope of this chapter.

Lessons from this study may serve as policy input to the implementation of a new ET system in a particular region. Lessons learned from the EU-ETS, for example, should not only address primarily economic issues, such as those discussed in Stavins (2007), with reference to a future ET in the US. They should also explicitly address aspects beyond economics, such as the ethical approaches discussed below.[2]

This chapter emphasises the dichotomy between economics and 'other social sciences' and humanities (viz., 'ethics' in this case). This dichotomy is often not nearly as pronounced as portrayed here, but emphasising it can serve pedagogical purposes. A mainstream view on economics is used in order to illustrate the degree to which mainstream economics can contribute, which is appropriate since much of the policy-relevant economic work on ET is mainstream.

As useful as the separation of efficiency from other aspects (such as distributional ones) can be in analysis, integration is crucial in practice. Far too often, the analytical separation of distributional aspects from efficiency leads to a neglect of distributional aspects in policy solutions based on economic approaches. This is then reflected in the lack of dialogue between proponents and critics of these

---

[2] Stavins (2007) briefly mentions the objection to ET that claims it is unethical to trade in pollution (p. 54). The way he rebuts this criticism shows that he does not take the issue seriously: 'However, few would agree that people are behaving immorally by cooking dinner, heating their homes, turning on a light, or using a computer. Yet all of these activities result in CO_2 emissions.'

approaches (cf. the discussion of ethical aspects in Stavins, 2007 and of economics' contributions in Lohmann, 2006, 2009).

8.3 Example: The European Union's Emissions Trading Scheme

ET systems are mainly established in the energy context but they have also been implemented in other areas such as water quality regulation. This section focuses on the European Union's Emissions Trading Scheme (EU-ETS) and later sections frequently refer to the EU-ETS for illustration. Details on many other ET examples can be found in Kraemer, Interwies, and Kampa (2002), OECD (1999, 2001), Stavins (2003), Sterner (2003), Freeman and Kolstad (2007), Tietenberg (2006) and IETA (2008).

The fundamentals of the EU-ETS, Europe's cap-and-trade system for CO_2 that started in 2005, will not be covered here. They are well described in Ellerman and Joskow (2008), Anderson (2008), Convery, Ellerman, and De Perthuis (2008), Ellerman and Buchner (2007), Ellerman (2008), Stavins (2007) and the official EU websites (EU, 2003, 2004, 2006, etc.). A wealth of empirical information is available from reports by Point Carbon and EEA (Point Carbon, 2008; EC, 2011; EEA, 2011, 2008 and previous years) and the issue of competitiveness is treated in Hourcade et al. (2007). Table 8.1 lists the basic features.

The EU-ETS is institutionally independent of the Kyoto Protocol (KP) insofar as it started operation with the trial period from 2005 to 2007 before the KP entered into force in February 2005 and as it is expected to remain in operation even if no agreement is reached on a successor for the KP for the post-2012 period. This is reflected in the rules for the subsequent trading period 2013–2020 (EU, 2009), which, besides institutional streamlining, increases the scope of the EU-ETS to include certain additional activities (for example, ammonia and aluminium production) and gases (such

Table 8.1 Basic characteristics of the EU-ETS

Starting date	1.1.2005
Periods	2005–2007/2008–2012/2012 onwards (details to be agreed)
Gases traded	CO_2 (maybe N_2O after 2012)
Industries covered	11,000 installations, power; cement; iron and steel; glass, ceramics and bricks; pulp and paper; refineries
Participating countries	EU-25, since 2006 EU-27
Initial allocation	Mainly grandfathering; increasing percentage of auctioning is planned in 2012
CDM credits	Can be used with some restrictions regarding volumes and types of projects
Volumes	50 percent of EU CO_2 emissions, 40 percent of EU total GHG emissions
Cap	2298 Mt CO_2 annually in 2005–2007 and 2028 Mt CO_2 annually in 2008–2012
Emissions	2125 Mt CO_2 annually in 2005–2007

8 Contributions of Economics and Ethics to an Assessment of Emissions Trading

as N_2O). Furthermore, the European Council has committed to at least 20 percent greenhouse gas (GHG) reductions by 2020, irrespective of other countries' actions or the presence of an international agreement (30 percent if other countries aim for comparable reductions).

8.3.1 Performance

Currently, assessment of the EU-ETS is only available for 2005–2007, as the second period runs till the end of 2012. The trial period showed that the EU succeeded in establishing a functioning emissions trading scheme. A single price for permits emerged in a liquid market and firms factor in this price in their business decisions: these are among the EU-ETS's main achievements. The trial period also provided important lessons on how the distributive and efficiency aspects of allocation could be separately treated. On the other hand, emissions were only slightly reduced (EC, 2008). This was to be expected, however, with a cap of only a few percent below the business-as-usual (BAU) scenario. In addition, a scarcity of accurate data on emissions meant that fluctuations (such as weather and macroeconomic effects) could easily result in overly generous caps. Furthermore, political pressure to get the system up and running quickly was high and the process involving all the member states was complex. Just how much abatement was actually undertaken is difficult to estimate, as deriving this would require the construction of a reliable counterfactual without the EU-ETS in place.

A decisive design flaw of the trial period was its separation from the subsequent periods, as no banking of permits from 2005–2007 to the next period was permissible. This led to a crash in the price of first-period permits in 2006 as an oversupply materialised that could not be used for compliance after 2007. However, the 2005–2007 period had always been planned as a trial period intended for developing the infrastructure and improving the design and implementation of the system for subsequent periods. The trial period was also important because of the lack of experience with ET in Europe. This learning phase has had some positive influence on the second period and the discussion on the EU-ETS after 2012, as seen in the national allocation plans and commission decisions on their acceptance, for example (lower caps, some harmonisation, etc.). In addition, the infrastructure for a functioning ET, such as market institutions, emissions registries, monitoring, reporting and verification procedures, is now in place. In general, the optimistic characterisation of the EU-ETS is reflected in this quote from Ellerman and Joskow (2008, p. iv): 'The initial challenge is simply to establish a system that will demonstrate the societal decision that GHG emissions shall have a price and to provide the signal of what constitutes appropriate short-term and long-term measures to limit GHG emissions. In this, the EU has done more with the ETS, despite all its faults, than any other nation or set of nations.'

However, Anderson (2008) adopts a more critical stance. In particular, he emphasises the importance of political economy aspects in the context of national allocation: '[...] allocation setting is a process fraught with technical difficulty and tough political choices, where industry holds an information asymmetry over regulators and national governments can produce projections of methodologies, designed to protect their industries' (p. 37). Here we may also mention the various frauds that have been perpetrated in the EU-ETS over the past few years (e.g., EC, 2010). However, these do not reflect fundamental problems in the ETS but rather certain design flaws and insufficient security in the accounting systems.

8.3.2 Criticism

A clear area for improvement is the *highly decentralised character* of the EU-ETS, which affects almost every aspect of the system, most prominently national caps and allocation procedures. There have been suggestions to change this drastically, which would result in a much more centralised system in which elements such as national allocation plans would no longer be necessary. This would lead to the emergence of a much more harmonised and homogeneous system and the avoidance of inequalities in the treatment of similar installations, closures and new entrants in different countries (EU, 2008). The system will also improve in the future because, unlike during the trial period, unrestricted banking (but no borrowing) is possible for 2008–2012 and subsequent periods. In addition, the added provision for unrestricted banking in the 2008–2012 period will dampen price volatility and, together with stricter price caps, will greatly reduce the likelihood of price crashes such as the one in 2006.

Another main point of criticism is directed towards over-allocation. This refers to the fact that the emissions cap in the 2005–2007 trial period was not binding, meaning that allocations to the units covered in the system were so generous that, overall, no scarcity for emission rights emerged. This became clear in 2006 when the data on 2005 emissions were made public. The problem stemmed partly from a lack of accurate data on previous emissions, the difficulty of defining reliable BAU scenarios and a moderate reduction goal of only a few percent. Under such conditions, fluctuations in weather and economic factors can turn a moderate scarcity into a surplus. This makes it difficult to set a binding cap. With more stringent reduction goals, such as those in place for the second period and those planned for the third period after 2012, the danger of over-allocation is minimal. Other, more critical voices have emphasised the role of industry interests and lobbying for high allocation (for example, Anger, Böhringer, & Oberndorfer, 2008). Over-allocation and corresponding low abatement in the covered sectors places a heavy compensatory burden on other sectors to meet national targets under the Kyoto Protocol. Whether over-allocation will persist as an issue will only become clear later in the 2008–2012 period. Since the end of 2008, however, carbon prices have dropped sharply and repeatedly. This can be attributed to decreased demand due to the worldwide financial crisis and, consequently, reflects over-allocation in light of these unforeseen circumstances (cf. CarbonPositive, 2009).

Windfall profits refer to specific issues regarding the interaction of the EU-ETS with the electricity industry. In a liberalised electricity market, the price for an emissions permit corresponds to the opportunity costs of using it for compliance with one's own cap rather than selling it. Electricity generators that receive permit allocations for free include the market permit price in their cost calculations, as the marginal generation costs include the costs of CO_2. Positively priced, freely-distributed permits will therefore lead to increased electricity prices and additional revenues for the electricity producers. This 'selling' of permits that were received for free generates 'windfall profits' for industry, which have been heavily criticised in the political process. From an economic standpoint, the increase in electricity prices in a liberalised market with an ET system in place is efficient. The situation would be similar with auctioned permits, as generators would be allowed to transfer these permit costs to the consumer. This discrepancy between economic efficiency, which prescribes pricing in opportunity costs of freely-allocated permits, and ethical considerations that highlight the injustice of selling permits received at zero cost, is notable.

8.3.3 Key Issues

Coverage is not complete in the EU-ETS. Affected agents include only units in the most highly-emitting industries and traffic, households and agriculture are not included. Such partial coverage can lead to uncertainties in emissions and reduction costs. This is of particular relevance in the context of the Kyoto Protocol, as the countries participating in the EU-ETS face national caps. Part of the task of keeping below the caps will be achieved within the sectors covered in the EU-ETS but any remaining gap in meeting the Kyoto goals has to be borne by the sectors that are not covered. Generous allocation in the EU-ETS may therefore result in large reduction requirements for the other sectors. This is not efficient, as marginal reduction costs can thus diverge significantly between sectors covered and those not covered. Partial coverage that focuses on large industrial units clearly has some organisational advantages regarding issues such as registries, monitoring and verification. However, the scope of the EU-ETS will increase from 2013 onwards; air traffic, for example, will be included in the system. Not all analysts are in favour of a more comprehensive extension of scope, though. Tilford (2008), for example, counsels against including the road transport or agriculture sectors because of the high monitoring costs involved and the expected low incentives for change if agents are small-scale emitters. The other risk of incomplete coverage in the EU is leakage, whereby emission-generating activities are relocated to unregulated countries (see, for example, Hourcade et al., 2007).

The *cap* set in an ET system should be determined by the damage costs of emissions in relation to abatement costs and benefits from emissions (such as economic growth). The EU has committed itself to caps that further the policy goal of keeping the global average temperature from increasing by more than 2°C. However, it is very difficult to translate temperature increases to an optimal emissions path and estimates for necessary reductions to achieve a specific goal are uncertain (e.g., den Elzen, Meinshausen, & van Vuuren, 2007). The EU-ETS started with a modest cap,

140 A. Muller

which can also be understood as a compromise designed to get the system running. Increasingly strict caps will be imposed in the future and the legislation for the post-2012 period (EU, 2009) includes a reduction path with a cap that decreases by 1.74 percent annually. This corresponds to the EU's reduction goal of 21 percent below 2005 levels by 2020 for the sectors covered.

With regard to the *initial permit allocation mechanism*, most permits in the EU-ETS were allocated for free (grandfathered), with levels depending on historical emissions. Although the rules allow for the possibility of auctioning part of the permits (5 percent in the trial period and 10 percent in 2008–2012), only a small fraction of the permits was or is planned to be auctioned in the first two periods (Ellerman & Joskow, 2008). From 2013 to 2020, a much larger share will be auctioned (EU, 2009): 100 percent for the power sector from 2013 onwards should be the rule and an increasing percentage for the other sectors, reaching 70 percent in 2020, with plans to reach 100 percent in 2027. As noted above, the main problem with free allocation is not so much the disadvantageous effects on efficiency if designed incorrectly as it is the huge wealth transfer to private firms (two billion permits annually generates a huge amount of money, even at a low permit price of 10 Euros), none of which accrues directly to the government. This wealth transfer has also triggered considerable rent-seeking behaviour in the covered industries. On the other hand, without grandfathering, industry opposition would likely have been much stronger and the EU-ETS would probably not have become operational as quickly as it did. Initial allocation is of particular relevance for new entrants and plant closures; the rules here are still diverse but there are plans to harmonise them.

The importance of distributive issues is acknowledged in the context of the EU-ETS, but they are usually seen as falling outside the scope of economics. Thus, Ellerman and Joskow (2008), for example, state that 'The key policy challenge is to ensure that the mechanisms used to achieve distributional goals and to resolve political economy challenges do not distort abatement behaviour or competition [i.e., efficiency challenges]' (p. 39).

8.4 The Economics of Emissions Trading

This section discusses the key economic challenges involved in the concrete implementation of emissions trading. The three issues that were the focus of the example in Section 8.3 (coverage and agents involved, cap, initial allocation) are analysed from the perspective of economics. Many other aspects that are relevant to this case fall within the core competence of economics and some of these are briefly reviewed in Section 8.4.4.[3]

---

[3] General references for Section 8.4 are Kosobud (2000), Stavins (2003), Sterner (2003), Schneider and Wagner (2003) and Tietenberg (2001, 2003, 2006) and, specifically for carbon regulation, Neuhoff (2008). For key issues and examples, see also OECD (1999, 2001). On the origins and history of ET, see also Gorman and Solomon (2002). Inspiring discussions of key topics from

8 Contributions of Economics and Ethics to an Assessment of Emissions Trading   141

Addressing the potential problems related to these key issues falls mainly within the competence of economics. At the same time, concrete implementation is governed by many non-economic considerations; for example, in the likely case that the choice of relevant agents for an ET is made through a process of political deliberation. Even here, however, economic arguments provide important insights in such areas as uncertainty, incomplete information, lobbying and market power.

Economics can often clarify issues and derive several possible chains of cause and effect but it may be difficult to decide on their validity for a specific situation. In these cases, complementing economic analysis with other disciplines can then provide additional insights.

8.4.1 Coverage/Relevant Agents

The first essential step in creating an ET system is to identify the relevant agents; that is, the emitters subjected to regulation. The EU-ETS, for example, has partial coverage, only including firms in the six most highly emitting sectors (cf. Section 8.3).

Agents are best chosen under the aspect of 'similarity'. An ET system works ideally with participants that are equal in all relevant aspects apart from their marginal abatement costs; all such agents should[4] then be included in the system. This ideal situation seldom prevails, however, and several other criteria can influence the choice of agents.

Properties of the Pollutant

Firstly, the physical properties of the pollutant to be regulated should be taken into account. Agents anywhere in the world can take part in an ET system for the reduction of CO_2 since it is a global pollutant. For a local pollutant, however, agents are dissimilar in terms of the geographical distribution of the adverse effects of their pollutant emissions and the geographical boundaries of the ET system should account for this (e.g., Krysiak & Schweitzer, 2006; Atkinson & Morton, 2004). Thus, sulphur dioxide (SO_2) emissions in the US can be traded by energy producers in the US but this trade should not be global, as that would lead to inefficient outcomes due to the possibility of geographic high pollution hot spots. Even within the US, certain wind patterns should be taken into account in order to achieve optimal solutions.

Similarly, flexibility in time ('borrowing and banking'), which can greatly reduce compliance costs (e.g., in the lead trade, Newell & Rogers, 2006), may be restricted

---

economics, legal issues and implementation are offered in Toman (2003). See also the introductory chapter in Muller and Sterner (2006) for a collection of important economic aspects to be considered when implementing market-based policy instruments.

[4] The prescriptions here are made with reference to the efficiency of their outcomes. An action 'should' or 'should not' be performed depending on whether it is more or less efficient, respectively.

142   A. Muller

in order to avoid temporal 'hot spots' (e.g., Kling & Rubin, 1997; Schennach, 2000). Long-term temporal aspects also matter. The dynamics of GHG concentrations and global warming, for example, suggest several emission reductions paths that limit borrowing on a larger time-scale, assuming a certain target is to be reached (cf. den Elzen et al., 2007).

Market Power and Lobbying

Apart from the similarity of agents with respect to the physical properties of their emissions, economics requires a 'level playing field' for the agents involved. Key characteristics such as market power must be sufficiently similar for a market to function. Several authors have addressed this in the context of ET (e.g., Misiolek & Elder, 1989; Liski & Montero, 2006; Hagem & Westskog, 2009; Maeda, 2003). A related issue is lobbying, which some observers believe resulted in an overly generous initial allocation and the dramatic permit price crash (cf. Section 8.3). Lobbying is typically directed towards influencing the choice of a policy instrument and keeping the costs to affected industries low (e.g., Anger et al., 2008).

Coverage

An ET system should not include agents that are highly dissimilar and dissimilarities may trigger action to reduce them. On the other hand, dissimilarities between largely similar agents often have political causes. The most extreme manifestation of this is in an ET system with partial coverage in which only a fraction of all emitting agents can participate, as in the EU-ETS (see Section 8.3). All countries in the EU-ETS have overall reduction goals but these goals cannot be met through the firm-specific goals for the sectors covered in the EU-ETS alone; other sectors must compensate for the gap. Due to generous initial allocation to the participating sectors, the burden on the other sectors may be higher than it should be according to efficiency criteria (that is, equalising marginal abatement costs; see also Stavins (2007) for further economic discussion of the disadvantages of partial coverage).[5]

A global instance of this politically created dissimilarity was generated by the Kyoto Protocol, under which developing countries do not take part in ET directly, as they do not have specified reduction goals. The Clean Development Mechanism (CDM, Boyd et al., 2009), however, is equivalent to ET in its use of purchases to shift abatement obligations to other parties. This view of the CDM pictures developing countries as emissions trading partners that are not subject to a cap but are nevertheless permitted to sell certain emission reductions, namely those realised under the CDM. In so doing, these countries meet a higher standard than they

---

[5] Full coverage is best achieved by 'upstream' regulation; that is, '[...] on fossil fuels at the point of extraction, processing, or distribution, not at the point of combustion' (Stavins, 2007, pp. 7, 17).

8 Contributions of Economics and Ethics to an Assessment of Emissions Trading 143

would under 'business as usual'.[6] One problem with this interaction is the potential exhaustion of cheap reduction options in the case of late-entry by developing countries that are then left only with expensive abatement options (Muller, 2007; Narain & t'Veld, 2008). This problem would not arise in a system in which all participants were covered from the beginning. On the other hand, establishing ET among firms from developing and developed countries can be problematic for the power asymmetries and differences in marginal utilities from consumption. Sheeran (2006) points out that equal marginal abatement costs that emerge in an ET system are not efficient when the involved agents are highly dissimilar. The participation of developing countries in ET can thus be judged positively or negatively depending on the focus of analysis. Section 8.5.4 discusses distributive aspects and justice in relation to ET and developing countries in greater detail. The fact that some sectors are not included in the EU-ETS and that some countries do not face caps under the Kyoto Protocol reflects concessions made in the policy process in order to avoid blocking implementation of the Protocol or the EU-ETS. Thus, while economics can inform the optimal (i.e., 'efficient') ET system, the final design is often shaped in the political process.

8.4.2 The Cap

The cap for each trading period determines the size of the market as well as the 'scarcity' of the good 'emissions rights'. The cap is usually set as a fixed limit to the total aggregate emissions of the relevant agents. This is the case in the EU-ETS, where the cap is set nationally as an annual average fixed emissions quantity for each trading period for the covered sectors and varies across countries (cf. Section 8.3). Other possibilities are 'flexible' caps, such as 'intensity caps' that set a limit on quantities such as emissions per GDP. Here, the flexibility is intended to avoid constraining economic growth (e.g., Dudek & Golub, 2003; Ellerman & Sue Wing, 2003; Kolstad, 2004; Jotzo & Pezzey, 2007). This may be helpful for integrating developing countries in a post-Kyoto agreement.

Ideally, the level of the cap should be chosen based on the characteristics of the pollutant. Knowledge of the damage function (i.e., a function linking emission levels to damages incurred) *and* the translation of these damages into a one-dimensional monetary dimension) are necessary in order to settle on a policy that is optimal (i.e., efficient), in that the gains from avoided damages equal the costs of avoidance measures at the margin (see, e.g., Sterner, 2003). However, uncertainty often makes it impossible to produce an accurate quantification of the damages from a pollutant and to derive a reliable damage function. Future damages from climate change, for example, are subject to ongoing controversy, and it is virtually impossible to reliably assess and monetise potential future damages from a large sea-level rise due

---

[6] This additionality is, however, a main point of criticism of the CDM, as it is often not met (cf., e.g., Michaelowa & Purohit, 2007; Michaelowa, 2007).

to melting polar ice-caps or disruption of the global food system due to increased water stress. Furthermore, any aggregate cap has to be translated into national emission goals. This necessitates decisions on distributive issues (cf. Section 8.5.4). The analysis is further complicated by the uncertainty of future abatement costs.

Economic theory suggests that where the damage function is uncertain, relatively steep damage functions (with respect to mitigation costs) are better regulated by fixing the pollutant's quantity (as with ET's quantity cap) than its price (e.g., through a tax). The thinking behind this is that steep damage functions may reflect stepwise, irreversible, catastrophic damages beyond a certain level of pollutant emissions. It is advantageous, therefore, to set a cap below this threshold and to avoid the uncertainty in aggregate emissions that comes with a price instrument (Weitzman, 1974; Krysiak, 2008).

8.4.3 Initial Allocation

A national emissions cap must be translated into sector- or firm-specific caps that reflect the amount of emissions they are allowed to emit for free (this amount can also be zero, as is the case with auctioned emission rights). The scheme that is usually favoured, on efficiency grounds, is an auction (for a recent review on auctions in ET, see Neuhoff & Matthes, 2008). Depending on the context, however, the advantage of auctioned permits over other allocation mechanisms is not always clear (e.g., Fischer, Parry, & Pitzer, 2003). Other mechanisms are free initial allocation based on historic emissions (grandfathering, as was largely done in the EU-ETS, see Section 8.3) or tied to historic or current production. These last mechanisms often involve 'updating', whereby the relevant baseline is adjusted after the passage of some years.

The choice of allocation mechanism has consequences for the distribution of costs. Emissions trading via auctioned permits follows the polluter-pays principle, while ET with grandfathered permits does not. When permits are grandfathered, costs related to polluting up to the level of the emissions cap are borne by society, not by the polluters. The concrete choice and design of the policy instrument, therefore, has significant ramifications for the bearers of these costs, even though the same environmental goal is achieved in either case (e.g., Sterner, 2003, chapter 15). Initial allocation involves huge sums of money and stimulates lobbying, which was seen as important in setting up both the EU-ETS and the SO_2 trade in the US (cf. Section 8.3; Joskow & Schmalensee, 1998). Economics can guide the choice of allocation mechanisms based on their efficiency and it can suggest which cost distribution is likely to emerge. However, a normative judgement on which cost distribution is most 'desirable' is beyond its scope.

Initial allocation also influences agents' strategic decisions. Linking allocation with factors that can be influenced by the agents has negative effects on efficiency. This occurs with updated allocation, for example, when firms anticipate that increased output will lead to increased allocation (e.g., Böhringer & Lange, 2005a, b, Mackenzie, Hanley, & Kornienko, 2008).

8.4.4 Further Aspects

This subsection briefly addresses further topics that lie within the key competence of microeconomic and macroeconomic analysis. Many topics could be considered, including effects on employment, the comparative efficiency of subsidies for research and development of energy-efficient or low-carbon technologies versus setting policy incentives for direct mitigation; the prospects for financial speculation within ET systems and potential adverse effects of it; the possibility of reaping a double dividend[7]; the effects of a 'safety valve' (i.e., a price cap to hedge against price spikes); and how transaction costs (e.g., Stavins, 1995) and differences in institutional frameworks between developing and developed countries may change the adequacy and operation of certain policy instruments (e.g., due to insecure property rights or problems with monitoring and enforcement, cf. Somanathan & Sterner, 2006). Another possible topic is firms' strategic responses to the introduction of an ET. As an example, it is sometimes claimed that ET reduces incentives for domestic innovation, since permits can be bought abroad to avoid the necessity of domestic reductions. It has been claimed that firms are more likely to relocate in response to the introduction of an ET system (the 'pollution haven hypothesis'; see, e.g., Millimet & List, 2004), while others assert that environmental regulation triggers efficiency increases and innovation that increase competitiveness (the Porter hypothesis). Although much has been written on all of these topics in the field of economics, space considerations have limited the following discussion to a brief treatment of linking separate ET systems and to the relationship between ET and other policy instruments, public good provision and regressivity.

It is important to assess how ET systems interact with other policy instruments, such as alternative climate change mitigation instruments, to market liberalisation policies in the energy sector and to policies to promote renewable energy. For example, what is the relationship between the price for emission permits and electricity prices in a liberalised market (Sijm, Hers, Lise, & Wetzelaer, 2008)? How can an EU-ETS with a fixed quantity cap be linked with other nascent ET systems, such as those whose caps are tied to GDP development (Fischer, 2003; Anger, Brouns, & Onigkeit, 2009; Anger, 2008; Flachsland, Marschinski, & Edenhofer, 2009; Jaffe & Stavins, 2008; JET-SET, 2008; Sterk, Mehlin, & Tuerk, 2009)? A range of literature deals with such topics, focusing especially on effects on efficiency and aggregate costs. Findings to date have tended to resist generalisation; they underscore the need to assess these questions in detail for each case anew.

Public good provision in the context of environmental policy and technological innovation poses some interesting challenges. Like taxes, emissions trading systems internalise external costs, but research and development (R&D) for abatement technologies often has the character of a public good. Thus, it is often suggested that

---

[7] A double dividend arises if welfare increases on account of an environmental policy at the same time that pollution is reduced. Revenues may be used to lower distortive taxes (e.g., on labour). See, for example, Goulder, Parry, Williams III, and Burtraw (1999).

146   A. Muller

subsidies for R&D or special technologies be combined with environmental policies like ET (e.g., Stern, 2007, chapter 16; Aldy & Stavins, 2008). Parry (2003) states that 'in cases where appropriability is weak, such as new technologies that are applicable to large numbers of firms and are easy to imitate even if patented, additional research stimulants may have a valuable role'.[8]

In internalising the external costs of certain types of consumption, environmental policy increases their price, which is disproportionately borne by the poorer members of society. Since revenue redistribution is the remedy for regressive policies, ET's potential regressivity cannot be remedied as long as allocation is done by grandfathering, since this does not generate income for the government. Taxes and ET with auctioned permits, however, may be compensated for their regressivity and are thus more favourable on this score (cf., e.g., Fullerton, 2008; West & Williams, 2004; Parry, 2004).

8.5 Fundamental Aspects Beyond Economics

We have seen that the range of topics that economics can treat – the issues and problems amenable to economic and technical reasoning – is astonishingly wide. However, there are many issues that economic approaches cannot adequately address. The remainder of the paper addresses these, with a focus on ethical aspects. It explores the issue of commodification of emissions (i.e., the design of hitherto non-marketed goods as goods to be traded), the allegation of indulgence trade, the theoretical need to monetise all decision-relevant aspects in economic policy instruments, and issues of justice. Although all four areas are highly interconnected, they are kept separate here for the purpose of this discussion.[9]

All of these issues form the basis for criticism of emissions trading, particularly that of justice (sometimes framed under 'carbon colonialism'; see, e.g., Lohmann, 2008, 2009; Bumpus & Liverman, 2008). In addition, critics often claim that the economics of ET will not work. They tend to highlight the lack of incentives for R&D and related issues, including the non-optimality of the resulting long-term reduction path, 'carbon lock-in' (i.e., the absence of strong incentives to change to

---

[8] For surveys of these and related issues, see Jaffe, Newell, and Stavins et al. (2003) and Requate (2005).

[9] Caney (2009) covers many of the arguments presented here. He organises and formulates his discussion very differently, however, which complements the present discussion. Goodin (1994) also describes a systematisation of ethical arguments related to market-based policy instruments (MBI) for environmental regulation. However, he partly addresses different aspects, often comparing the MBIs to a complete ban. In this light, some of his arguments against MBIs do not have the intended result, since they are targeted against any use of the environment, per se, not against some special type of regulation of such use. His other arguments largely correspond to those discussed below, although neither the neo-colonialism argument nor certain other ET-specific aspects are raised in his discussion.

8 Contributions of Economics and Ethics to an Assessment of Emissions Trading 147

a renewable energy-based economy) and the lack of co-benefits of reducing fossil fuel use (such as a reduction of local pollution from coal power plants).[10] The field of economics itself acknowledges these potential pitfalls (cf. the discussion in Section 8.4).[11] A related criticism is that ET induces false complacency about its effectiveness in curtailing climate change (cf. Smith, 2007). However, such a criticism can be levelled at any policy instrument designed to mitigate climate change that has the confidence of the public and policymakers, regardless of whether it is warranted. However, the CDM and non-permanent reductions abroad make this concern particularly relevant. Additionality is controversial because it is often not met (cf. footnote 6 above) and sequestration of carbon in forests that may later be burned is arguably a shaky basis for emissions permits. Finally, ET is subject to the general critique that, by its very design, it fails to challenge the capitalist structures that are allegedly at the root of the problem and, according to this reasoning, cannot therefore contribute to its solution (e.g., Altvater, 2008).

8.5.1 Commodification – New Goods in New Markets

ET creates a new good – a permit to pollute – and a new market in which these permits are traded. The process of creating a new good that can be exchanged for money has been called 'commodification'. Economically, the creation of new markets does not pose particularly difficult *conceptual* problems. The only peculiar aspect in this case is the need to artificially create demand for permits by setting a cap on overall emissions. In practice, though, carbon markets pose immense *practical* challenges.

Some Clarifications

Many objections to the commodification of pollution are not based on economic reasoning. Formulated somewhat polemically, this critique of emissions trading as a means of GHG regulation raises the fundamental issue of whether it should be permissible to privatise 'the atmosphere'. This is an erroneous but common interpretation of emissions trading (e.g., Lohmann, 2006; Bachram, 2004). A more accurate formulation would be that certain use rights to the 'greenhouse gas absorption capacity of the atmosphere' can be privately owned and traded.

The good traded in ET is not, therefore, private property rights on hitherto publicly owned resources. Instead it is use rights for the emissions absorption capacity of some natural systems (e.g., greenhouse gas absorption capacity of the atmosphere). The ownership status of these use rights is usually unspecified before emissions trading systems are established, and these sinks tend to be overused as common-pool resources. Because ownership is unclear beforehand, distributing permits to pollute does not necessarily entail a redistribution of use rights from the

---

[10] See, e.g., Lohmann (2008, 2009) and their references.

[11] See also Mansfield and Boyd (2007).

public to the private domain. An ET system assigns use rights to the polluter or the polluted depending on who pays. When permits are auctioned, a (private) firm buys the right to pollute from the public, whereas with grandfathered permits these rights are assigned to the (private) firm and the public must pay the firm to retrieve them (to prevent the firm from using them to pollute, for example).

Under ideal conditions, these two ways to assign rights are equivalent in terms of the efficiency of the outcome (the Coase theorem; see, e.g., Perman, Ma, McGilvray, & Common, 2003). At the same time, they have very different implications in terms of other criteria, especially justice. Furthermore, in practice, power differences muddy this equivalence and differences in ability and willingness to pay typically mean that the outcome depends greatly on whether rights are assigned to the polluting firm or left in the public domain.

It is also worth noting that there is nothing novel about establishing clear ownership status for such use rights and allowing them to be traded. Many resources have been managed (successfully, if done correctly) by assigning private and sometimes tradable use rights to them (e.g., individual transferable quotas in fisheries or grazing rights on land; cf. Sterner, 2003). Assigning use rights to the sink capacity of the atmosphere is conceptually no different.

Commodification – Special Characteristics

In light of these clarifications, three separate arguments against commodification can be discerned. The first argument hinges on special characteristics. Accordingly, entities can have symbolic value, sanctity or religious connotations, values that are important to social interaction or related to personhood (such as friendship, care), values connected to the history of a certain entity or its use, or intrinsic values that are independent of human beings (cf. O'Neill, Holland, & Light, 2008; Gilbert, 2008; Pellegrino, 1999; Arneson, 1992). It has been argued that it is morally wrong to *trade* entities with such characteristics in exchange for money, when trade is even possible. This argument has been used against the privatisation of drinking water (as a gift from the earth; see, e.g., Dalton, 2001), trade in human organs (based on the 'principle of totality' of the human body, cf. the discussion in Borna, 1987) and trade in health care services (Pellegrino, 1999).

Opposition to the commodification of entities on the basis of their special characteristics need not appeal directly to spiritual aspects, nature's intrinsic value or human dignity. The argument can also be premised on 'basic needs' and the resources necessary to satisfy them, such as a minimum quantity of clean water. Some observers have argued that commodification should only be allowed on the portion of resources that exceeds the amount necessary to satisfy basic needs. This line of argument is discussed below under the heading of 'Justice'. A third type of special characteristic refers to entities with a special function in society that cannot be fulfilled if they are traded. An example is voting rights in a democracy. This is also best framed in the context of justice, as the inalienability of such basic entities in society protects against injustice stemming from differences in the distribution of wealth.

Basing the argument against emissions trading on the special characteristics of the good traded raises the question of why it should apply to such a specific issue as rights to the atmosphere's capacity as a sink for GHG. Such an argument may instead reflect an aversion to using the atmospheric sink so heavily or to using the atmosphere as a sink at all, perhaps due to its sanctity, symbolic character or other intrinsic value. However, this is inherently a criticism of the lifestyle that fuels our (over)use of the atmosphere as a sink, not a criticism of ET as an instrument that seeks to regulate this use.

A 'Slippery-Slope' Argument Against Commodification

Some opponents of use rights to the atmosphere wish to end what they see as the ever-expanding commodification of objects and activities and the domination of society by markets and competition. They draw the line at the commodification of hitherto non-commodified entities. According to these critics, excessive commodification increases the risk of egoism and of undermining ethical behaviour in society.[12] This is probably not relevant to trading GHG emissions, since many negative aspects of commodification – such as extremity of outcomes, lack of information, externalities and inequality (Kanbur, 2001) – can be avoided in this case. In addition, this line of argument has broader unsettling implications for the use of many standard policy instruments to regulate GHG emissions. Taxes involve a similar commodification, in principle,[13] since they can also be regarded as a payment to pollute. Whether they are actually perceived as such, however, would need to be investigated empirically. Only a command and control regulation, such as prescribing specific emission levels for various activities, can entirely avoid this criticism.

Secondly, some claim that commodification has detracted from mystery, spirituality and romanticism in the world and has made life more superficial. It seems, however, that emission rights as a highly abstract good are not tied to people's perception of the world and the danger of their detracting from spirituality seems minimal. Similarly, public perception of this would need to be investigated empirically.

The arguments presented thus far against commodification and markets for certain entities can also be seen as opposition to utilitarian thinking or cost-benefit analysis as the dominant basis for decision making in environmental policy.[14]

---

[12] Cf. examples in Shleifer (2004), Roth (2007), Kulik, O'Fallon, and Salimath (2008); see also footnote 39 in Nussbaum (2000) and references in Goodin (1994, p. 581). For a contrarian view, see Suchanek (1997).

[13] This is not the case where the payment is understood to have a *punitive* dimension.

[14] For an accessible in-depth discussion of the problems of utilitarian thinking in environmental policy and some alternatives, see, e.g., O'Neill et al. (2008).

Commodification and Power Asymmetries

The third argument against commodification focuses on power relations between actors in a (potential) market. If market participants' power is expected to be highly asymmetrical, commodification and the corresponding setup of the market becomes problematic since participation in the market may not be voluntary and exploitation of weak market participants is likely. The classic illustration is organ trade. In purely economic terms, it is a win-win situation when a poor, healthy Indian peasant sells one of his kidneys to a rich, ailing Westerner. It is doubtful, though, whether the Indian's decision to sell his kidney is a truly free decision uninfluenced by his state of poverty (also cf. Roth, 2007). In the context of emissions trading, power asymmetries may mean that only those who can afford it will buy their freedom from making domestic reductions, while the less wealthy actors may be forced to sell their permits.

Economics recognises the problem of market power asymmetries where these asymmetries reflect market failures (see Section 8.4). Here, however, the case of asymmetries in the absence of market failures is of greater relevance. The most important asymmetries are those that stem from the restriction of personal freedom inherent in individuals' 'ability to pay', a restriction that is widely accepted in market societies. In theory, the ability to pay ideally coincides with a willingness to pay. In reality, this is clearly not the case. Wealthier people have a higher ability to pay and therefore a larger choice set than the less well-off. Because these differences in wealth are largely (e.g., on the global level) *not* due to individual decisions to be more or less wealthy, these differences in ability to pay can be unjust. This injustice is carried through in every situation where goods and services are exchanged for money and, therefore, especially in markets. A striking example of this is the adverse effects that come with changes in competitive structures, such as the food vs. bioenergy debate. Demand for crops for biofuel production has contributed to price increases for agricultural commodities, adding to the stress on the poor (Evans, 2009). One way to avoid injustices due to asymmetric market power relations is to organise allocation of scarce goods using non market-based mechanisms such as command-and-control instruments. This presents other disadvantages, though, as it imposes other restrictions of personal freedom and poor implementation may produce the adverse effects of a planned economy.

Power asymmetries are relevant to all forms of ET. Where firms are the market participants, it is less problematic, since any power asymmetries do not directly impinge on individuals. Indirectly, however, if power asymmetries force certain firms to lay off workers or to close, individuals bear the costs of their effects. If a global ET system were implemented, asymmetries could arise between players from industrialised and developing countries (cf. Section 8.4). Were ET to be established for individuals, this criticism would also apply, as the less wealthy might be forced to sell more of their emission rights than they would otherwise freely choose to do.

8.5.2 Is Emissions Trading Indulgence Trade?

This criticism addresses aspects of responsibility. For some critics of ET, the problem is not the commodification of the atmosphere but the opportunity ET offers to avoid 'cleaning up one's own waste' by buying emission permits. This is seen as shirking responsibility (e.g., Smith, 2007). The core issue addressed by this criticism is the opportunity ET offers to postpone or entirely avoid changing a company's or country's own current and future emission-generating activities. Clearly, this comes at some cost but, according to the logic of ET, it is determined by the cost of abatement measures. As long as it is cheaper to buy emission rights from abroad than to abate domestically, domestic abatement will not take place. This is economically efficient, at least in the simplest cases.[15]

This criticism is based on the assumption of a *moral* difference between one's own emissions and those of others (as global pollutants, GHG emissions do not show any physical differences). This view has been advanced by the G77 and NGOs, which 'expect conversion of the sinner, not just payment for damages ... For them, the causes of maldevelopment have to be removed, not just its effects contained' (Ott & Sachs, 2000, p. 17). Such a formulation is at odds with economic (consequentialist utilitarian) thinking, where ultimately only the consequences matter, irrespective of the mindset behind the actions.

Depending on how the boundaries for the set of ET participants are drawn, some emission trade can be legitimate, even in light of this critique. It can be argued, for example, that the relevant actor that should reduce its *own* emissions is a country. Thus, one might insist that an industrialised country reduce its own national emissions domestically rather than through CDM-type purchases, while still agreeing that emitting firms within the industrialised country may trade emissions with one another. Similarly, a natural boundary for an ET system is in fact the group of industrialised countries. They are, in the aggregate, responsible for the bulk of past and (for the time being) current emissions. It could then be argued that, as a group, they should reduce emissions internally without buying emission reduction credits in developing countries. Hence, this argument is not necessarily applicable to the EU-ETS, for example. This depends on how strictly the notion of 'taking responsibility' is used in this non-consequentialist approach, as this governs the degree of flexibility in substituting one's own abatement with the financing of abatement elsewhere.

8.5.3 Monetisation and Market-Based Policy Instruments

This subsection addresses problems related to monetisation. In principle, this is relevant to any market-based policy instrument since, in an optimal situation, the

---

[15] If aspects such as R&D are included, for example, ET without measures targeted at increased R&D may lead to inefficient outcomes (cf. Section 8.4). This is, however, a different argument against ET than the critique of avoiding responsibility.

152   A. Muller

price or quantity cap is based on the costs of pollution as the tax and the permit price equal the marginal cost of pollution in this optimum. In order to determine these costs, various physical and societal effects of the pollutant have to be valued in monetary terms. These values are then summed, rendering the multidimensional effects of pollution in a one-dimensional quantity. In reality, the quantity cap (or the tax level) is often based on political reasoning rather than on a thorough physical assessment of the damages. The result is the pragmatic 'price-standard-approach' of Baumol and Oates (1971), in which a certain cap or tax level is chosen, and the role of economics is to inform its implementation at minimal cost, not to inform the choice of the cap level per se. Although monetisation is necessary in principle, it cannot stand behind each concrete implementation of an emissions trading system.

Practical Impossibility of Monetisation

There are several lines of criticism related to monetisation. The first is based on the view that it is impossible to aggregate relevant aspects into one dimension. It is not only the complexity of physical reality that limits monetisation. In addition, there are the problems with capturing individual preferences – especially when these are changing over time – and the evolution of monetary values in the wider economy (Hirsch-Hadorn & Brun, 2007; see also the discussion in O'Neill et al., 2008).[16] This can easily be illustrated with examples: what is the proper procedure for aggregating the loss of species or land due to a rise in sea-level, the loss of livelihoods due to droughts, increased heat deaths and many more factors in order to arrive at an estimate of the impacts of climate change over the next hundred years? The difficulties of such monetisation are reflected in the wide range of estimates that have recently been generated for the costs of climate change (cf. Stern, 2007; Nordhaus, 2007; Weitzman, 2007; Sterner & Persson, 2008). A major point of contention is the discount rate; that is, how to weigh costs incurred by future generations.

Although the argument concerning the practical difficulties of monetisation is valid, it is not relevant here because it is not an ethical argument.

Intrinsic Values

A second line of criticism of monetisation is based on the claim that some things cannot have monetary value. These have intrinsic value, not merely instrumental value (cf. O'Neill et al., 2008, p 114). This is tightly linked to the first critique of commodification, presented above under special characteristics. A compromise solution is to apply multi-criteria analysis (e.g., Roy, 1996) instead of cost-benefit analysis. This explicitly acknowledges the multidimensionality of criteria and their incommensurability and avoids aggregation of relevant aspects into one dimension. Multi-criteria analysis does not provide definite guidance on the best course of

---

[16] Bromley and Paavola (2002) provide an inspiring discussion of problematic issues related to values, monetisation and contested choices in environmental economics and policy.

8 Contributions of Economics and Ethics to an Assessment of Emissions Trading   153

action but it does offer a valuable way to reduce complexity to a tractable level. Clearly, multi-criteria analysis is also an option to deal with the practical impossibility of monetisation described above. The objection to monetisation based on intrinsic values is similar to the criticism of commodification based on intrinsic values (cf. Section 8.5.1.2 above) and it can be questioned on similar grounds.

Monetisation and Injustice

Closely related to commodification and its third line of criticism (cf. Section 8.5.1.4 above) is the claim that monetisation is unjust. It is first asserted that the measurement unit that monetisation imposes – money – is also a unit against which disparate things can be exchanged and which is distributed unequally among individuals according to their ability to pay.[17] A second line of reasoning claims that monetisation may be unjust in cases where it only provides a unique metric to compare valuations of things that are not actually exchanged. In monetising mitigation costs for future damages from climate change, the choice of weighting, discount factors and purchasing power parity factors can devalue costs to future generations and/or developing countries and thereby be unjust. More generally, in any case where conflicting options are valued via monetisation, certain sides may be inadequately represented and given correspondingly lower weights in the final accounting, potentially leading to unjust outcomes. Unlike the earlier argument, this one is not of direct relevance as an argument against ET, since here an exchange of goods actually takes place.

8.5.4 Emissions Trading and Justice

The previous discussion of both economic and ethical aspects of ET has shown that questions of justice arise everywhere. This subsection explores some of the most important ones, namely, the charge of (neo-)colonialism, basic needs, indulgence trade and initial allocation, in further detail. The discussion collects several of the themes discussed above, complements them with additional points and rearranges them under the focus of justice. It is limited largely to justice in the context of ET and does not address the general question of justice and climate change.[18]

(Neo-)colonialism

Many critics have voiced fears that ET will become an additional showcase of the North's exploitation of the South (e.g., Agarwal & Narain, 1991; Lohmann, 2008,

---

[17] For more details, cf. the corresponding discussion in Section 8.5.1.4 above.

[18] There is a considerable body of research on questions related to justice and climate change and climate policy in general (e.g., Shue, 1993; Jamieson, 2001; Gardiner, 2004; Page, 2006; Vanderheiden, 2008a, b; Gardiner et al., 2009) but there is much less specific work on justice and emissions trading in particular (e.g., Caney, 2009; partly Goodin, 1994; Sagoff, 1999).

2009; Bumpus & Liverman, 2008; Liverman, 2004 and also NGOs such as FERN, 2005; Sinks Watch, 2007 or Bond & Dada, 2004). This concern is sometimes framed under the charge of '(neo-)colonialism'. Their rationale is that players in a global ET system are too unequal to allow just outcomes and that, in particular, the developing countries will not gain from ET because it is unlikely to lead to broader and faster development and technology transfer, for example. In addition, these critics claim that it is often unclear who reaps the benefits of the reduction projects in developing countries.

These criticisms of ET mainly refer to the potential inequality of market participants. This problem is also acknowledged in economics (cf. Section 8.4) but more out of concern for adverse effects on efficiency than for potential inequity or injustice. The degree to which these inequalities hinder just outcomes is the subject of ongoing debate. Miranda, Dieperink, and Glasbergen (2002), for example, suggested that positive development effects are possible if the market is enhanced with content beyond economics; that is, if it is layered with a 'social meaning'. Many economists are optimistic that these potential problems can be resolved, while other social scientists, such as social geographers, tend to be more sceptical (Bailey, 2008; Liverman, 2009 and references therein).

The neo-colonialism claim is mainly, if often implicitly, geared towards the CDM. By the very structure of the CDM, developed and developing countries have a different status, which may foster the interpretation of 'colonialism' (in the sense of the North exploiting resources in the South for their own benefit without leaving an adequate share in the country of origin). CDM also presents the risk that the host country does not share greatly in the financial or development benefits (e.g., Muller, 2007).

However, the CDM is formally independent of ET since support for ET does not necessarily imply advocacy for the CDM. ET does not necessarily coincide with 'reductions abroad' or in developing countries in particular. Still, ET and the CDM are, admittedly, closely related, and a certain number of CDM credits can be used for compliance in the EU-ETS. As a result, ET may well lead to increased CDM activities. Currently, emissions trading does not yet take place between developing and developed countries. If future agreements implement emissions trading globally, trade between developed and developing countries could bring all the above-mentioned problems connected to power inequities to the fore (cf. Section 8.4). However, the colonialism critique would only be relevant to ET if market inequalities showed specific enduring patterns.

Justice and Basic Needs

This subsection directly continues the discussion of basic needs and commodification from Section 8.5.1.2 above. Many people consider trade in things that contribute to satisfying basic needs to be unjust, since unequal distribution of wealth may be so extreme as to compel the poor to trade amounts that are necessary for their own basic needs. In order to forestall such an outcome, a certain quantity of the resource in question must be provided at zero or low cost, while 'luxury

8 Contributions of Economics and Ethics to an Assessment of Emissions Trading 155

consumption' beyond this may be priced according to market forces and may therefore take on the characteristics of an ordinary good. A striking depiction of the spectrum of emissions' 'social quality' characteristics places methane emissions from rice paddies of subsistence farmers on one end and CO_2 emissions from gasoline combustion of sport utility vehicles in cities on the other (e.g., Agarwal & Narain, 1991; cf. also Shue, 1993).

Justice and the Allegation of Indulgence Trade

A third aspect relates to the allegation of indulgence trade, emphasising aspects of (in)justice. This criticism does not focus on necessary 'self-correction' regarding maldevelopment (cf. Section 8.5.2) but on the fact that ET allows wealthy people, with their high ability to pay, to enjoy full freedom of action, while the actions of others may be restricted because of their lower ability to pay for more emission rights. Poorer people may even be forced to agree to trade their emission rights for money. Again, based on similar arguments as those above, this aspect is relevant where developing and developed countries interact but less applicable to firms in the context of the EU-ETS, for example. This potential problem therefore reflects injustice embedded in all markets, where individuals' freedom is restricted according to their ability to pay. This also relates to the argument against ET based on basic needs.

Justice and Initial Allocation

Sections 8.3 and 8.4 discussed potential problems with initial allocation. The main issue is the distribution of the wealth embodied in the emissions permits. Accordingly, grandfathering is often seen as unjust, since the most emission-intensive companies get the largest share of permits for free and may even capture additional windfall profits by passing on the opportunity costs of the permits to the end-consumers (cf. Section 8.3). The degree of coverage in broader regulatory schemes also involves issues of justice. Energy-intensive industries could be freely allocated permits under the EU-ETS, while households might have to meet any extra-EU-ETS reduction burdens through other means. The recognised problems of power asymmetries, vested interests and lobbying play an important role in initial allocation (cf. Section 8.4).

Aspects of justice connected to initial allocation have broader relevance, as the question of who has the right to emit how much can be framed as one of initial allocation in a global context (cf., e.g., Meyer & Roser, 2008). An even global distribution of emissions rights has been suggested as the initial allocation in a hypothetical global ET system. For example, the one ton of CO_2 per capita proposal (ESC, 2008) could be used to calculate national caps for each nation. This would solve the problem of initial allocation among nations but it would not resolve how to implement the necessary regulation to achieve this cap within nations.

Table 8.2 The main ethics-based arguments against carbon emissions trading (ET) systems

Argument based on	Further specification	ET should not be allowed because ...	Judgment: This argument...
Commodification	*Special characteristics* – Special values	– ... entities with special values must not be traded.	– ... is problematic, as it is questionable whether the atmosphere's GHG absorption capacity (AC) has such special value. In order to be less problematic, the argument needs to be reformulated. It then would not be directed against commodification but against over-use of the atmosphere.
	– Basic needs	– ... amounts necessary to meet basic needs must not be allocated by a market.	– ... is discussed under 'justice'.
	Slippery slope	... it leads to commodification of life in all its aspects. This increases egoism and undermines ethical behaviour.	... is problematic because it is questionable whether this applies to commodification of the atmosphere's GHG AC, which is a non-problematic good in many respects.
	Power asymmetries	... power asymmetries between market participants are unavoidably too large for equitable solutions.	... applies in general, in particular if trade between individuals were established. It may not apply for wisely regulated trade between firms that are largely equally powerful.
Indulgence trade	Avoid maldevelopment	... there is a moral difference between my own action and that of someone else.	... applies in general; it may not apply if the boundaries comprising 'own emissions' are chosen wisely (e.g., firms in EU-countries only).
Monetisation	Practical impossibility	... it is practically impossible to collect the necessary information to set up a workable ET system.	... applies, but it is not an ethical argument.
	Intrinsic values	... things possessing intrinsic values should not be monetised.	... is similar to the argument against commodification based on special value.

Table 8.2 (continued)

Argument based on	Further specification	ET should not be allowed because ...	Judgment: This argument...
	Justice		
	– Market exchange situations	– Monetisation is unjust mainly due to the restriction of liberty linked to differences in ability to pay.	– ... is similar to the argument against commodification based on power asymmetries.
	– Valuation	– Monetisation is unjust due to inadequate representation of certain aspects (typically those more important to less powerful stakeholders) when valuing conflicting options.	– ... is not of direct relevance because an exchange of goods actually takes place under ET.
Justice	Neo-colonialism	It exploits the South.	... may apply, depending on how the ET system is implemented.
	Basic needs	Amounts necessary to meet basic needs must not be allocated by a market, as the unequal distribution of ability to pay can lead to unjust outcomes.	... may apply, depending on how the ET system is implemented.
	Indulgence trade	It is unjust, because the rich can pay and increase their liberty while the poor cannot.	... applies in general, unless market participants are largely equal in terms of their ability to pay.
	Initial allocation	As it leads to unjust distribution of emission rights/wealth.	... need not apply, as initial allocation can be designed to avoid such injustice.

158   A. Muller

A domestic ET system only re-introduces the problem of initial allocation for various industries.[19]

In addition, the distribution of the proceeds earned through emissions trading raises issues of justice. The wealth transfer in developing countries could be considerable if a global initial allocation of one ton of CO_2 per capita is agreed upon in future international emissions conventions. The proper beneficiary of these money flows – individual citizens, municipalities, regional or national governments – would have to be resolved. Depending on governance structures, such large wealth inflows could lead to 'resource curse' or 'rentier state' effects in which government finances are largely independent of taxpayers and lack accountability, and in which small groups and elites can easily appropriate revenues.

8.5.5 Synthesis

Table 8.2 summarises the main ethics-based arguments against carbon emissions trading presented in this chapter. The table shows that almost all sound ethical arguments against ET are based on problems of injustice: power asymmetries, neo-colonialism or basic needs. The conclusion from this is that arguments related to special characteristics of the good traded (atmospheric pollution absorption capacity) do not apply, since the good lacks the special characteristics – such as intrinsic values – necessary for this argument to hold. The only other argument that can be consistently adduced is indulgence trade in carbon reduction; that is, that ethically it matters who does the reducing. However, this argument assumes that avoiding individual maldevelopment, and not aggregate emissions reductions, is the ultimate goal.

8.6 Conclusions

This chapter has presented a wealth of information on emissions trading as a policy instrument for internalising external costs and it has addressed aspects critical for its successful implementation. Although economics has much to contribute on the issue, other disciplines – here exemplified by ethics – need to complement economics in order to conduct an encompassing, comprehensive assessment of ET's merits. For example, the chapter discussed what economics has to say on the potential for diverse market participation in an ET system without compromising its successful implementation. This was complemented with perspectives from ethics, bringing in aspects of justice and responsibility. The contributions of both disciplines to the issue of ET exemplify the fact that, in order to grapple successfully with

---

[19] See also Caney (2009). The distribution of the rights to emit or the costs of reducing emissions is implicitly a key challenge for any carbon regulation. Sagoff (1999) discusses this in the context of GHG emissions trading.

the complex problems of today, disciplines need to complement one another, mutually conceding the others' expertise in their core competences, while also accepting their own limitations.

Economics is capable of raising very clear questions on a wide range of topics, thus greatly clarifying many of the issues involved. Although it can be more difficult to provide answers, particularly for surefire policy guidance, economics can often at least describe what will *not* work and thereby help avoid ineffective, costly policies. For the EU-ETS, for example, economics would have counselled against implementing a system that allows for so much diversity of rules across countries. Economics' strength in detailed formalised analysis, however, comes at the cost of a far-reaching simplification, which is necessary to make reality tractable to its formalisms.

Ethics, on the other hand, can clarify in detail the values, attitudes, convictions and arguments behind 'gut feelings of common morality'. This means it can help clarify ethical arguments raised against or made in favour of ET, identifying where those are flawed and where they may legitimately apply. Ethics addresses a number of fundamental questions to which economics is blind, and it deals in practical issues that go beyond economics' core competences. Like economics, however, ethics often cannot provide single clear-cut solutions. Rather, it provides the methods and concepts with which to clarify the 'ethical' aspects of ET in detail, aspects on which economics must remain silent.

Where real-world complexity prevents economics from pinpointing a clear solution to a problem, ethics may provide a short-cut to engaging in effective action. Instead of endlessly debating the best implementation of a global emissions trading system, ethical arguments may suggest that industrialised countries accept responsibility and take the lead in reducing their own emissions. This reduces the burden on economic analysis, which can then focus on implementing an ET system that extends to industrialised countries only.

Furthermore, ethical analysis may pave the way for implementing solutions that economics proposes by clarifying the ethical content of arguments raised against such solutions. For critics of ET, who cite the atmosphere's intrinsic value, for example, ethical analysis can clarify that this reflects an opposition to the (over-)use of the atmosphere as a sink for anthropogenic emissions rather than opposition to concrete regulation of this use by means of emissions trading.

Is emissions trading a viable option for carbon regulation? If so, how should an ET system be designed? All in all, this study has shown that ET can play an important role. Economics' most important basic design recommendation for an ET system is to avoid large power asymmetries between participants. Further, there should be high coverage, for instance in an upstream system that targets fossil fuel importers. The system should be harmonised across different national jurisdictions and initial allocation should be done through auctioning. Inclusion of gases other than CO_2 and inclusion of different sectors (such as agriculture) in a downstream system need to be handled very carefully. Similarly, interaction with other policy instruments needs careful design. For its part, ethics suggests that a global emissions trading system would be problematic, especially if both industrialised and developing countries were to take part. Therefore, the issues of power asymmetries (including ability to pay) and basic needs suggest restricting the scope of participating countries. If all the problems identified in Sections 8.3 and 8.4 are forestalled or remedied, the resulting ET system is not objectionable, as long as it is not extended to a global system. Furthermore, the CDM must be regulated in order to avoid the problems of power asymmetries, asymmetric abilities to pay and conflicts with basic needs.

This chapter concludes with a much farther-reaching issue. A valid question is whether more radical ideas are needed to solve the pressing problems of our time. Are policy instruments like ET enough to pave the way to a wholly different world that has mitigated climate change and eliminated poverty? Is it sufficient to realise the goal of a truly sustainable energy system? ET is a pragmatic mechanism that can play an important role, but it is not a substitute for a more fundamental reflection on our ultimate values and goals and the means to realise them.

Acknowledgments Many thanks to Jürg Minsch, Dominic Roser, Markus Huppenbauer, Carsten Köllmann, Daniel Spreng, David Goldblatt and Thomas Flüeler for inspiring discussion and helpful comments. Special thanks to David Goldblatt, whose editorial input greatly improved the text. Financial support from the University Priority Program in Ethics, University of Zurich, and the 'Stiftung Mercator Schweiz' is gratefully acknowledged.

References

Agarwal, A., & Narain, S. (1991). *Global warming in an unequal world*. New Delhi: Center for Science and Environment.

Aldy, J., & Stavins, R. (2008). *The role of technology policies in an international climate agreement*. Issue Paper. The Harvard Project on International Climate Agreements. All web links accessed November 16, 2011, http://belfercenter.ksg.harvard.edu/files/The%20Role%20of% 20Technology%20Policies%20--%20Harvard.pdf

Altvater, E. (2008). Für ein neues Energieregime. Mit Emissionshandel gegen Treibhauseffekte? *Widerspruch, 54*, 5–17.

Anderson, J. (2008). Emission trading in Europe. In *Climate change and sustainable energy policies in Europe and the United States*. The Transatlantic Platform for Action on the Global Environment (T-PAGE). Washington, DC: Natural Resources Defense Council (NRDC); Brussels: Institute for European Environmental Policy (IEEP).

Anger, N. (2008). Emissions trading beyond Europe: Linking schemes in a Post-Kyoto world. *Energy Economics, 30*(4), 2028–2049.

Anger, N., Böhringer, C., & Oberndorfer, U. (2008). *Public interest vs. interest groups: Allowance allocation in the EU Emissions Trading Scheme*. ZEW Discussion Paper, No. 08-023. ftp://ftp. zew.de/pub/zew-docs/dp/dp08023.pdf

Anger, N., Brouns, B., & Onigkeit, J. (2009). Linking the EU Emissions Trading Scheme: Economic implications of allowance allocation and global carbon constraints. *Mitigation and Adaptation Strategies for Global Change, 14*, 329–398.

Arneson, R. (1992). Commodification and commercial surrogacy. *Philosophy and Public Affairs, 21*(2), 132–164.

Atkinson, S., & Morton, B. (2004). Determining the cost-effective size of an emission trading region for achieving an ambient standard. *Resource and Energy Economics, 26*, 295–315.

Bachram, H. (2004). Climate fraud and carbon colonialism: The new trade in greenhouse gases. *Capitalism Nature Socialism, 15*(4), 5–20.

8 Contributions of Economics and Ethics to an Assessment of Emissions Trading 161

Bailey, I. (2008). Geographical work at the boundaries of climate policy: A commentary and complement to Mike Hulme. *Transactions of the Institute of British Geographers, NS33,* 420–423.

Baumol, W., & Oates, W. (1971). The use of standards and prices for protection of the environment. *Swedish Journal of Economics, 73,* 42–54. (Similar: Chapter 11 in W. Baumol & W. Oates (1988). *The theory of environmental policy, 2nd edition.* Cambridge: Cambridge University Press)

Böhringer, C., & Lange, A. (2005a). On the design of optimal grandfathering schemes for emission allowances. *European Economic Review, 49,* 2041–2055.

Böhringer, C., & Lange, A. (2005b). Economic implications of alternative allocation schemes for emission allowances. *Scandinavian Journal of Economics, 107*(3), 563–581.

Bond, P., & Dada, R. (2004). *Trouble in the air: Global warming and the privatised atmosphere.* Durban, South Africa: Center for Civil Society.

Borna, S. (1987). Morality and marketing of human organs. *Journal of Business Ethics, 6,* 37–44.

Boyd, E., Hultman, N., Roberts, J. T., Corbera, E., Cole, J., Bozmoski, A., et al. (2009). Reforming the CDM for sustainable development: Lessons learned and policy futures. *Environmental Science and Policy, 12*(7), 820–831.

Bromley, D., & Paavola, J. (Eds.) (2002). *Economics, ethics, and environmental policy.* Malden, MA: Blackwell.

Bumpus, A., & Liverman, D. (2008). Accumulation by decarbonization and the governance of carbon offsets. *Economic Geography, 84*(2), 127–155.

Caney, S. (2009). Justice, morality and carbon trading. *Ragion Pratica, 32.*

CarbonPositive. (2009, February 16). Carbon price lows test market faith. *Carbon News and Info.*

Convery, F., Ellerman, D., & De Perthuis, C. (2008). *The European carbon market in action: Lessons from the first trading period.* Interim report. Joint Program on the Science and Policy of Global Change, Massachusetts Institute of Technology (MIT). Cambridge, MA: MIT.

Dalton, G. (2001). *Private sector finance for water sector infrastructure: What does Cochabamba tell us about using this instrument?* Occasional Paper, No. 37. Water Issues Study Group, School of Oriental and African Studies. London: University of London.

den Elzen, M., Meinshausen, M., & van Vuuren, D. (2007), Multi-gas emission envelopes to meet greenhouse gas concentration targets: Costs versus certainty of limiting temperature increase. *Global Environmental Change, 17*(2), 260–280.

Dudek, D., & Golub, A. (2003). 'Intensity' targets: Pathway or roadblock to preventing climate change while enhancing economic growth? *Climate Policy, 3*(S2), 21–28.

EC. (2008). *Greenhouse gas monitoring and reporting, progress report and appendices, several years.* European Commission. http://ec.europa.eu/environment/climat/gge_progress.htm

EC. (2010). *Communication from the Commission to the European Parliament and the Council towards an enhanced market oversight framework for the EU Emissions Trading Scheme, European Commission, COM (2010) yyy final.* http://ec.europa.eu/clima/news/docs/communication_en.pdf

EC. (2011). *Emissions Trading System (EU ETS).* European Commission Climate action. http://ec.europa.eu/clima/policies/ets/index_en.htm

EEA, European Environment Agency. (2008). *Application of the Emissions Trading Directive by EU Member States.* EEA Technical report, No. 13/2008. http://reports.eea.europa.eu

EEA. (2011). European Union Emissions Trading Scheme (EU ETS) data from CITL. http://www.eea.europa.eu/data-and-maps/data/european-union-emissions-trading-scheme-eu-ets-data-from-citl-3

Ellerman, D. (2008). *The EU Emission Trading Scheme: Prototype of a global system?* The Harvard Project on international climate agreements. Discussion Paper 08-02.

Ellerman, D., & Buchner, B. (2007). The European Union Emissions Trading Scheme: Origins, allocation, and early results. *Review of Environmental Economics and Policy, 1*(1), 66–87.

Ellerman, D., & Joskow, P. (2008). *The European Union's Emissions Trading Scheme in perspective.* Arlington, VA: Pew Center on Global Climate Change.

Ellerman, D., & Sue Wing, I. (2003). Absolute versus intensity-based emission caps. *Climate Policy, 3*(S2), 7–20.

ESC. (2008). *Energy Strategy for ETH Zurich*. http://www.esc.ethz.ch/box_feeder/StrategyE.pdf

EU, European Union. (2003). *Directive 2003/87/EC of the European Parliament and of the Council of 13 October 2003 establishing a scheme for greenhouse gas emission allowance trading within the Community and amending Council Directive 96/61/EC*. http://eur-lex.europa.eu/LexUriServ/LexUriServ.do?uri=CELEX:32003L0087:EN:HTML

EU. (2004). *Directive 2004/101/EC of the European Parliament and of the Council of 27 October 2004 amending Directive 2003/87/EC establishing a scheme for greenhouse gas emission allowance trading within the Community, in respect of the Kyoto Protocol's project mechanisms*. http://eur-lex.europa.eu/LexUriServ/LexUriServ.do?uri=CELEX:32004L0101:EN:HTML

EU. (2006). *Communication from the Commission to the Council and to the European Parliament on the assessment of national allocation plans for the allocation of greenhouse gas emission allowances in the second period of the EU Emissions Trading Scheme, COM (2006) 725 final*. http://eur-lex.europa.eu/LexUriServ/site/en/com/2006/com2006_0725en01.pdf

EU. (2008). *EU-ETS post 2012, in particular the Proposal for a Directive of the European Parliament and of the Council amending Directive 2003/87/EC so as to improve and extend the greenhouse gas emission allowance trading system of the Community*. http://ec.europa.eu/clima/policies/ets/index_en.htm

EU. (2009). *Directive 2009/29/EC of the European Parliament and of the Council of 23 April 2009 amending Directive 2003/87/EC so as to improve and extend the greenhouse gas emission allowance trading scheme of the Community*. http://eur-lex.europa.eu

Evans, A. (2009). *The feeding of the nine billion, global food security in the 21st century*. London: The Royal Institute of International Affairs, Chatham House.

FERN. (2005). *Carbon 'offset' – no magic solution to 'neutralise' fossil fuel emissions*. FERN Briefing Note, June 2005. The Forests and the European Union Network FERN, www.fern.org

Fischer, C. (2003). Combining rate-based and cap-and-trade emissions policies. *Climate Policy, 3*(S2), S89–S109.

Fischer, C., Parry, I., & Pitzer, W. (2003). Instrument choice for environmental protection when technological innovation is endogenous. *Journal of Environmental Economics and Management, 45*, 523–545.

Flachsland, C., Marschinski, R., & Edenhofer, O. (2009). Global trading versus linking. Architectures for international emissions trading. *Energy Policy, 37*(5), 1637–1647.

Freeman, J., & Kolstad, C. (Eds.) (2007). *Moving to markets in environmental regulation: Lessons from twenty years of experience*. Oxford: Oxford University Press.

Fullerton, D. (2008). *Distributional effects of environmental and energy policy: An introduction*. NBER Working Paper, 14241. http://www.nber.org/papers/w14241

Gardiner, S. (2004). Ethics and global climate change. *Ethics, 114*, 555–600.

Gardiner, S., Caney, S., Jamieson, D., & Shue, H. (2009). *The ethics of climate change*. Oxford: Oxford University Press.

Gilbert, J. (2008). Against the commodification of everything. *Cultural Studies, 22*(5), 551–566.

Goodin, R. (1994). Selling environmental indulgences. *Kyklos, 47*(4), 573–596.

Gorman, H., & Solomon, B. (2002). The origins and practice of emissions trading. *The Journal of Political History, 14*(3), 293–320.

Goulder, L., Parry, I., Williams III, R., & Burtraw, D. (1999). The cost-effectiveness of alternative instruments for environmental protection in a second-best setting. *Journal of Public Economics, 72*(3), 329–360.

Hagem, C., & Westskog, H. (2009). Allocating tradable permits on the basis of market price to achieve cost effectiveness. *Environmental and Resource Economics, 42*(2), 139–149.

Hirsch Hadorn, G., & Brun, G. (2007). Ethische Probleme nachhaltiger Entwicklung. In R. Kaufmann, P. Burger, & M. Stoffel (Eds.), *Nachhaltigkeitsforschung – Perspektiven der*

8 Contributions of Economics and Ethics to an Assessment of Emissions Trading 163

Sozial- und Geisteswissenschaften. Bern, Switzerland: Schweizerische Akademie der Geistes- und Sozialwissenschaften.

Hourcade, J.-C., Demailly, D., Neuhoff, K., Sato, M., Grubb, M., Matthes, F., et al. (2007). *Differentiation and dynamics of EU-ETS Industrial Competitiveness Impacts*. Cambridge, UK: Climate Strategies.

IETA. (2008). *Online library of the International Emissions Trading Association IETA*. http://www.ieta.org

Jaffe, A., Newell, R., & Stavins, R. (2003). Technological change and the environment. In K.-G. Mähler & J. Vincent (Eds.), *Handbook of environmental economics, Vol. 1* (pp. 461–516). Amsterdam: North-Holland.

Jaffe, J., & Stavins, R. (2008). *Linkage of tradable permit systems in international climate policy architecture*. The Harvard Project on International Climate Agreements Discussion Paper, 08-07. John F. Kennedy School of Government, Harvard University. Cambridge, MA: Harvard University.

Jamieson, D. (2001). Climate change and global environmental justice. In P. Edwards & C. Miller (Eds.), *Changing the atmosphere: Expert knowledge and global environmental governance* (pp. 287–307). Cambridge, MA: MIT Press.

JET-SET. (2008). *Joint emission trading as socio-ecological transformation*. http://www.sozial-oekologische-forschung.org/en/104.php?LM=list&%20MODE=detail&PAG=2&PN=28&PUB=363. Homepage: http://www.wupperinst.org/en/projects/proj/index.html?projekt_id=97&bid=137

Joskow, P., & Schmalensee, R. (1998). The political economy of market-based environmental policy: The U.S. Acid Rain Program. *Journal of Law and Economics, 41*(1), 37–85.

Jotzo, F., & Pezzey, J. (2007). Optimal intensity targets for greenhouse gas emissions trading under uncertainty. *Environmental and Resource Economics, 38*, 259–284.

Kanbur, R. (2001, 2004). On obnoxious markets. http://www.arts.cornell.edu/poverty/kanbur/Obnoxious%20Markets.pdf. Revised version. In S. Cullenberg & P. Pattanaik (Eds.,). (2004). *Globalization, culture and the limits of the market: Essays in economics and philosophy*. Oxford: Oxford University Press.

Kling, C. L., & Rubin, J. (1997). Bankable permits for the control of environmental pollution. *Journal of Public Economics, 64*(1), 99–113.

Kolstad, C. (2004). The simple analytics of greenhouse gas emission intensity reduction targets. *Energy Policy, 33*(17), 2231–2236.

Kosobud, R. (Ed.). (2000). *Emissions trading*. New York: John Wiley.

Kraemer, A., Interwies, E., & Kampa, E. (2002). *Tradeable permits in water resource protection and management* (Paper presented at the OECD Expert Workshop on Implementing Domestic Tradeable Permits, September 2001, Paris).

Krysiak, F. (2008). Prices vs. quantities: The effects on technology choice. *Journal of Public Economics, 92*, 1275–1287.

Krysiak, F., & Schweitzer, P. (2006). *The optimal size of a permit market, Working Paper*. http://wcms-neu1.urz.uni-halle.de/download.php?down=1488&elem=1040373

Kulik, B., O'Fallon, M., & Salimath, M. (2008). Do competitive environments lead to the rise and spread of unethical behavior? Parallels from Enron. *Journal of Business Ethics, 83*, 703–723.

Liski, M., & Montero, J.-P. (2006). On pollution permit banking and market power. *Journal of Regulatory Economics, 29*(3), 283–302.

Liverman, D. (2004). Who governs, at what scale and at what price? Geography, environmental governance, and the commodification of Nature. *Annals of the Association of American Geographers, 94*(4), 734–738.

Liverman, D. (2009). Conventions of climate change: Constructions of danger and the dispossession of the atmosphere. *Journal of Historical Geography, 35*(2), 279–296.

Lohmann, L. (Ed.). (2006). *Carbon trading: A critical conversation on climate change, privatization and power*. Development Dialogue Nr 48. Uppsala: Dag Hammarskjold Foundation.

Lohmann, L. (2008). Carbon trading, climate justice and the production of ignorance: Ten examples. *Development, 51*(3), 359–365.

Lohmann, L. (2009). Climate crisis: Social science crisis. In M. Voss (Ed.), *Der Klimawandel: Sozialwissenschaftliche Perspektiven.* Wiesbaden: VS-Verlag.

Mackenzie, I., Hanley, N., & Kornienko, T. (2008). The optimal initial allocation of pollution permits: A relative performance approach. *Environmental and Resource Economics, 39,* 265–282.

Maeda, A. (2003). The emergence of market power in emission rights markets: The role of initial permit distribution. *Journal of Regulatory Economics, 24*(3), 293–314.

Mansfield, M., & Boyd, E. (Eds.) (2007). *Commodifying carbon – the ethics of markets in Nature.* ECI Workshop Report, July 16, 2007. http://www.eci.ox.ac.uk/publications/downloads/commodifycarb-report.pdf

Meyer, L., & Roser, D. (2008). Climate justice: Past emissions and the present allocation of emission rights. In P. Cobben, P. Coppens, A. Gosseries, & U. Marti (Eds.), *Distributive justice.* Berlin: Springer.

Michaelowa, A. (2007). Untergräbt der Clean Development Mechanism den internationalen Klimaschutz? *Die Volkswirtschaft, 9,* 20–23. Bern, Switzerland: Federal Department of Economics.

Michaelowa, A., & Purohit, P. (2007). *Additionality determination of Indian CDM projects.* Cambridge, UK: Climate Strategies.

Millimet, D., & List, J. (2004). The case of the missing pollution haven hypothesis. *Journal of Regulatory Economics, 26*(3), 239–262.

Miranda, M., Dieperink, C., & Glasbergen, P. (2002). The social meaning of carbon dioxide emission trading. *Environment, Development and Sustainability, 4,* 69–86.

Misiolek, W., & Elder, H. (1989). Exclusionary manipulation of markets for pollution rights. *Journal of Environmental Economics and Management, 16*(2), 156–166.

Muller, A. (2007). How to make the CDM more sustainable – the potential of rent extraction. *Energy Policy, 35*(6), 3203–3212.

Muller, A., & Sterner, T. (Eds.) (2006). *Environmental taxation in practice.* Aldershot, UK: Ashgate.

Narain, U., & van t'Veld, K. (2008). The Clean Development Mechanism's low-hanging fruit problem: When might it arise, and how might it be solved? *Environmental and Resource Economics, 40*(3), 445–465.

Neuhoff, K. (2008). *Tackling carbon – how to price carbon for climate policy.* http://www.climatestrategies.org/research/our-reports/category/32.html

Neuhoff, K., & Matthes, F. (Eds.) (2008). *The role of auctions for emissions trading.* Climate Strategies, UK. Cambridge, UK: University of Cambridge. http://www.climatestrategies.org

Newell, R.G., & Rogers, K. (2006). The market-based lead phasedown. In J. Freeman & C. Kolstad (Eds.), *Moving to markets in environmental regulation: Lessons from twenty years of experience* (pp. 173–193). New York: Oxford University Press.

Nordhaus, W. (2007). A review of the Stern Review on the economics of climate change. *Journal of Economic Literature, 45*(3), 686.

Nussbaum, M. (2000). The costs of tragedy: Some moral limits to cost-benefit analysis. *The Journal of Legal Studies, 29*(2), 1005–1036.

OECD. (1999). *Implementing domestic tradable permits for environmental protection.* OECD Proceedings. Paris: OECD.

OECD. (2001). *Implementing domestic tradeable permits.* OECD Proceedings. Paris: OECD.

O'Neill, J., Holland, A., & Light, A. (2008). *Environmental values.* London: Routledge.

Ott, H., & Sachs, W. (2000). *Ethical aspects of emissions trading.* Wuppertal Papers 110. http://www.wupperinst.org/uploads/tx_wibeitrag/WP110.pdf. Assumingly similar to Ott, H. & Sachs, W. (2002). The ethics of international emissions trading. In L. Pinguelli-Rosa & M. Monasinghe (Eds.), *Ethics, equity and international negotiations on climate change.* Northampton: Edward Elgar.

Page, E. (2006). *Climate change, justice and future generations.* Cheltenham: Edward Elgar.

8 Contributions of Economics and Ethics to an Assessment of Emissions Trading

Parry, I. (2003). On the implications of technological innovation for environmental policy. *Environment and Development Economics, 8*, 57–76.

Parry, I. (2004). Are emission permits regressive? *Journal of Environmental Economics and Management, 47*, 364–387.

Pellegrino, E. (1999). The commodification of medical and health care: The moral consequences of a paradigm shift from a professional to a market ethic. *Journal of Medicine and Philosophy, 24*(3), 243–266.

Perman, R., Ma, Y., McGilvray, J., & Common, M. (2003). *Natural resource and environmental economics*, 3rd edition. New York: Addison Wesley.

Point Carbon. (2008). Carbon 2008 – Post-2012 is now. Røine, K., Tvinnereim, E., & Hasselknippe, H. (Eds.). http://www.pointcarbon.com/research/carbonmarketresearch/analyst/1.912721

Requate, T. (2005). Dynamic incentives by environmental policy instruments. *Ecological Economics, 54*, 175–195.

Roth, A. (2007). Repugnance as a constraint on markets. *Journal of Economic Perspectives, 21(3)*, 37–58.

Roy, B. (1996). *Multicriteria methodology for decision aiding*. Dordrecht: Kluwer.

Sagoff, M. (1999). Controlling global climate: The debate over pollution trading. *Philosophy & Public Policy, 1*(1), 1–6.

Schennach, S. M. (2000). The economics of pollution permit banking in the context of Title IV of the 1990 Clean Air Act Amendments. *Journal of Environmental Economics and Management, 40*, 189–210.

Schneider, F., & Wagner, A. (2003). Tradable permits – ten key design issues. *CESifo Forum 1/2003*, 15–22.

Sheeran, K. (2006). Who should abate carbon emissions? A note. *Environmental and Resource Economics, 35*, 89–98.

Shleifer, A. (2004). Does competition destroy ethical behaviour? *AEA Papers and Proceedings, 94*(2), 414–418.

Shue, H. (1993). Subsistence emissions and luxury emissions. *Law and Policy, 15*(1), 39–53.

Sijm, J., Hers, S., Lise, W., & Wetzelaer, B. (2008). *The impact of the EU ETS on electricity prices*. Final report to the DG Environment of the European Commission. Petten: Energy Research Centre of the Netherlands ECN.

Sinks Watch. (2007). *SinksWatch – an initiative to track and scrutinize carbon sink projects*. http://www.sinkswatch.org

Smith, K. (2007). *The carbon neutral myth: Offset indulgences for your climate sins*. Carbon Trade Watch, The Transnational Institute. www.tni.org, http://www.carbontradewatch.org/pubs/carbon_neutral_myth.pdf

Somanathan, E., & Sterner, T. (2006). Environmental policy instruments and institutions in developing countries. In M. Toman & R. Lopez (Eds.), *New options for sustainable development*. Oxford: Oxford University Press.

Stavins, R. (1995). Transaction costs and tradeable permits. *Journal of Environmental Economics and Management, 29*(2), 133–148.

Stavins, R. (2003). Experience with market-based environmental policy instruments. In K.-G. Mähler, & Vincent, J. (Eds), *Handbook of environmental economics, Vol. 1*. Amsterdam: Elsevier. (Stavins' list of ET examples is also reported in Schneider & Wagner, 2003 above).

Stavins, R. (2007). *A U.S. cap-and-trade system to address global climate change*. Brookings Institution Discussion Paper, 2007–2013. John F. Kennedy School of Government, Harvard University. Cambridge, MA: Harvard University.

Sterk, W., Mehlin, M., & Tuerk, A. (2009). *Prospects of linking EU and US Emission Trading Schemes: Comparing the Western Climate Initiative, the Waxman-Markey and the Lieberman-Warner proposals*. May 2009. Cambridge, UK: Climate Strategies.

Stern, N. (2007). *Stern Review on the economics of climate change*. HM Treasury (UK Economics and Finance Ministry). http://www.hm-treasury.gov.uk/sternreview_index.htm

Sterner, T. (2003). *Policy instruments for environmental and resource management.* RFF Press. London: Earthscan.

Sterner, T., & Persson, M. (2008). An even Sterner Report: Introducing relative prices into the discounting debate. *Review of Environmental Economics and Policy, 2*(1).

Suchanek, A. (1997). Verdirbt der Homo Oeconomicus die Moral? In K. Lohmann & B. Priddat (Eds.), *Ökonomie und Moral – Beiträge zur Theorie ökonomischer Rationalität* (pp. 65–84). Munich: Oldenbourg.

Tietenberg, T. (Ed.) (2001). *Emissions trading programs.* 2 Volumes. Aldershot: Ashgate.

Tietenberg, T. (2003). The tradable-permits approach to protecting the commons: Lessons for climate change. *Oxford Review of Economic Policy, 19*(3), 400–419.

Tietenberg, T. (2006). *Emissions trading – principles and practice.* RFF Press. London: Earthscan.

Tilford, S. (2008). *How to make EU emissions trading a success.* London, UK: Centre for European Reform.

Toman, M. (2003). *Understanding the design and performance of emissions trading system for greenhouse gas emissions.* Proceedings of an Experts' Workshop to Identify Research Needs and Priorities. *RFF Discussion Paper 03–33.* Washington DC: Resources for the Future.

Vanderheiden, S. (2008a). *Atmospheric justice. A political theory of climate justice.* Oxford: Oxford University Press.

Vanderheiden, S. (Ed.) (2008b). *Political theory and global climate change.* Cambridge, MA: MIT Press.

Weitzman, M. (1974). Prices vs. quantities. *Review of Economic Studies, 41*, 477–491.

Weitzman, M. (2007). A review of the Stern Review on the economics of climate change. *Journal of Economic Literature, 45*(3), 703.

West, S., & Williams, R. (2004). Estimates from a consumer demand system: Implications for the incidence of environmental taxes. *Journal of Environmental Economics and Management, 47*, 535–558.

Chapter 9
No Smooth, Managed Pathway to Sustainable Energy Systems – Politics, Materiality and Visions for Wind Turbine and Biogas Technology

Ulrik Jørgensen

Abstract Wind energy and biogas are considered to be economically viable renewable energy solutions to the increasingly acknowledged climate challenge. Political controversy over long-term and short-term goals, changing economic criteria and performance assessments, as well as technical constraints related to design, systems embedding and material agency of the technologies have contributed to the 'roller coaster' character of assessments and development. This contribution explores how conflicting relationships between renewable energy technologies based on wind and biomass, as well as institutional and regulatory changes, have shaped the Danish energy system. Important lessons can be learned from case studies on the assessment of technology and its anticipated properties, the role of alliances and predictions for change, as well as the need for the continued modification of energy innovation strategies and regulatory policies in a climate of continued controversy over means and ends.

Keywords Biogas · Changing actor perceptions · Constructivist approach · Political controversy · Wind energy

9.1 Wind and Biogas – Today's Success, Yesterday's Failure?

Wind turbines and biogas facilities, among other new current energy technologies, are considered to be fully functional and well-established energy-producing technologies. They are core contributors to renewable energy programmes, with high rates of growth and penetration in contemporary energy plans in Europe and other parts of the world. They are even often seen as icons of renewable energy, illustrating the existence of alternatives to fossil fuel-based technologies and some actors also consider them to be an alternative to nuclear power. Compared to other

---

U. Jørgensen (✉)
Department of Management Engineering, Innovation and Sustainability, Technical University of Denmark, Lyngby, Denmark
e-mail: uj@man.dtu.dk

D. Spreng et al. (eds.), *Tackling Long-Term Global Energy Problems*,
Environment & Policy 52, DOI 10.1007/978-94-007-2333-7_9,
© Springer Science+Business Media B.V. 2012

energy-producing technologies such as solar cells or wave- and tidal-based power generation, wind turbines and biogas are even seen as almost mature technologies and well-defined investment opportunities. This does not imply, for example, that all actors expect these technologies to become the sole or dominant provider of electricity. The old and well-established technology of hydraulic power will continue to provide renewable energy in some countries, as will nuclear power and other technologies that provide relatively improved energy efficiency, such as gas. Other compensating technologies that reduce energy efficiency are not yet feasible and safe, such as CO_2 capture and storage, but have also received much attention as a way to prolong the utilisation of fossil fuels, especially coal. While nuclear power has re-entered the energy arena as a potential carbon-neutral technology, it is still greatly contested because of its potential environmental impact. Wind and biomass have taken the lead in the area of truly renewable energy sources and have become priority technologies, though there are still open questions as to their scalability and integration into the general energy system, which must operate continuously and independent of weather conditions and other variables. The effects of biomass on land use and other environmental impacts have also raised questions about the constraints on energy provision from this source.

Wind and biogas technologies have repeatedly been discounted as unsuccessful and without a future within energy systems. The present investigation was triggered by the varying assignments of meaning and properties (qualities) to these technologies throughout their history, since these changes cannot simply be related to an idea of linear technological improvements resulting in social changes (Bijker, 1995). In this respect, the history of these technologies is related equally to the social visions and economic interests of the actors involved, to the willingness to invest in piecemeal improvements, to policy support and to the priorities put on social and environmental problems. In some theoretical contexts, these changing expectations and conditions can be seen as a reflection of the uncertainties involved. However, by taking the priorities of the various actors seriously, they can be seen as the result of contradictory assessments of the technologies involved and conflicting objectives for societal change. This chapter presents a critical assessment of the history of these two energy technologies, based not simply on their technical improvements but on their socially embedded and interwoven history as 'socio-technologies' (for a broader discussion, see Bijker et al., 1987).

9.2 Why Such a Turbulent History and Conflicting Assessments?

Defining technology as a seamless web of technical entities and social meaning and processes has radical and important consequences for how meaning and properties (qualities) are assigned to a technology (Hughes, 1987). Otherwise, how would it be possible to explain that a technology is sometimes viewed as a future solution to climate problems and at other times discarded as a poor and inefficient energy provider? Basic qualities assigned to technologies are not intrinsic properties but instead depend on the network of relations to actors specifying these qualities and reflect the perspectives and context in which this technology operates. Technical, environmental and economic qualities are assigned to technologies by different groups of actors based on the specific roles they are given within economic institutions, social structures and broader energy systems (Bijker, 1995, Callon, Méadel, & Rabeharisoa, 2002).

Engineering and economic (ex-post analytical) reconstructions of technology may highlight some of a technology's features and properties and view them as more important. This usually provides a relatively specific perspective on the technology and sets the conditions for its use without explaining its full historic role or the qualities assigned to it (Lundvall et al., 2002). If the sociotechnical view from constructivist theory is taken at the outset, the changing and provisional views on technologies and their role throughout history can be explained by the associations, or mediations (Callon, 1987), made by the actors engaged in the use and development of a technology. These mediations also represent actors' institutional and material policies, highlighting the role technologies have as bearers of social, economic and technical visions (Rip, 2009).

9.3 Energy Systems in Transition

New energy technologies have generated controversy and conflict among actors who have attempted to give them a role within existing energy systems. Established energy systems are characterised by their sociotechnical regime, resulting from the development of technologies, the supply of energy sources, distribution infrastructure and practices of energy usage as well as from the relevant institutions maintaining and regulating their operations and boundaries. Such technological regimes are difficult to change and resist new technologies and new institutions that challenge the dominant practice and interest structures (Hughes, 1983, 1987; Shot & Geels, 2008). This not only entails a strong path dependency of the technological regimes related to economic and technical priorities but also usually involves a discursive, conceptual framing of what should be considered 'good solutions' (Garud & Karnøe, 2001).

Finding a place and a role for new energy technologies within existing regimes involves a longer period of transition. This could, for a time, be bypassed by creating institutional (social and economic) niches in which innovations can be developed, tested and nurtured, leading to improvements and lessons being learned without having to compete with existing technologies and their supporting institutions (Elzen, Geels, & Green, 2004; Geels & Raven, 2007). Eventually, however, the relationship of new energy technologies to the broader institutional and technical infrastructure has to be identified and stabilised. This often leads to conflicts of interests over the reconstruction, or sometimes even the substitution and destruction, of existing regimes. Such identifiable transition periods have occurred throughout the history of society's energy systems; they also illustrate the multi-faceted process of change, in

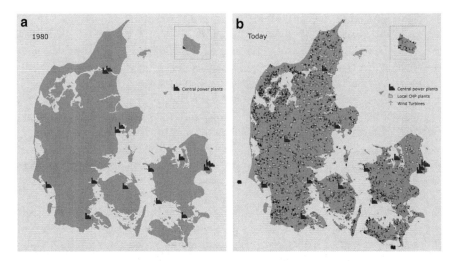

Fig. 9.1 The transition from central power utilities in 1980 (**a**) to distributed energy production today (**b**)
Source: Energienet, from http//www.energienet.dk

which the active engagement of actors is crucial for any outcome (van der Vleuten & Raven, 2006). Existing regimes may be able to resist change, even if the need for change appears obvious from some actors' perspectives, and new technical and institutional solutions have to fight hard to become part of society's energy solutions (Geels, 2004). The impact of the ongoing transition can be illustrated by the radical change in the distribution and scale of power producing units (Fig. 9.1).

In economic theory, the concepts of competition and of optimal solutions are often used to argue for a market-based solution to technological change. The approach to socio-technology and the existence of institutionally embedded regimes, described above, sets the stage rather differently for the economic processes involved. Market mechanisms and economic choices are framed by actors' expectations and the institutional constructs involved in regimes, which include oligopolies and policy-dependent forms of restricted competition and choices of technical solutions.

9.4 Different Storylines and Creations

It may have already become clear that the 'stories' (versions of history) that have been told about technical solutions and social change are the product of the promotional activities of established institutions and the stakeholders involved. In particular, these 'stories' are intrinsically woven into the social, economic and political perspectives of the actors involved, who are committed to defining and giving meaning to their preferred solutions and choices. This implies that the history of social and technological change comes in a variety of often conflicting versions of the assessments of technologies and their impact, which are directly related to the interpretations resulting from specific actors' interests and disciplinary references (Staudenmaier, 1989). The assignment of qualities to a technology at a given time by the actors involved may differ and thereby provide different accounts of the technology. Consequently, it is questionable whether these relate to the same technology; from a sociotechnical perspective, technologies are not stable and depend on use and context and, therefore, do not appear as single entities. Historical accounts differ in terms of how qualities are assigned to social change and technologies, depending on how the accounts are framed and the context in which they are made.

Stories told about wind turbines and biogas facilities have conflicted and changed dramatically over the lifetime of these technologies, depending on society's and actors' changing discourses and how qualities relating to issues of, for example, 'growth', 'security', 'local supply', 'alternative energy', 'renewable sources', 'climate solution' and 'energy efficiency' have been assigned. The storylines are important in the construction of the bigger picture for describing interests and perspectives and for mapping present relationships and power structures, as well as for defining and preparing for future changes. Stories about technology are usually not simply neutral accounts or historical reconstructions of conflicts over choice and the use of technology in society. They also embed themselves deeply in the creation of contemporary choices by creating scenarios for future development and change (Rip, 2009).

The present discussion of the role and conditions of 'story'-telling and creation covers accounts of grass-root movements and environmental organisations, as well as those of leading research institutions, public authorities and companies. All of these actors are deeply engaged in framing policy and defining the basis of contemporary energy systems, as well as trying to shape their future development. The point here is not that these often different historical accounts are false, per se, or part of a conspiracy to deliberately mislead, although this may well be the case sometimes. It is much more important to understand that these differing historical accounts are the product of different ideas and interpretations because the actors disagree on the qualities assigned to socio-technologies and have different ideas and visions about future change.

9.5 The Long History of Wind: From Rural Electrification to Local Initiatives

The use of wind energy for electricity production was first developed in the early 20th century, when the newly discovered potential for using electricity as a way to power machinery and light was considered an important contribution to the development of factory production and farming. In Denmark, the folk high school movement conducted experiments in rural electrification and in 1893 the engineer Poul La Cour began experimenting with wind turbines to produce electricity and using hydrogen for energy storage and as an energy carrier. This led to the Danish

Wind Electricity Enterprise building a series of wind turbines for farmers and villages. Their design was simple but efficient in delivering electricity for local farm production and was closely connected to the technical improvements in machinery and farming techniques of the time, including improvements in husbandry and the quality of farm produce, and to the education and training of people in the folk high schools (Klitmøller & Thorndahl, 2008).

In the 1920s, local and individual rural wind turbines were outperformed and overtaken by petrol-driven power generators. This resulted from a fall in fossil fuel prices, as well as the lack of success in storing power by methods such as hydrogen separation or batteries, which made local electricity production dependent on wind and weather conditions.

A new wave of experimentation occurred in the 1940s due to the scarcity of fuel during World War II, followed by post-war demand for electricity for industrial growth and a temporary slowdown in the construction of new fossil fuel power plants and steam turbines. During the war, wind turbines had again been considered as a potential source of electricity. F. L. Schmidt, the large cement and machine factory, developed and produced a series of large-scale (approximately 50 kW) wind turbines, with 21 turbines operating around 1950. After the war, the newly established, government-funded Wind Energy Commission used data from the turbines as the basis for constructing a 200 kW test turbine in Gedser and for analysing the potential for further technical improvements (Christensen, 2008).

The turbines were seen as small power plants that supplied electricity to local electricity grids; as such, they had to compete with other larger power generators driven by steam turbines and fuelled by oil. Even though wind technology was improved in terms of design and size and grew more efficient over the next two decades, the reduced price of fossil fuels made these experiments with wind turbines less attractive. Wind turbines were again abandoned as part of the official energy policy and the technology was deemed useless. This is illustrated by the Energy Commission's 1962 report, which stated that 'Wind energy will never play an important role in Danish energy supply'. This was based on a calculation that the price per kWh of electricity from wind turbines was double that from coal-fired power plants (Jørgensen & Karnøe, 1995a; Thorndahl, 2005).

Although none of the turbines from WWII exist today, the test turbine in Gedser has become world-famous – at least within energy movements and among wind turbine engineers – as has Johannes Juul, the chief engineer who worked on its design and data. This turbine provided solid information for the next period of experimentation with wind turbines, which started in the mid-1970s, just a little more than a decade after the technology was judged useless. With US funding, the Gedser turbine was even re-started for two years of further testing in 1976. Juul never accepted the Energy Commission's conclusions and he made a public statement that 'If we could construct new turbines based on the knowledge gained from experiments with the Gedser wind turbine they would be much more efficient' (Hvidtfelt, 2001).

9.6 The Oil Crisis and Controversy over Nuclear Power

With the first oil crisis in the early 1970s, a new political and social situation that was linked to the controversy over the risks of nuclear power and the potential future shortage of fossil fuels created a new and urgent need to seek alternative energy solutions. The dramatic increase in the consumption of fossil fuels after WWII and the growing dependency on oil led to an effort to find other large-scale power production technologies. During the Cold War, there was increased public distrust of government information on the impact of nuclear war (as demonstrated by the large public participation in the Easter Marches in Denmark). The strategic effort to develop a 'peaceful utilisation of nuclear power' was met with scepticism and the conflict over pollution from nuclear facilities led to violent confrontations in Germany and the United Kingdom, for example. In Denmark, the experience of non-violent resistance to nuclear warfare was taken up by the anti-nuclear power movement (OOA – 'Organisation for Information on Nuclear Power'). This added to political tensions in the 1970s over society's energy strategy but it also set the stage for the development of alternative solutions to the energy crisis.

The criticism focused on the social impact of centralised energy systems and institutions and was often associated with police action and security measures, in opposition to local supply; this led to wind turbines and biogas facilities being seen as a logical small-scale and preferable alternative (OOA, 1980; Jungk, 1977). The critique not only attacked fossil fuels and nuclear power plants but also included ideas on how technology was part of the general concept of localising 'power' within society. Nuclear power was part of a technocratic dream of continuous growth in an industrial consumer society, which turned this technology into the logical next step in the development of still larger power utilities to satisfy the endless need for electricity. Small-scale energy alternatives benefited from being locally produced and managed.

From the perspective of large power plants, wind turbines were too small. This opinion restated the established conclusion that wind turbines were useless and an inferior alternative in any realistic future strategy for supplying society's power needs. These perspectives and visions were held by most government officials and power utilities, including the Nuclear Energy Commission as it laid the ground for the utilisation of nuclear power in Denmark. In the grass-roots movements, on the other hand, the alternative social vision that 'small is beautiful' and the idea of having thousands of small, locally-owned energy facilities satisfied the concept of a renewed focus on more local responsibility and less material consumption (as advanced by Schumacher, 1973). In both these points of view, social visions set the stage for anticipated sociotechnical developments following different, preferred regimes and each defined distinct research and innovation agendas for the actors involved. The anti-nuclear movement soon created a separate branch (the OVE or 'Organisation for Renewable Energy'), which focused on alternative technologies that were developed outside the existing power supply system and institutional regime and built instead on local and self-organised experiments with wind turbines and biogas facilities. To this end, the data and experience gathered by Johannes

Juul proved very useful and helped build the new functional designs, overcoming some of the earlier technical problems in wind turbine design (Jørgensen & Karnøe, 1995a).

9.7 Wind Turbine Entrepreneurs

Environmental movements and wind energy entrepreneurs crossed the barriers of earlier experience by taking the small-scale wind turbine seriously as a power-producing unit. They were inspired by the idea of developing a local, independent energy supply (as illustrated in Fig. 9.1). Without this sociotechnical vision, the willingness to run the risk of failure – in contrast to the risk-averse behaviour of investors in power grids – would not have been fed by a similarly strong belief in the potential of the technology (Jørgensen & Karnøe, 1995a). The movement was supported by local investors, who believed in the vision and often collectively invested in one or more turbines. In this regard, there is nothing special about wind turbine technology when it is seen as just another technology. As the history of technology has shown, many new technical ideas have been developed over time and in a series of piecemeal improvements before becoming sound, generally accepted technologies. In practical terms, nurturing and supporting the radical transformation of a society's energy systems requires a lot of hard work from the entrepreneurs involved. The disagreements along the way stemmed from the lack of common ground between the different ideas and visions of the involved actors. However, these disagreements have been crucial for facilitating change and the learning process.

In the case of wind turbine development, the 'bottom-up' practices of the local and small industrial entrepreneurs involved, who based their design on a series of experiments and the slow process of scaling up and optimising their operational designs, were the most important source of new technological know-how. Besides focusing on a scaling-up process, Danish entrepreneurs also tested a large variety of different designs before selecting today's predominant design of a three-bladed, first stall- and later pitch-regulated, transmission-based and grid-coupled wind turbine. In contrast, the energy crisis of the early 1970s also led to 'top-down' approaches by governments investing in large-scale testing facilities. These approaches, such as the Growian turbine in Germany and the even older WWII Putnam turbine in the US, made no significant contribution to the development of today's functional wind turbines (Karnøe, 1991). Even the considerable investment in research into engineering large-scale turbines did not provide a great deal of relevant information in the early years of development and researchers later adopted alternative solutions and small-scale technology as objects of study (Jørgensen & Karnøe, 1995b). This can be partly explained by the reliance on established aerodynamics technology based on existing airplane design not taking into account the specific problems of designing wind turbines.

This point illustrates the importance of innovation strategies and active anticipation of problems when new technologies are developed. Even though classic engineering experience tells us that scale matters and that new problems and issues emerge when construction methods are transferred from one use to another, research programmes were scaled up to support wind energy development without taking this into account. An important reason for this was the complexity of the aerodynamics of airflows along turbine blades operating on the edge of turbulence. Another reason is related to the structural dynamics of tower and blade design, which have to withstand vibrations over long periods of time. The degree of challenge was underestimated and basing the design of large turbines on research and translated technology from the aircraft and space industries led to a series of failures (Gipe, 1995).

Whereas the United States viewed engineering design departments as the main contributor to technological change, the Danish turbines were developed based on experiences and more direct contact with users – often with the grass-root innovators using the turbines themselves. The industry involved in the next stage of commercialisation and up-scaling evolved from the agricultural machinery industry, which had experience working with large machines under tough conditions. This experience and knowledge of vibration stress on materials turned out to be crucial to the success of the early wind turbine industry in Denmark. It also resulted in steady improvements in energy efficiency as shown in Fig. 9.2.

While research input turned out to be of secondary importance, the continuous exchange of knowledge and experience between users, entrepreneurs, innovators and industry engineers at annual wind meetings, organised by OVE, among others,

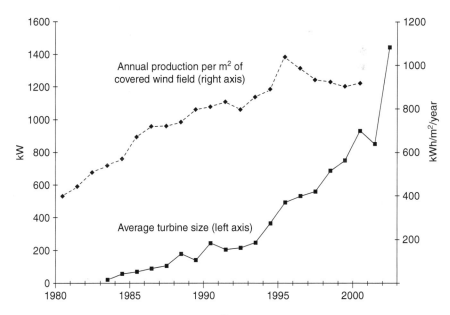

Fig. 9.2 Energy efficiency by production per m^2 and average size of wind turbines
Source: Skytte, Jensen, Morthorst, and Olsen (2004)

was important for problem-solving and the growth of the wind turbine sector. In addition to these meetings, a test station was established at the government's energy research facility at Risø (which had hitherto been involved in research on nuclear power), where new designs could be tested and verified, thus giving customers better data for their investment decisions (Karnøe, 1991). In addition, the test station became part of the knowledge and exchange network that is so important for a system of innovation to grow. This was the case when Danish industry started to export to the growing and much larger markets in Germany and California.

9.8 Public Policy Support

The controversy over Danish energy policy led to the crafting of an alternative energy plan in 1976, written by several energy researchers who supported the energy movement. The plan resulted in a greater focus on renewable energy and the creation of government support schemes for energy installations. This followed several years of public controversies and the engagement of different groups of actors in the policy debate over energy visions and measures allowing experimentation. Most important, though, was the fact that the government forced energy grid companies to accept turbines' connection to the grid and to buy surplus power, based on defined feed-in tariffs, from wind turbine owners. In this first phase of investment, almost all single turbines and turbine parks were owned by cooperatives and local communities, which meant that the feed-in tariffs created a predictable investment horizon for wind turbines. At the same time, these policy measures supported market demands and kept investors pushing for energy efficiency, which created a competitive environment for continuous improvements in wind turbines as long as the feed-in tariffs were also adjusted according to the improved energy efficiency of the technology. It also resulted in a situation where wind energy together with a number of other renewable energy sources accounts for an increasing share of power production (Fig. 9.3).

These policies also helped overcome the resistance of the dominant energy regime, which viewed wind turbines as too small and inefficient, by lowering the entry barriers for the new technology and protecting the industry in its 'infancy' period. Important elements in the effectiveness of this 'market-oriented' policy were continual adjustment of the feed-in tariffs, following efficiency improvements in the technology, and the maintenance of a competitive edge for investors and industry (Jørgensen & Strunge, 2002).

9.9 Why Was the Danish Turbine Industry Successful?

Comparisons have already been made with developments in the US and Germany, where the government invested heavily in research programmes and industrial innovation. In most of these cases, the piecemeal engineering and bottom-up approach

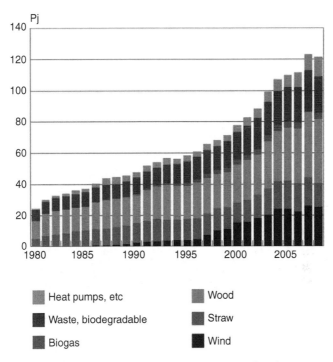

Fig. 9.3 The quantitative change in electricity production by type of producer
Source: Danish Energy Agency

that emerged from the involvement of grass-root entrepreneurs, energy movements and local industry ended up producing more efficient and reliable wind turbines. The engaged actor constellations behind these experiments were committed to finding solutions, while the research and industry contracts set up by US and German governments, for example, did not produce the same urgency and search for solutions. The exception has been the German Enercon company, with its multi-polar generators, which has avoided the heavy transmissions by connecting the turbine blades directly to the generators.

Comparison with countries such as Sweden and Holland reveals interesting differences in the role of the alternative energy movements and their impact on the birth and growth of the wind turbine industry. While the new industry's managers in Denmark may not have shared the sociotechnical visions and ideas of the early grass-roots movements, they were influenced by the entrepreneurial spirit and basic scepticism regarding arguments against the potential of wind turbine technology. The energy movements in other countries did not engage in the creation of alternative solutions in the same way. The context in which the alternative energy movement operated and their members' perception of their potential additional roles as inventors and entrepreneurs, as well as their involvement in controversies over specific political support measures, are all important factors that have typically been absent from contemporary historical research on renewable energy development.

9.10 Changing Priorities and New Locations

During the late 20th century, several energy plans were created that focused on how the energy supply system was composed and on how to integrate new energy supply technologies into the electricity grid, partly for the purpose of maintaining the security of supply. Not least of the problems was the need to design new system control strategies, both at the technical and institutional level. This made it possible to increase the contribution of wind turbines from 15 percent to a goal of 25 percent or more (Fig. 9.4). Achieving such goals increases the need for either a backup capacity in other parts of the general electricity system in case wind energy supply fails or increased use of systems to manage energy consumption and storage. It also requires better grid connections and general management of the energy system. The first batch of wind turbines have been connected relatively close to consumption locations. However, plans for a wind-based energy supply capacity above 15–20 percent require wind turbine parks to be moved offshore, which changes the properties of turbine technology and assigns features to them that are similar to other large power plants that supply the grid and consumers over longer distances. Planning for

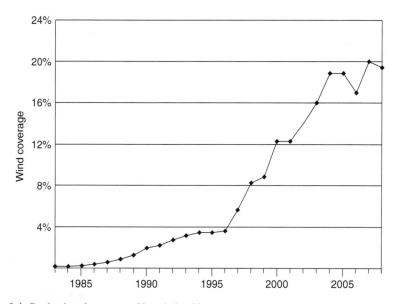

Fig. 9.4 Production share covered by wind turbines
Source: Danmarks Vindmølleforening

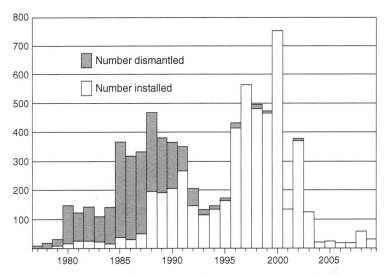

Fig. 9.5 Number of turbines installed and dismantled based on their year of installation
Source: Danmarks Vindmølleforening

the required backup capacity also involves connecting to other energy sources, such as hydroelectric power plants, and connecting over long-distance power networks.

With the increase in the number of wind turbines and their size, which now exceeds the original estimated size for efficient turbines, the networks of supporting actors have changed, as have their perspectives. This has included the substitution of earlier installed wind turbines of lower capacity with new and bigger turbines (Fig. 9.5). The scene has changed since early alternative energy entrepreneurs challenged the utilities and local investors installed wind turbines. Utilities now represent the core investments in wind turbine parks, which are often sited offshore because of their size. A new constellation of actors has taken over and local investors are now rare, while institutional investors are showing growing interest in wind turbine-based energy production.

9.11 CO$_2$ Emissions and Climate Challenges

The Kyoto Protocol and the European Emissions Trading System have created new mechanisms for the development of energy systems in Europe. These do not necessarily benefit renewable energy systems or promote climate improvement. The EU emission trading scheme favours heavy 'polluters' in compensating for existing emissions by establishing a completely new market for CO$_2$ quotas. This has resulted in a focus on the short-term optimisation of energy supply systems, such as by favouring combined cycle gas turbines for their short payback time and their CO$_2$ emissions, which are lower than conventional fossil fuel power plants. Because of CO_2 reduction policies, wind energy has now been given a steady, if perhaps temporary, role in the general power supply system but the qualities assigned to these technologies have changed. Yet the EU energy policies have delayed more radical adjustments to the energy supply systems and, at the same time, given the impression that power companies are willing to act in favour of climate mitigation. The inflexible adjustment of quota frameworks and their distribution has even produced lower CO_2 prices than the promoters of the trading scheme anticipated. These mechanisms have resulted in very little additional investment in renewable energy and, especially, in the development of new energy technologies (Jørgensen, 2005).

At a more strategic level, European liberalisation policies and the narrow focus on CO_2 trading schemes have had a radical impact on the distribution of actors in the power arena. The new market construction favours utilities and proven conventional technologies, resulting in a major shift from local investors and entrepreneurs towards conventional financial markets and large investors, which has led to short-term optimisation (Jørgensen & Strunge, 2002). Modern wind turbine technology is considered to have made an important contribution to the general energy supply system. Nevertheless, other renewable energy technologies based on, for example, solar, wave or geothermal power, have still not reached a stage at which they can be developed and proven in the much-needed process of implementation and continued pressure for piecemeal adjustments and improvements. The centralised focus on single policies and market mechanisms has changed the context for innovation but this new context is not readily responsive to changing conditions in a process that is inevitably characterised by constant and mutable conflicts of political interests.

9.12 Local Biogas Experiments

The energy crises of the 1970s prompted experimentation with several other alternative renewable energy technologies in addition to wind turbines. Heat pumps, wave-based technology, solar cells and solar heating were explored but, along with wind turbines, biogas proved to be one of the most promising technologies adopted in local experiments. Excess manure production from Danish pig farms, in particular, made this technology potentially useful for heat production on farms. In biogas production from anaerobic digestion, the natural fermentation process transforms manure into methane, CO_2 and a residue that can be used as fertiliser.

Like wind energy, biogas has a long history, with early experimentation having begun in the 1880s using waste from larger cities. The growing emphasis on hygiene and the treatment of wastewater meant that these experiments did not have a major impact on energy supplies. The next phases of experimentation took place in the 1930s, in World War II and again in the 1950s, motivated in each case by energy supply shortages. Access to affordable fossil fuels in Europe led to the abandonment of these early experiments in Europe, while they continued, evidently with some success, in countries such as India (Lybæk & Møller, 1999).

9 No Smooth, Managed Pathway to Sustainable Energy Systems – Politics, ...

Biogas has developed differently from wind technology and it illustrates the importance of the vision and risk-taking of the actors involved, as well as the need to learn from technological drawbacks and failures. The actors involved in the modern biogas experiments of the 1970s were local farmers, some of whom were also involved in the energy movement. The technology depended on the supply of manure and other residue products, which made it difficult to involve a larger community beyond the farming sector in the experiments. When some of the initial problems with stabilising the fermentation process emerged, engineering firms became involved in improving and proliferating facilities. A programme for biogas innovation and testing was funded in 1978, which established a consortium to improve the efficiency of biogas plants and publicised the results in a special newsletter. In addition, a test facility for biogas plants was established at a government research facility in Horsens (Geels & Raven, 2007; Lybæk & Møller, 1999).

All of the established facilities presented a series of problems, as the processing of manure was found to be unstable and required specialised knowledge of the operation of biochemical plants. This made it difficult for these small facilities to function efficiently and most local experiments were abandoned in the 1980s. Despite improvements in biogas yields, especially with the use of supplementary organic waste in the process, the initial problems made a generation of young farmers sceptical about the usefulness of biogas technologies (Lybæk & Møller, 1999). Since these farmers were both financial investors and participants in the learning process, their capacity for continued experimentation was limited when the first experiments were not successful and the overall visions did not demand further experimentation. At the same time, energy prices dropped again and attempts to create an industry with specialised producers was not successful, partly because of the absence of an international market that could legitimise further investments and overcome failures. The decline was not absolute, however. Some local experiments were continued, supported by an experimental facility established by members of the 'Nordisk Folkecenter for Vedvarende Energi' energy movement (the Nordic Peoples' Centre for Renewable Energy). This movement kept local initiatives and strategies alive, in parallel with the efforts made to sustain similar ideas and experiments with local wind turbines. In both cases, the centre was involved in designing a 'blacksmith' version of a wind turbine and a biogas plant, with the latter actually becoming a stable small-scale producer of biogas.

9.13 Communal Biogas Facilities

The problems with stabilising the biochemical fermentation process caused the focus to shift in the latter part of the 1980s, with communal biogas installations adopted instead of small-scale local farm units. Because of the larger scale and the possibility of investing in more advanced monitoring equipment, professional firms started to replace local farmers in the oversight of these installations and their operations. The new centralised biogas plants seemed to be more promising, although early experimental plants experienced serious problems due to the lack of expertise on how to build large processing units (Raven, 2005). The construction of these plants was subsidised by government grants and loans and was coordinated within the 1988 Biogas Action Plan. The Danish government intervened because of the need to reduce the levels of nitrate and other pollutants from agriculture and to handle the surplus of manure at individual farms. The Danish Environmental Protection Agency drafted regulations governing when manure could be spread as a fertiliser in the Water Environment Action Plans of 1987. These regulations required every pig farmer to create storage facilities for manure or to join collective facilities like the ones established at communal biogas facilities.

The network of actors changed to include new professional institutions and several ministries. The qualities assigned to biogas plants also changed, with such plants helping to solve problems of storage and groundwater pollution as well as producing energy. As such, they were no longer primarily part of an alternative energy vision but a practical means to sustain large-scale pig farming in Danish agriculture – activities that were often criticised as being intrinsically unsustainable. Even though energy prices fell in the 1990s, combined demand sustained the construction of communal biogas plants. To stabilise the process in the biogas reactors, manure was mixed with other biomass residue, resulting in more stable and efficient biogas production. Dependency on government economic support for investment in facilities hindered the spread of centralised biogas facilities when the new liberalisation-oriented energy policy was introduced at the turn of the century and the fixed price feed-in tariffs were removed, as this created uncertainty over energy prices. Instead of pushing for new biogas plants, the government changed course and argued that future energy investment should be based on market principles. With the change of government in 2001, most funding of renewable energy investment and research was discontinued, leaving investment in new centralised biogas plants in limbo.

9.14 Next Stages: Integrated Solutions on Farms

Figure 9.6 illustrates the overall change of the direction of development and technological paradigms. The figure also shows the renewed interest in farm biogas production facilities. This was linked to energy use but also reflected the demand for a comprehensive solution to improving the quality of the groundwater and energy efficiency and reducing the use of manure and odour from large-scale pig farms.

In recent years, biogas facilities at the farm scale have become a stable technology, which can be run and maintained without special attention and can be integrated into the new types of high-tech production facilities that characterise modern pig farming. Consequently, investment in farm-scale plants has risen sharply since the late 1990s and some of the technology appears fairly sophisticated, even though it is dependent on local initiative and the investment capacity of the farmers. In

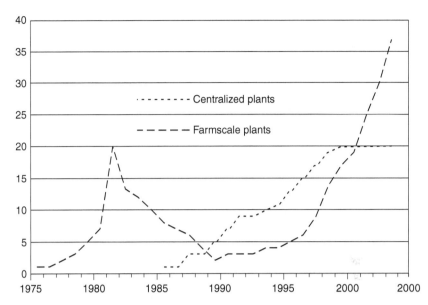

Fig. 9.6 Farm-scale and centralised biogas plants in Denmark
Source: Geels and Raven (2007)

this respect, the continued activities of farmers and researchers involved in the first phase, as well as the lessons learned about how to regulate the biochemical process, have been crucial. At the same time, the motivation for investing in biogas plants has changed from being part of an energy vision to becoming part of a strategy to make pig farming slightly more sustainable, even though such meat production may be unsustainable when viewed from a total lifecycle perspective.

Government policies now tend to set high standards for waste and energy efficiency for new pig farms. Consequently, regulators have forced a new perspective on environmental technology policy as a condition for obtaining a local licence to operate. This has turned original local ideals and commitments upside down, solidifying the farms' status as industrialised production facilities, not only in size but also in the way they are planned and regulated, with an integrated approach to production yields, waste handling, energy efficiency and open land policy.

9.15 Economic Mantras and Master Plans

Knowledge of technologies, their properties and their use from within scientific disciplines depends heavily on the perspective, problem sensitivity and object framing of these disciplines. Each individual discipline (such as economics, environmental science, life cycle assessment, energy planning, political science and market analysis) has a specific focus and a normative set of models for representing its core subjects. Consequently, the practitioner's choice of disciplinary expertise determines the kind of analysis and advice to be expected. This is also true of the selection of core mechanisms in processes of social change and in the stylised models used to explain change. Disciplines and the professional knowledge they support also produce and select the specific properties associated with a technology (Bijker, 1995). This critical perspective is also relevant in the study of energy systems and their development.

The foregoing analysis is primarily based on approaches from the sociology of technology and the economics of innovation, with some contributions from the history of technology. These approaches emphasise the importance of the specific actors involved in technological change, the properties assigned to emerging technologies and the sociotechnical regimes of which they become part. The advantage of this perspective is that these approaches allow a critical analysis of the framing of political and economic decisions and of professional knowledge. Although they are not immune to the problem of disciplinary blinders described above, their critical and reflexive components may leave them in a better position to compensate for it. The approaches also reduce the risk of writing a simplified history based on hindsight and only telling the story of the 'winners' while overlooking the specific involvement of actors in constellations that have been crucial for developing technologies, for learning processes and for transforming what may start as a local vision and experiment into a global and stabilised solution.

Energy policies at the European Union level and, to a large extent, national Danish energy policies have been dominated by economic rationales. While the choice of a local versus a centralised power system and the choice of energy technology, for example, are secondary to economic models, market access, price competition and ownership are core issues. CO_2 quotas and the liberalisation of the EU power and energy markets were introduced to secure optimal – and short-term – solutions for investment in new energy technologies, which would lower emissions at minimal cost. In Europe, these policies have primarily led to power company mergers, which have created strategic ownership of power utilities, grid facilities and customer access but have had little impact on the creation of a more sustainable energy supply structure (Jørgensen, 2005). Compared to conventional command-and-control policies, including specific support policies for renewable energy technologies, this economic rationale has had virtually no impact to date. Instead of encouraging the development of efficient means of transitioning to new energy systems, these rationales have helped to sustain existing power regimes and fossil-fuel-based energy technologies, which has led to delayed transitions. Much higher prices for carbon and fossil fuels are required before economic rationales will favour sustainable solutions. Rather than being neutral instruments that help identify optimal solutions, these economic rationales have become politicised tools in the continuous fight over the direction of change.

Technology- or system-based models used in energy planning often take the idea of an economically optimal solution as a starting point and depend on relatively stable assumptions about the efficiency of energy technologies and prices for fossil fuels. Such models cannot easily reflect the impact of innovations, behavioural changes based on energy savings, and changing political priorities. Even as recently as the mid-1980s, climate change was not prominent on the agenda of most countries' energy planning and most of today's economic models for energy systems still only operate on the basis of feeble assumptions about the need for change. Other models operate on policy-based scenarios for the creation of energy systems and include the need for energy savings and increased efficiency. However, these models do not usually focus on how to enable actors to provide these changes at the conceptual and behaviour levels. The lessons learned from such scenarios are, at best, that alternative energy futures exist and that sustainable energy solutions based on renewable sources are feasible, yet they do not show how the political process can be harnessed to make these futures a reality.

9.16 Innovation Contexts

The two case studies presented in this chapter have demonstrated the importance of specific social and political contexts for the success of innovations. The actors involved have shown themselves to be crucial, in their breadth of sociotechnical vision, willingness to experiment and perseverance in their quest for solutions and technical improvements. Expectations and assigned qualities – or lack thereof – are crucial for an actor's focus and choice of specific energy technologies, as well as the technologies' eventual combination and integration into existing institutional settings and distribution networks. Power utilities were unlikely to agree to experiment seriously with wind turbines as long as they considered them to be basically inefficient and unstable small power units. Later, when they could assign positive properties to them within CO_2 emission reduction initiatives, they were willing to overcome problems of grid integration and to formulate backup strategies to stabilise supplies. With the inclusion of tariff differentiation and other complementary policies and technologies, wind turbines may eventually even replace a significant share of (and, in many recent scenarios, in combination with biomass, even 100 percent of) the coal-fuelled, large-scale steam turbine generators as the backbone of the power system, under pressure from climate mitigation policies. As for biogas units, individual farmers need a better way to integrate pollution, energy production and large-scale facilities to achieve renewed interest in this technology. Even then, questions remain concerning the sustainability of the overall meat production in which biogas plays a part.

One lesson here is that while actors in the dominant sociotechnical regime often have easy access to both political power and investment funds, they are not always the greatest innovators or risk-takers, or especially willing to experiment and develop new technologies. Such processes usually require the bottom-up involvement of actors who see the benefits of new technologies, experimentation, learning processes and commitment to the distribution of results as core elements of their activity. The history of wind turbine and biogas has demonstrated that the innovative commitment of 'grass-roots' entrepreneurs, NGOs and public support has been crucial to their development. Without this commitment, wind turbines would not have become such a well-established contributor to the renewable energy system.

A second lesson concerns the relationship between technological change and economic processes in the long term. Actors' specific sociotechnical visions often change and since the learning process for new technologies depends on these visions and the qualities assigned to the constellations of technologies, the process cannot be seen as stable and prescribed but as something that will have to be adjusted and changed during the development process. Specific path dependencies may be established along the way and specific market relations may tend to stabilise actor relations. In addition, technical difficulties may materialise, perhaps as reverse salients, as described by Hughes (1987). It counts whether these dependencies and challenges are seen as definite obstacles or as hurdles, as shown in the wind turbine and biogas cases. The actors involved must constantly analyse whether difficulties should be ascribed to technical problems, institutional factors, insufficient learning or combinations of these, and they must adjust their innovative efforts accordingly. While wind turbines became a practicable technology for large groups of public investors after only a decade, biogas facilities were not adopted with the same enthusiasm because of the technical problems involved in turning these facilities into workable solutions within the constellation of actors involved at that time.

9.17 Governance of Transition – In Need of an Interactive and Conflict-Based Approach

Policy models in economic theory and in political science often anticipate a control centre that sets standards and guides development, at least as something intrinsic to the idea of governance. Central government has been responsible for certain parts of the cases described above that have contributed to the processes of change. However, the cases demonstrate that these situations are not just the result of established powers within these centres but constitute the co-construction of power relationships and policy discussion that legitimised specific interventions. Whether these work for or against sustainable energy solutions depends on the specific historic setting and the ideas and actors involved. The liberalisation of EU power markets and the construction of a European carbon trade system are examples of such top-down interventions that have resulted in some important transformations in the energy system but that have not met the billed expectations for change in the dominant regime and its unsustainable performance. Instead of furthering investment in renewable energy, they have stimulated carbon offset activity in which countries buy CO_2 quotas and install technologies with a short payback time. Although the technologies were introduced based on decisions made by a powerful central authority, the actors involved in implementing the new governance structures were influential in setting the stage with their own interpretations.

The notion of governance has been used in political science to emphasise the institutional and networked character of specific areas in which agency is distributed across actors and the knowledge required for action is not available to a central governing body but rests with other actors in the network. Governance also takes into account the need for negotiation of objectives and means in circumstances where translation from general political goals to specific implementation targets is crucial to the effectiveness and legitimacy of the regulation (Smith & Stirling, 2007). Specific forms of governance are required for energy system transformations and, for example, complex objectives like CO_2 reductions, environmental concerns, international dependencies and the security of supply in order to handle the conflicting interests of the parties involved.

The historical case studies demonstrate that eschewing conflict and focusing on only one policy measure or one central economic agent to provide expected results can easily destroy the general capacity to adjust to change in terms of how energy supply is configured, how energy is used and in the role of the dominant sociotechnical regime. Acting within a context of conflict and constant negotiation is part of the challenge, since trying to solve problems fully at every stage of development may not provide long-term satisfactory solutions – as seen from the perspective of society's need for the sustainable transformation of its energy system. Policy-makers, analysts and entrepreneurs see competing developmental directions and technical paradigms but a simple top-down choice cannot be made since a variety of actors and ideas is crucial to the process. The leap to a definitive solution by seeking the most convenient or efficient technology may lock out committed innovators and entrepreneurs, as well as processes in which learning and piecemeal engineering can be practised.

In Denmark, for example, if the utility companies had been made the only providers of power and given a central role in decision-making on the introduction of energy technologies based on renewable sources, the continuing process of improvement would not have occurred and the viability of new system compositions would probably have been lower. Alternatively, if wind turbines and biogas facilities had had access to investment funding and support that was not constantly adjusted according to the increasing efficiency of the technology, far too many early designs might have been installed and not renewed once the next generation of wind and biogas technology came along – their sunk costs would have been wasted. These have been controversial issues during the decades of change; balancing interests and facilitating innovation and experimentation turned out to be an important aspect of governing transition. When this de facto governance of feed-in prices was terminated in the late 1990s in favour of specific, albeit inadequate, market forces, the transition came to a halt. Only recently, thanks to a sense of urgency on climate change, steps towards a continued transition have been considered again, albeit in a very top-down policy manner. Still, the impact of these steps may not live up to the political and corporate rhetoric on mitigating climate change.

The dominant top-down technology approach has shown its weakness in contemporary climate-based policies, scenarios and plans. Politicians and companies have criticised consumers and citizens for not taking serious steps while they themselves, afraid of backlash from constituents and clients, have avoided committing to radical

188   U. Jørgensen

regulatory measures. There is a clear need for bottom-up experimenters and committed grass-root movements to take up the challenge and become an innovative force in energy system transition.

References

Bijker, W. (1995). *Of bicycles, bakelites and bulbs – Towards a theory of sociotechnical change.* Cambridge: MIT Press.

Bijker, W., Hughes, T. P., & Pinch, T. (1987). *The social construction of technological systems.* Cambridge: MIT Press.

Callon, M. (1987). Society in the making: The study of technology as a tool for sociological analysis. In W. Bijker, et al. (Eds.), *The social construction of technological systems* (pp. 83–103). Cambridge: MIT Press.

Callon, M., Méadel, C., & Rabeharisoa, V. (2002). The economy of qualities. *Economy and Society, 31*(2), 194–217.

Christensen, B. (2008). FLS 'Aeromotoren' – en dansk pionermølle. In *Kapitler af vindkraftens historie i Danmark*, 4. årgang, Elmuseet a.o. 11–19.

Elzen, B., Geels, F., & Green, K. (2004). *System innovation and the transition to sustainability. Theory, evidence and policy.* Cheltenham: Edward Elgar.

Garud, R., & Karnøe, P. (2001). Path creation as a process of mindful deviation. In R. Garud & P. Karnøe (Eds.), *Path dependence and creation* (pp. 1–38). Philadelphia: Lawrence Erlbaum.

Geels, F. W. (2004). From sectoral systems of innovation to socio-technical systems. Insights about dynamics and change from sociology and institutional theory. *Research Policy, 33*, 897–920.

Geels, F. W., & Raven, R. P. J. M. (2007). Socio-cognitive evolution and co-evolution in competing technological trajectories: Biogas development in Denmark (1970–2002). *International Journal of Sustainable Development & World Ecology, 14*, 63–77.

Gipe, P. (1995). *Wind energy comes of age.* New York: John Wiley.

Hughes, T. P. (1983). *Networks of power – Electrification in western societies, 1880–1930.* Baltimore: Johns Hopkins University Press.

Hughes, T. P. (1987). Evolution of large technological systems. In W. Bijker, T. P. Hughes, & T. Pinch (Eds.), *The social construction of technological systems.* Cambridge: MIT Press.

Hvidtfelt, K. (2001). *Tilting at windmills: On actor-worlds, socio-logics, and techno-economic networks of wind power, 1974–1999.* PhD thesis, Department of Science Studies, Aarhus Universitet, Aarhus.

Jørgensen, U. (2005). Energy sector in transition – Technologies and regulatory policies in flux. *Technology Forecasting and Social Change, 72*, 719–731.

Jørgensen, U., & Karnøe, P. (1995a). The Danish wind-turbine story: Technical solutions to political visions? In A. Rip, T. J. Misa, & J. Schot (Eds.), *Managing technology in society – The approach of constructive technology management* (pp. 57–82). London: Pinter Publishers.

Jørgensen, U., & Karnøe, P. (1995b). Samfundsmæssig værdi af vindkraft, Report no. 4 in *Dansk vindmølleindustris internationale position og udviklingsbetingelser.* Copenhagen: AKF's forlag.

Jørgensen, U., & Strunge, L. (2002). Restructuring the power arena in Denmark: Shaping markets, technologies and environmental priorities. In K. Sørensen & R. Williams (Eds.), *Shaping technology, guiding policy: Concepts, spaces and tools* (pp. 286–318). Cheltenham: Edward Elgar.

Jungk, R. (1977). *Der Atom-Staat.* Munich: Kindler.

Karnøe, P. (1991). *Dansk vindmølleindustri – en overraskende international success.* PhD thesis, Samfundslitteratur, København.

Klitmøller, L., & Thorndahl, J. (2008). Dansk Vind Elektricitets Selskab – den første forening til fremme af vindkraften i Danmark. In *Kapitler af vindkraftens historie i Danmark.* 4. årgang, Elmuseet a.o., 2–10.

9 No Smooth, Managed Pathway to Sustainable Energy Systems – Politics, ...

Lundvall, B.-Å., Johnson, B., Andersen, E. S., & Dalun, B. (2002). National systems of production, innovation and competence building. *Research Policy, 31*, 213–231.

Lybæk, R., & Møller, P. (1999). *Teknologisk innovative og markedsunderstøttende strategier for fremme af gårdbiogasanlæg.* Masters thesis, Institut for Miljø, Teknologi og Samfund, Roskilder Universitetscenter.

OOA (1980). *Fra atomkraft til solenergi – en indføring i energidebatten.* Copenhagen: Organisationen til Oplysning om Atomkraft.

Raven, R. P. J. M. (2005). *Strategic niche management for biomass.* PhD thesis, Technische Universiteit Eindhoven, Eindhoven.

Rip, A. (2009). Technology as prospective antology. *Synthese, 168*(3), 405–422.

Schumacher, E. F. (1973). *Small is beautiful: Economics as if people matter.* London: Blond and Briggs.

Shot, J., & Geels, F. W. (2008). Strategic niche management and sustainable innovation journeys: Theory, findings, research agenda, and policy. *Technology Analysis & Strategic Management, 20*(5), 537–554.

Skytte, K., Jensen, S. G., Morthorst, P. E., & Olsen, O. J. (2004). *Støtte til vedvarende energi?* København: Jurist- og Økonomforbundets Forlag.

Smith, A., & Stirling, A. (2007). Moving outside or inside? Objectification and reflexivity in the governance of socio-technical systems. *Journal of Environmental Policy & Planning, 9*(3–4), 351–373.

Staudenmaier, J.M. (1989). *Technology's storytellers – reweaving the human fabric*, Cambridge, MA: MIT Press.

Thorndahl, J. (2005). Gedsermølle – banebrydende dansk vindmølleteknologi I 1957. In *Kapitler af vindkraftens historie i Danmark*, 2. årgang, Elmuseet a.o, 15–21.

van der Vleuten, E. B. A., & Raven, R. P. J. M. (2006). Lock-in and change: Distributed generation in Denmark in a long-term perspective. *Energy Policy, 34*(18), 3739–3748.

Chapter 10
Technical Fixes Under Surveillance – CCS and Lessons Learned from the Governance of Long-Term Radioactive Waste Management

Thomas Flüeler

Abstract Carbon dioxide (CO_2) capture and storage (CCS) is a technical option for avoiding higher CO_2 concentrations in the Earth's atmosphere while still using carbon-intensive technologies in power production and other industries. It highlights the tension between the advantage of a short-term 'quick fix' and the disadvantage posed by the risk of long-term leakage and, from a technology policy perspective, the danger of perpetuating carbon lock-in. This chapter assesses CCS against criteria taken from the controversial and long-lasting governance of radioactive waste. As the dimensions covered by this issue are manifold and intertwined, there is no 'one' methodology with which to analyse it (such as a technology assessment of the n-th order). Instead, cross-disciplinary investigations make it possible to draw lessons from contentious long-term environmental issues and social science research, which is necessary before embarking on this route on a large scale.

Keywords Carbon Capture and Storage (CCS) · Climate change · Lock-in · Radioactive waste · Safety assessment · Sociotechnical system · Technical fix · Technology assessment · Total-system analysis

Global climate change due to human activity has been well documented, as has the need for concerted action (Table 10.1). Over 100 countries have thus far pledged their support for the 2°C target, which refers to the maximum global temperature increase allowed in order to avoid unduly harmful climatic effects (Pachauri & Reisinger, 2007). This goal can only be reached if CO_2 emissions are at least halved by 2050. The 400-ppm CO_2-equivalents target, which is approximately the current concentration of greenhouse gases (GHG), is estimated to provide a 75 percent chance of keeping global warming below 2°C (Meinshausen et al., 2009; Stern in Richardson et al., 2009).[1]

---

[1] The long-term 2-degree target can only be reached at a climate sensitivity (CS) of 3°C, the most likely value according to the IPCC (Clarke et al., 2009, p. S67). CS denotes the global mean equilibrium temperature response to a doubling of CO_2-eq concentrations. There is great uncertainty about the true value of CS.

T. Flüeler (✉)
Institute for Environmental Decisions (IED), ETH Zurich, 8092 Zurich, Switzerland
e-mail: thomas.flueeler@env.ethz.ch

D. Spreng et al. (eds.), *Tackling Long-Term Global Energy Problems*,
Environment & Policy 52, DOI 10.1007/978-94-007-2333-7_10,
© Springer Science+Business Media B.V. 2012

Table 10.1 Climate stabilisation scenarios and resulting long-term global average temperature

Stabilisation level (ppm CO_2-eq)	Global mean temperature increase at equilibrium (°C)	Global mean sea level rise[a] (m)	Year CO_2 needs to peak	Year CO_2 emissions back at 2000 level	Reduction needed in 2050 CO_2 emissions compared to 2000
445–490	2.0–2.4	0.4–1.4	2000–2015	2000–2030	−85 to −50
490–535	2.4–2.8	0.5–1.7	2000–2020	2000–2040	−60 to −30
535–590	2.8–3.2	0.6–1.9	2010–2030	2020–2060	−30 to +5
590–710	3.2–4.0	0.6–2.4	2020–2060	2050–2100	+10 to +60
710–855	4.0–4.9	0.8–2.9	2050–2080		+25 to +85

[a] Above pre-industrial level at equilibrium from thermal expansion only (no melting of ice sheets, glaciers and ice caps)
Sources: after Pachauri and Reisinger (2007, p. 67), and presentation by Pachauri at UN HQ, New York, 24 September, 2007, Richardson et al. (2009, p. 19)

The general consensus is that a multitude of technologies and approaches must be deployed in order to stay within the ambitious target of 2°C warming (Figs. 10.1, 10.2, and 10.3). Up to 2007, almost 29 billion tonnes (Gt) CO_2 were emitted from fossil fuel combustion, which is the source of over two-thirds of total greenhouse gas emissions (IEA, 2009c). Out of a variety of geoengineering options,[2] CCS dominates technical and policy discussions. Indeed, the potential overall

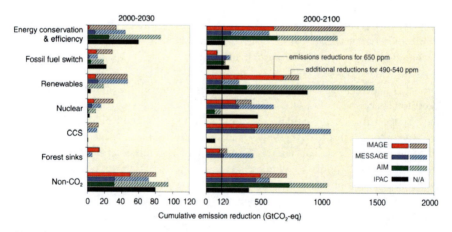

Fig. 10.1 Total contribution of CO_2 reduction options according to various scenario models (*colours*), concentration targets (*solid*: 650 ppm, hatched: 490–540 ppm) and timelines (2030, 2100). CCS ranges from 0 to well over 1,100 Gt of CO_2 mitigation
Source: Pachauri and Reisinger (2007, p. 68, IPCC AR4)

---

[2] See Royal Society (2009), Crabbe (2009), controversial comments, e.g., Schneider (2008), Schiermeier (2009), Victor, Morgan, Apt, Steinbruner, and Ricke (2009). In this context, geoengineering is the intentional large-scale modification of the Earth's environment to combat climate change. CCS is a CO_2 mitigation measure, ocean iron fertilisation an indirect CO_2 one. An antiwarming technique is the enhancement of cloud reflectivity, such as seeding the stratosphere with

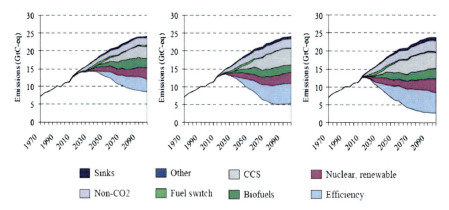

Fig. 10.2 Contribution of reduction options over time (by 2100) and according to climate targets (CO_2 concentrations in ppm). The mitigation option of CCS comes just after efficiency measures in its capacity to reduce CO_2 emissions
Source: van Vuuren et al. (2007)

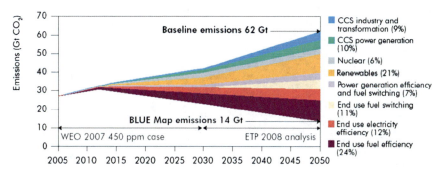

Fig. 10.3 Contribution of CO_2 reduction options over time, as forecasted by the World Energy Outlook 2007 (2030) and ETP (2050, 'BLUE' scenarios describe least-cost pathways to reach a 50 percent reduction by 2050 compared to 2005). CCS is said to contribute almost one-fifth to mitigation by 2050
Source: Goldstein and Tosato (2008, p. 4)

contribution by CCS to climate stabilisation is significant (Figs. 10.1, 10.2, and 10.3). According to prospective implementers, 3–7 Gt CO_2 might be injected annually into suitable geological formations (Rohner, 2008). However, policy options for effectively and efficiently reaching sustainable climate conditions must be thoroughly assessed in order to avoid misallocating resources that might be better used otherwise. The question is whether CCS as represented in the wedge in Fig. 10.3 holds promise as a climate change solution or instead will prove ineffective, a sideshow, or even a red herring.

---

sulphur aerosols (Pollard et al., 2009; with possible ozone damage: Tilmes, Müller, & Salawitch, 2008).

10.1 Carbon Capture and Storage (CCS): The Concept

In principle, there are three technical approaches to reducing CO_2 emissions. The first is to increase energy efficiency and the second is to substitute 'dirty' systems with CO_2-reduced or even CO_2-free systems (coal to gas, renewables, possibly nuclear). The third option is to seize CO_2 loads, without major changes to the basic technical process, right at the 'dirty' point sources (fossil-fuelled power plants, oil refineries, cement kilns, iron and steel factories, etc.), move them to suitable stacks and bury them, hopefully forever. This third option is the basic notion of CCS.

Although the idea is compelling, the associated process chain is long (Fig. 10.4): It takes six coordinated and well-established steps, from mining of the resource and transport to conversion in the factory, CO_2 capture/compression, transportation (via pipelines or tankers) and finally storage.

Capture in power plants[3] can be executed along three processes: post-combustion, pre-combustion or oxyfuel (Fig. 10.5). Post-combustion uses a solvent to capture the CO_2 from the flue gas, which requires low levels of nitrogen and sulphur oxides and, therefore, requires energy-intensive de-NO_x and desulphurisation facilities. Pre-combustion involves reacting a fuel with air or oxygen that

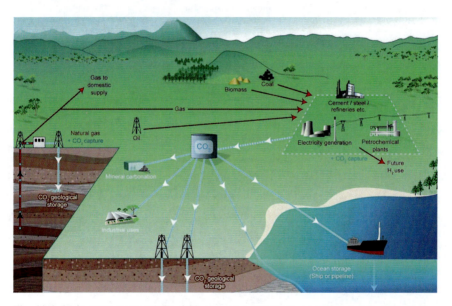

Fig. 10.4 Main components of the CCS system: 1. Resource extraction (mining) or production (biomass), 2. Transport, 3. Conversion (power plants, cement works, etc.), 4. CO_2 capture, 5. Transportation, 6. Storage (oil or gas fields, saline formations or coal beds; mineral carbonation; ocean). Reuse is possible but negligible
Source: IPCC (2005)

---

[3] Other available and emerging capture techniques are listed in Plasynski et al. (2009, pp. 27–30).

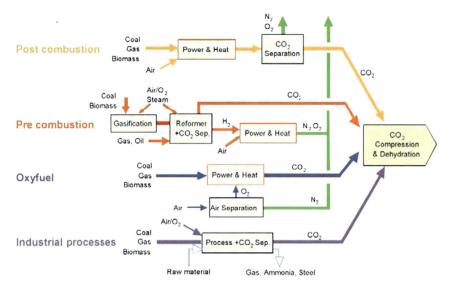

Fig. 10.5 Process chains associated with CO_2 capture. As industrial processes do not intrinsically depend on fossil fuels, they might theoretically be designed 'fossil-free'
Source: IPCC (2005)

contains CO and H_2. CO_2 is separated and H_2 is used as a fuel. In oxy-combustion, the fossil fuel is burnt with almost pure oxygen instead of air. The air separation is energy-intensive. Only post-combustion is current technology in niches whereas gasification and oxyfuel are both at early stages of development. Through all processes, the thermal efficiency of the plants drops to between 34.5 percent (minimum) and 55.6 percent (maximum) (Adams & Davison, 2007). There are even proposals to extract CO_2 directly from dispersed sources (Lackner in IOP, 2008) or to produce methanol from CO_2 (Stucki, Schuler, & Constantinescu, 1995), although both of these methods are even more energy-intensive than 'conventional' CCS.

CO_2 storage is planned as injection into geological formations (declining oil fields, unminable coal seams or saline aquifers) or into the deep ocean. A more stable fixation of CO_2 would be direct mineral carbonation; that is, converting CO_2 gas to solid carbonates such as calcite or magnesite (Seifritz, 1990). This occurs naturally by weathering but at a slow pace. Researchers have suggested relatively rapid in situ carbonation of peridotite (Kelemen & Matter, 2008). The IPCC estimates that this would require 60–180 percent more energy than a power plant without CCS (IPCC, 2005, pp. 321, 330). As for ocean storage, the ocean naturally sequesters CO_2 from the atmosphere at a rate of 2 Gt C/year. Injecting liquid CO_2 into seawater causes a reduction in pH that is potentially harmful to marine life. According to Ormerod, Freund, and Smith (2002), near-field modelling studies suggest that this impact would be reduced if techniques disperse the CO_2 and minimise transient increases in CO_2 concentrations. More data are necessary to understand sub-lethal impacts over longer periods of time.

10.2 Plethora of Challenges to Reduction Policy Without Delay

Although both science and society have recognised climate change as a serious problem (EC, 2009), mankind has continued along a high-emissions path. The signatory states of the Kyoto Protocol show a mixed record, with a slightly increasing emission balance overall.[4] The reference scenario of the World Energy Outlook 2009 still foresees a 40 percent global increase of primary energy demand by 2030, resulting in energy-related CO_2 emissions of over 40 Gt, compared to less than 29 Gt in 2007 (IEA, 2009c). Barriers to effective reductions exist at political, institutional and individual levels. Different parts of the world have diverging views on the responsibilities of 'first movers' (historical polluters such as North America, Australia or Europe versus 'new' polluters like China, India and developing countries). Incentives, trading and enforcement mechanisms are weak or not in place in these latter countries and large-scale lifestyle changes towards sustainable development are rare.

In such a tangled situation, the characteristics of CCS seem attractive. It is a quick and technical, narrowly located but high-potential solution that has no need for widespread and extensive efficiency improvement in dispersed facilities, equipment, appliances or in 'software' such as institutions and behaviour. Is it an emergency brake until more durable measures become effective or does it in fact delay such measures (Hawkins, 2003)? Is it a 'horn of plenty' or a Trojan horse (de Coninck, 2008)? Or is it another Faustian Bargain (Spreng, Marland, & Weinberg, 2007), in addition to the nuclear bargain that Weinberg (1972) cautioned against almost 40 years ago? Given the presence of systemic resemblances of nuclear and fossil-based technologies, especially regarding their waste, CCS (Celia, Bachu, Nordbotten, Gasda, & Dahle, 2005; Bowden & Rigg, 2005) is assessed here against the template of – or, rather, the strenuous experience with – the governance of radioactive waste management.[5] This field is also confronted with difficult safety demonstrations around long-term leakage issues in a charged political environment.

10.3 Six Criteria for Assessing CCS

Whether or not it is in fact 'one of the most intractable policy issues' (North, 1999) in countries with nuclear power plants, the long-term governance of high-level radioactive waste remains unsolved in most regions of the world and is a major stumbling block for the nuclear industry. Based on the 50-year history and continual

---

[4] http://unfccc.int/ghg_data/ghg_data_unfccc/items/4146.php. All web links accessed November 16, 2011.

[5] Governance is more than management and denotes, according to a European White paper on the subject: 'rules, processes and behaviour that affect the way in which powers are exercised ... particularly as regards openness, participation, accountability, effectiveness and coherence' (CEC, 2001).

evaluations of how industry, governments and society deal with nuclear waste (e.g., IAEA, 1999; CEC, 2002; NEA, 2008a; Colglazier, 1982; Kasperson, 1983; Carter, 1987; Blowers, Lowry, & Solomon, 1991; Freudenburg, 2004; Flüeler, 2006), it is possible to draw some key lessons for other long-term sociotechnical environmental issues such as CCS. The approach is to assess the issue using a combination of disciplines and perspectives. These can include systems theory, integrated risk assessment, social sciences, technology assessment and management (implementation, compliance) (Flüeler, 2006). The six criteria listed below address issues that have proven to be crucial in technology policy debates (Ropohl, 1978/1999; Ravetz, 1980; Vlek & Stallén, 1980; Wynne, 1983; Morone & Woodhouse, 1986; Kasperson, Golding, & Tuler, 1992; approach in Flüeler, 2006).

1. Need for deployment and comparative benefits
2. Total-system analysis and safety concept
3. Dedicated internationally harmonised regulation and control
4. Economic aspects (costs and incentives)
5. Implementation
6. Societal issues

It is important to note that rather than simply being items that are checked off on a list (first 'technical', then social 'add-ons'), the criteria should relate to one another such that numbers 1 and 6 above are interlinked and influence the others all along. CCS is not 'just' a technical innovation but a sociotechnical system that is eventually to be implemented on a large scale. As the CCS concept is 'in the making' (Latour, 1987), approaches in science and technology studies (STS), such as 'interpretative flexibility', are useful for scrutinising arguments for their consistency (Collins, 1981; Bijker, Hughes, & Pinch, 1989). Although by no means a comprehensive study, the following is more a suggestion to carry out such an attempt at integral appraisal (going beyond additive assessments such as in Gough and Shackley (2006) or in the otherwise notable Toth, 2011).

10.3.1 Criterion #1: Need for Deployment and Comparative Benefits vis-à-vis Competing Technological Options

Radioactive waste is a technological and political constraint. It results from politically desired and officially promoted activity (the choice of nuclear power in national energy policy as well as from medicine, industry and research) and it is a physical by-product that must be dealt with in a sustainable way. Waste is the inevitable 'back end' of the nuclear fuel 'cycle' and literally 'downstream' in the controversial nuclear debate. It is an intrinsic characteristic of the nuclear system and must be dealt with accordingly.

With the climate debate having recognised that CO_2 is a pollutant, this otherwise harmless gas has become waste and part of the back end of fossil-fuel systems.

198 T. Flüeler

As such, it is as inseparable from coal utilisation as radwaste is from nuclear fission utilisation. There has also been mention of closing the carbon cycle (Powicki, 2007) just as the nuclear cycle was closed (NEA, 2000). Unlike nuclear waste, however, CO_2 is not tied to a specific technology but is produced in other conversion chains (like deforestation) and can be reduced in such chains (such as afforestation with entrained CO_2 sequestration) or in processes (such as building insulation with associated avoided CO_2 emissions). Consequently, CCS is in competition with other technologies with regard to other criteria, such as potential, efficiency and cost and it must therefore perform uncontrovertibly well. At this stage of development (R&D, few pilots, concept not established, see below), the performance of CCS has yet to be determined. Estimates of the storage potential (in terrestrial geological formations) cover a wide range:

- 1,678–10,000 Gt worldwide, according to IPCC (2005) (p. 34, lower estimate for all reservoir types to upper estimate for deep saline formations)
- 'Likely at least 2,000 Gt' worldwide in geological formations according to IPCC (2005) (p. 33, likely = 66–90 percent probability)
- 0–1,100 Gt worldwide as implemented in scenarios according to Fig. 7.1 (Pachauri & Reisinger, 2007)
- 800 Gt worldwide, according to the International Energy Agency Greenhouse Gas R&D Programme 2007 (IEA, 2007)
- 1,250–12,250 Gt worldwide, according to the IEA Greenhouse Gas R&D Programme homepage[6]
- 8,090–15,500 Gt worldwide, according to the IEA Roadmap (IEA, 2008b, 2009d)
- 2,000–10,000 Gt worldwide, according to Bradshaw et al. (2007)
- 3,600–12,920 Gt in the USA, as determined within the Regional Carbon Sequestration Partnerships (RCSP) initiative by the US Department of Energy[7]
- 14–250 Gt for the UK alone, with the range depending on assumptions of pore volume (according to Gough et al., 2006)

In addition to the climate target as such, the timing of mitigation measures is also relevant (Fig. 10.7 below). Krey and Riahi (2009) concluded from their scenario analysis that 'early abatement ... might be as important as the long-term concentration target itself in order to reduce the risk of exceeding specific temperature thresholds' (ib., p. S106).

10.3.2 Criterion #2: Total-System Analysis and Safety Concept

Even though the approach sounds thrilling as a technical solution, the Carbon Capture and Storage concept is not yet mature (Fig. 10.6) and its technical potential

---

[6] http://www.ieaghg.org, based on 20 USD/t CO_2 stored.

[7] http://www.netl.doe.gov/technologies/carbon_seq/core_rd/storage.html

10 Technical Fixes Under Surveillance – CCS and Lessons Learned from . . . 199

CCS component	CCS technology	Research phase	Demonstration phase	Economically feasible under specific conditions	Mature market
Capture	Post-combustion			X	
	Pre-combustion			X	
	Oxyfuel combustion		X		
	Industrial separation (natural gas processing, ammonia production)				X
Transportation	Pipeline				X
	Shipping			X	
Geological storage	Enhanced Oil Recovery (EOR)				X[a]
	Gas or oil fields			X	
	Saline formations			X	
	Enhanced Coal Bed Methane recovery (ECBM)		X		
Ocean storage	Direct injection (dissolution type)	X			
	Direct injection (lake type)	X			
Mineral carbonation	Natural silicate minerals	X			
	Waste materials		X		
Industrial uses of CO_2					X

[a]CO_2 injection for EOR is a mature market technology, but when this technology is used for CO_2 storage, it is only 'economically feasible under specific conditions'

Fig. 10.6 Current (2005) maturity of CCS system components. An X indicates the highest level of maturity for each component. There are also less mature technologies for most components
Source: IPCC (2005, p. 8)

cannot be reliably defined. Furthermore, its performance cannot be assessed because overall conditions, including its embeddedness in economic, technological and environmental policies, must be set beforehand. It is essential to start at the beginning with a total-system analysis.

System Approach

CCS is a potentially powerful but also ambiguous lever in the transition away from carbon-intensive economies and, therefore, if it is ever deployed it must be done so very prudently. This must start with a careful system analysis in order to understand all possible structural system changes its full-scale implementation might induce. For all energetic technological lines of CCS (pre-combustion, post-combustion, and oxyfuel, Fig. 10.5), a comprehensive life cycle analysis[8] along the process chain of

---

[8] This goes far beyond the 'life cycle of a CO_2 storage project' notion defined in IPCC (2005, p. 226) or the comparison of mere power plant characteristics (Hadjipaschalis, Kourtis, & Poullikkas, 2009; Mondol, McIlveen-Wright, Rezvani, Huang, & Hewitt, 2009).

extraction of the energy carrier, transport, conversion, CO_2 capture, compression, transportation and storage phases shows a 10–40 percent increase in resource use (e.g., coal and mining) if a 'low-carbon' use of fossil energy is targeted (IPCC, 2005; Wuppertal Institute, DLR, ZSW, PIK, 2008a; Grünwald, 2008). This implies a decrease of system efficiency to approximately 65 percent, currently even 45 percent[9] and an increase in classically toxic environmental impact (NO_x, heavy metals, fine particles; see Koornneef et al., 2008).

The net greenhouse gas reduction amounts to 78 percent for a German brown-coal plant (Wuppertal Institute et al., 2008a). Constraints on transportation distance and means are required (such as power plants as close as possible to storage sites, pipelines or large tankers). In a long-term management plan, all components have to be set in place in a timely manner so that coal provisioning, plant lifetime and storage volume and facility are synchronised. If power plants reach their foreseen termination before 2020 (when full-scale CCS might be in operation) and must be replaced without interruption, a smooth transition to low carbon with capture-ready plants will not be possible (as is the case with Germany, ib.). 'Technical lessons learned' is not enough (EU CO_2 Network, 2004) – after all, it is the maturity and deployment of the overall system, not just its components, that counts (Fig. 10.6). The definition of system boundaries and the choice of decisive energy penalty values are crucial, especially given the slim publicly available database (Page, Williamson, & Mason, 2009), as the plant efficiency sinks with carbon capture.

It is widely recognised that CCS can only have a bridging function (de Coninck, 2008). In other words, on the way to sustainable energy systems it must not substitute or hamper the development of renewables and efficiency measures because fossil fuels are finite, as is geological storage, despite the possible existence of huge reservoirs. The crucial question is the length of the transition period that CCS could or should be given (or, rather, that the world energy system should be given with the CCS tool) until the combination of efficiency measures and renewables takes effect. Arguably, it takes less time and effort to 'add' CCS to the existing energy supply infrastructure than it does to change the overall energy system (provision, distribution, consumption) to be based on renewables and energy efficiency. Some components, drivers and interrelations are noted in Fig. 10.7.

All in all, the system analysis must consider material and energy fluxes, the link of CCS with other energy (sub)systems, integration into the market and the interplay of actors (this is a very intricate issue, as alluded to in Praetorius & Schumacher, 2009, for example). CCS is 'more than a strategy for "clean coal"' (IEA, 2009d, p. 4; see also Vallentin, 2007; Viebahn et al., 2007; Van Alphen, Hekkert, & Turkenburg, 2009); its technology must also be adopted by biomass and gas power plants, and 'mixed' portfolios (namely coal/biomass/hydrogen) are possible. Consequently, 'negative emission scenarios' that provide a net reduction of CO_2 must be taken into account.

---

[9] http://www.co2captureandstorage.info/what_is_co2.php. Current CO_2 capture rates range between 80 and 90 percent; techniques are foreseen that would make it possible to reach 98 percent (Gibbins & Chalmers, 2008).

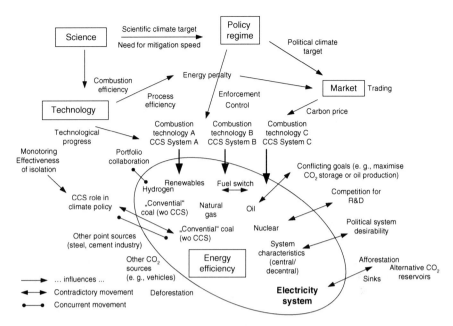

Fig. 10.7 System components, drivers, actors and their interrelationships. Issues are covered in the text

In order to increase transparency and comparability, it might be worthwhile including CCS into a global warming and environmental production efficiency ranking (a metric proposed by Feroz, Raab, Ulleberg, & Alsharif, 2009). The system approach reveals how its compartments, functions and dynamics are related, whether this is optimised and, if so, to what goal. One outcome of a system analysis is the revealing of unintended consequences. An example from the radioactive waste field is the dysfunctions (leakage) of Asse, the German waste facility.[10] To a large extent, this leakage can be traced back to the fact that this site was not designed for waste disposal but was misused as such after it had been exploited for rock salt. This process has had major repercussions on the German disposal programme, incidentally building on the Gorleben site with the same host rock as found in Asse. Consequently, abandoned oil and gas fields have to be carefully evaluated as potential CCS deposit sites, particularly as their deployment is very attractive to exploiters

---

[10] http://www.endlager-asse.de/EN. Conventional salt mining from 1909 to 1964. From 1965, research laboratory under the auspices of the Federal Ministry for Scientific Research and Technology. From 1967 to 1978, trial emplacement, then final disposal of low- and intermediate-level radioactive waste into the abandoned cavities under the relatively weak Mining Law. In 2009, operations were transferred to the Federal Office of Radiation Protection under the Atomic Energy Act, which requires a decommissioning concept and a long-term performance assessment. It is planned to retrieve all waste and dispose of it at the dedicated and licensed Konrad facility.

in the form of profits from extraction, higher yield due to CO_2 injection (up to three times), CO_2 credits and disclaimers in the case of short-term liabilities.[11]

Systemic Features

Some observers have argued that CCS is by no means a novel technological concept as CO_2, among other things, has long been injected into oil formations to recover additional oil (Benson, 2005, EOR in Fig. 7.6). What is radically different are the respective goals and scales of CCS and enhanced oil recovery. The new intention with CCS is to seclude CO_2 for the common good (a sustainable climate, as it is harmless to individuals, even at high concentrations), at an ample scale and for a very long time. This requires the implementation of an internationally effective system,[12] tested long-term safety performance of facilities and the overall system (including an adequate control mechanism) and, above all, the recognition of longevity as a factor to be dealt with.

When CO_2 is emitted, much of it remains in the atmosphere for around 100 years. The carbon cycle of the biosphere takes a long time to neutralise and sequester anthropogenic CO_2. According to Archer 2005, the lifetime of fossil-fuel CO_2 is 30,000–35,000 years (Archer, 2005) if averaged out over its entire distribution. The term 'lifetime' refers to the period required to restore equilibrium following an increase in the gas's concentration in the atmosphere. If one includes the long tail, Archer suggests that this figure is '300 years, plus 25% that lasts forever' (Archer, p. 6). This, along with the heat capacity of the oceans, means that climate warming is expected to continue beyond 2100, even if CO_2 emissions cease (see Fig. 10.8).

Figure 10.9 depicts the well-known long-term dimension of nuclear waste, with a drop in radioactivity for low-level waste after 1,000 years.

Comparing radioactive wastes with CO_2 reveals both differences and similarities, both in systemic and risk aspects (Tables 10.2 and 10.3, respectively). It is less about the sheer size of the release into the environment but has more to do with the nature of it (Benson, 2005). Disposal and storage systems are both associated with a similar risk mechanism: a low-level but long-term chronic release of the pollutants into the environment along with a slow degradation/alteration of an open system (geological formations and geological formations/climate, respectively) with concurrent large uncertainties. With the exception of some scenarios of human intrusion, potential impacts are hard to detect with respect to location and time.

Safety Concept

Waste and risk management for CO_2 case both exhibit an early stage of research and no maturity of the safety concept (Table 10.3), even though progress has been made (Maul, Metcalfe, Pearce, Savage, & West, 2007; Stenhouse, Galeb, & Zhou, 2009) and shortcomings have been recognised (Bachu, 2000).

---

[11] According to the April 2011 draft for a German CCS act, an operator is permitted to request the transfer of responsibility 30 years after closure of a storage site at the earliest (BMU, 2011).

[12] Apart from this, the electricity system integration is decisive, establishing which components are chosen and how effectively (see the exemplary combination given by Davison, 2009).

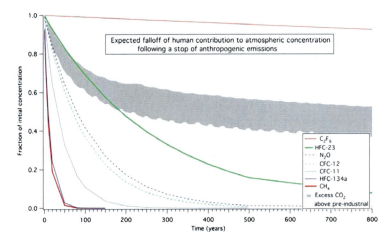

Fig. 10.8 Falloff of concentrations expected for various greenhouse gases, from a peak value following cessation of emissions. The decline of CO_2 to a high and stagnant level is notable
Source: Solomon, Plattner, Knutti, and Friedlingstein (2009), supplementary material

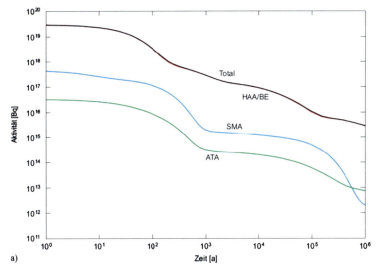

Fig. 10.9 Comparison of radioactive waste streams with respect to their radioactivity (measured in Becquerel, Bq). The reference is the Swiss nuclear programme with 50 years' history of operating five nuclear power plants with 3,000 MW_{el} power and waste from medicine, industry and research accumulated over 70 years (until 2050). Zeit [a] = time (in years), HAA/BE: high-level waste and spent fuel (3,000 metric tons), SMA: low-level and short-lived intermediate-level waste, ATA: (intermediate) alpha-toxic waste
Source: Nagra (2008, p. 27)

Table 10.2 System and institutional properties of radioactive waste disposal and fossil CO_2 storage. The large family of technologies is an <u>asset of CO_2</u> (<u>underlined</u>) whereas the absence of economical incentives and disposal funds, the high number of actors (polluters and potential implementers), the uncertain regulation and classification, the high volume and the non-dedicated type of facilities are **burdens to sustainable CO_2 storage (bold)**

System / Characteristic	Radioactive waste disposal	(Fossil) CO_2 CCS
Origin	Nuclear electricity generation; medicine, industry, research	Fossil-fuelled energy generation; oil refinery; cement, iron, steel industries
Sources	Comparatively few	Few large, innumerable small
Economy	Polluter-pays fee on each kWh	Not regulated to date
Resources (R&D, money)	Substantive	Sparse
Actors	Few state-by-state implementers; national regulatory agencies	Market-driven international companies; regulation undetermined
Volume	Small (Mt/60 yrs)	High (Gt/yr)
Waste package	Concentrated, consolidated waste stream	Dispersed
Nature of substance	Contaminant, waste	Greenhouse gas, commodity (waste)
Regulatory classification	Internationally classified	No standardised classification
Facilities	Dedicated repository with drifted galleries, abandoned mines substandard	Exploited reservoirs
Range of options/flexibility	Small	Large (from alternatives (e.g., renewables) to variants (e.g., ocean storage, mineralisation, etc.))

In the radioactive waste community, evidence, analysis and arguments for the safety of radioactive waste disposal sites must be gathered in a so-called 'safety case', which involves a stepwise and iterative procedure to provide risk assessments and appraisals, as well as confidence statements from site selection to closure and post-operational monitoring of a specific facility (NEA, 2008b). As specified below, the safety case for radioactive waste disposal envisions the following overall safety functions: isolation of the waste from the human environment, long-term confinement within the disposal system and attenuation of releases from the environment (IAEA, 2003, p. 5; IAEA, 2006, pp. 19passim). The key idea of compliance is the multiple-barrier concept, according to which myriad technical and natural barriers take effect in case one single barrier breaches (Fig. 10.10).

This defence-in-depth paradigm, which is central to nuclear installations, is missing in the CO_2 safety concept for geological storage (Fig. 10.11). The basic idea here is that caprocks, such as low-permeable aquitards, must prevent CO_2 from leaking back through sealed abandoned or dedicated injection wells or other pathways into the atmosphere. The dependence on what is basically one barrier is risky even if

10 Technical Fixes Under Surveillance – CCS and Lessons Learned from . . .

Table 10.3 Risk-related properties of radioactive waste disposal and fossil CO_2 storage. The longevity of pollutants, leakage as evasion mechanism, the low but chronic risk level and the long-term barrier mechanism are *common* characteristics. The relatively simple hazard system, good traceability and the wide range of options may be <u>assets of CO_2 (underlined)</u>, whereas most other waste and risk management aspects are at an initial phase and are therefore currently **burdens to sustainable CO_2 storage (bold).** See the text for explanations

Characteristic	Radioactive waste	(Fossil) CO_2
Pollutants		
Waste package	Concentrated, consolidated waste stream	None, dispersed
Toxicity	High (and low with short-lived low-level waste)	(Low but **dispersed**)
Duration of toxicity (residence time in systems)	Very long (millions of years)	Long (thousands of years)
(Long-term) hazard system	Complex (ionising radiation)	Simple (greenhouse gas effect)
Risk level	Low, chronic	Low, chronic
Protection limits	Radiation dose (mSv)	None (work safety: yes)
Indicators	Defined for safety functions	None (no thresholds)
Traceability	Bad	Better
Transport mechanism	Groundwater, air (gas)	Air, groundwater
Evasion mechanism	Leakage	Leakage, seepage
Risk appraisal		
Concept	Concentrate and confine	(Confine)
Barrier mechanism	Sorption, diffusion	Mineralisation, dispersion
System understanding	Existent, developed	Under construction
Site characterisation	Good, dedicated investigations (seismics, boreholes, etc.)	Medium (lost records of abandoned reservoirs)
Barriers:	Pellet, canister/container,	None;
Engineered (technical)	bentonite;	Sealed wells
Natural	Sealed galleries and shafts Host rock and other confining geological units	Caprock(s)
Assessment methodology	Relatively well-established performance assessment ('safety case')	Embryonic
Other relevant issues		
Use conflicts	Partly existent (can be avoided)	Often intrinsic (abandoned resource reservoirs)
Process of safety case	Stepwise, iterative	Undefined
In situ monitoring	Accessible shafts and galleries	Basically inaccessible (single-barrier seals)
Range of options/flexibility	Small	Large (variants, e.g., ocean storage, mineralisation)

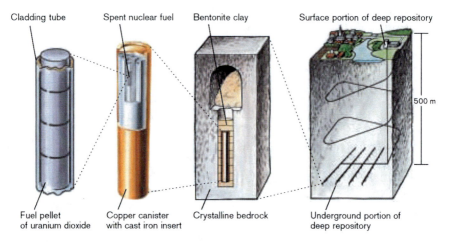

Fig. 10.10 Multiple-barrier concept in radioactive waste management: The fuel pellets, the canister, the bentonite-clay buffer and a primary rock (such as the crystalline basement in the Swedish case) are designed to prevent the radioactive substances in spent fuel from spreading into the environment
Source: SKB, from http://www.skb.se

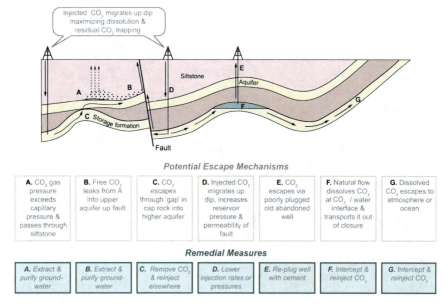

Fig. 10.11 Safety concept of CO_2 storage: CO_2 gas is injected into a 'storage formation' as the only barrier
Source: IPCC (2005, p. 35)

potential storage locations, such as – relatively few – depleted oil and gas reservoirs, are proven long-term traps of gas.

It is symptomatic that the term used is 'storage' rather than 'disposal'. In the nuclear community, the term 'storage' always refers to interim storage, whereas disposal is meant to be final in the sense that there is no intention of retrieval and that, if successful, (most) radionuclides will be kept out of the biosphere (IAEA, 2006, p. 1). Retention is also possible by building CO_2 into minerals (mineralisation, carbonation) but this has only been conceptualised in research, let alone implemented as state of the art (Fig. 10.6). CO_2 is expected to spread at a rapid pace: 8.6 Mt injected annually would produce a plume of 18 km^2 over 30 years (Benson, 2004).

The wide range of options – from ocean storage to final trapping in minerals – is ambiguous for the appraisal of the 'soundness' of the CO_2 safety conception. On one hand, it demonstrates the flexibility of waste management while on the other it indicates a lack of maturity in the discourse of options. Following lengthy expert and public debate in the radioactive waste area over a period of several decades (NEA, 2008a; CoRWM, 2006), final disposal in geological formations has emerged as the best – or the least bad – solution. Due to a rejection of the 'disperse and dilute' concept, sea dumping was prohibited by the London Dumping Convention and, later by the OSPAR Convention (LDC, 1972; OSPAR, 2007). Space disposition was dismissed and, likewise, sub-seabed disposal was abandoned for safety reasons, prohibition by maritime law and lack of public acceptance (NEA, 2000, 2008c; OSPAR, 2007). Although this is far from a 'closed issue', the scientific and non-scientific debate is more advanced in the radioactive waste discourse than in CCS (Tables 10.2 and 10.3). According to Bijker et al. (1989), 'closure' in science occurs 'when a consensus emerges that the "truth" has been winnowed from the various interpretations' (p. 12). The differences in advancement are most striking in the safety assessment methodology and procedure, as shown below.[13]

Safety Assessment

Safety analyses, which are usually called 'performance assessments' in the context of radwaste, include scenario development, conceptual and mathematical models development, consequence analysis, uncertainty and sensitivity analysis and confidence building. There is international agreement that it is 'not intended to imply a rigorous proof of safety, in a mathematical sense, but rather a convincing set of arguments that support a case for safety' (NEA, 1999, p. 11; NEA/IAEA/CEC, 1991, pp. 10, 13). Due to the long-term character of the repository system, the 'aim of the performance assessment is not to predict the behaviour of the system in the long term, but rather to test the robustness of the concept as regards safety criteria' (IAEA, 1999, p. 245fpassim). For this purpose, a refined repository philosophy and design was developed with technical barriers, monitoring cycles and quality assessment following one another (NEA/IAEA/CEC, 1991; ICRP, 1993). The instrument for this is the development of a so-called 'safety case'. According to NEA (1999,

---

[13] Adapted from Junker, Flüeler, Stauffacher, and Scholz (2008).

p. 22), 'A safety case is a collection of arguments, at a given stage of repository development, in support of the long-term safety of the repository. A safety case comprises the findings of a safety assessment and a statement of confidence in these findings. It should acknowledge the existence of any unresolved issues and provide guidance for work to resolve these issues in future development stages'. This implies that the point is not to start out by planning a repository in a manner that is as detailed as possible but to corroborate the safety case in steps that must be understood by all actors at every stage (NEA, p. 15).

As for the barrier structure, the safety case distinguishes between the near-field and the far-field. The near-field is defined as the engineered barrier systems (EBS), 'as well as the region of rock immediately around waste emplacement tunnels (extending a few metres from the tunnels) [the Excavation Damaged Zone, EDZ] that is significantly affected by thermal, hydraulic, chemical and mechanical changes induced by the presence of the waste and excavations. The far field is considered the [surrounding] host rock and the geosphere beyond [that is, other confining geological units], in which such effects [supposedly] are substantially smaller or negligible' (Nagra, 2002, p. 111). System considerations include the evolution of climatic, surface environmental and geological boundary conditions and they also involve 'long-term processes such as uplift/erosion, climatic changes, neotectonic events, etc. The evolution of the near field and far field, driven by the presence of the waste and influenced by the boundary conditions, are then discussed [in Nagra, 2002], including all aspects relevant to radionuclide transport' (Nagra, 2002, p. 111). The consequence analysis is executed in a subsequent step; that is, within the site-specific safety analysis (for low-level radioactive waste, LLW, for example, Nagra, 1994) and considers so-called 'critical groups' that might be affected by release of radioactivity within a defined scenario (Nagra, 1994, p. 170).

The realisation, that is, the method of the safety analysis, consists of the following steps (as carried out by the Swiss implementer Nagra for the host rock Opalinus Clay and spent fuel, SF, high-level radioactive waste, HLW, as well as long-lived intermediate-level waste, ILW, according to Nagra, 2002, p. IVf.):

1. The choice of a disposal system, via a flexible repository development strategy, that is guided by the results of earlier studies, including studies of long-term safety
2. The derivation of the system concept, based on current understanding of the features, events and processes (FEPs) that characterise and may influence the disposal system and its evolution
3. The derivation of the safety concept, based on well-understood and effective pillars of safety
4. The illustration of the radiological consequences of the disposal system through the definition and analysis of a wide range of assessment cases
5. The compilation of the arguments and analyses that constitute the safety case, as well as guidance for future stages of the repository programme.

The safety case, however, includes more information than just models and calculations of expected doses or risks. Additionally, it is based on the following broad lines of argument (Nagra, 2002, excerpts from pp. V, VIIIf. and chapter 8):

- Strength of geological disposal as a waste management option. This is supported by (i) the internationally recognised fact that a well-chosen disposal system located at a well-chosen site fulfils the requirement of ensuring the safety and protection of people and the environment, as well as security from malicious intervention, now and in the future, (ii) the existence of suitable rock formations in Switzerland and elsewhere, (iii) other safety assessments conducted world-wide, (iv) observations of natural systems and (v) the relative advantage of geological disposal over other options.
- Safety and robustness of the chosen disposal system. This is ensured by (i) a set of passive barriers with multiple phenomena contributing to the safety functions, (ii) the avoidance of uncertainties and detrimental phenomena through an appropriate choice of site and design and (iii) the long-term stability of the Opalinus Clay host rock and the repository due to a suitable geological situation.
- Reduced likelihood and consequences of human intrusion. This is supported by (i) the preservation of information about the repository, (ii) the avoidance of resource conflicts (that is, the absence of viable natural resources in the area proposed for the repository) and (iii) the compartmentalisation of the repository and the solidification of the wastes.
- Strength of the stepwise repository implementation process. This is supported by (i) the fact that, at the current project stage, not all the details of the repository system need to be determined and, therefore, the information basis only needs to be adequate for that particular stage, (ii) the reliance on understood and reliably characterised components (site, EBS), (iii) the involvement of stakeholders and the opportunities for feedback and improvements, (iv) the flexibility of the project, allowing new findings to be taken into account (such as findings regarding the detailed allocation of emplacement tunnels, choice of design options, placement of surface facilities and even siting; i.e., there are other possible sites for the Opalinus Clay host rock option as well as for other host rock options) and (v) the possibilities for monitoring and reversal (including retrieval of the wastes).
- Good scientific understanding that is available and relevant to the chosen disposal system and its evolution. This is supported by (i) the results from regional and local field investigation programmes, key elements of which include an extensive 3-D seismic campaign and an exploratory borehole in the potential siting area and the availability of information from other boreholes in the region, as well as complementary studies performed in the Mont Terri underground rock laboratory and in other laboratories, as well as observations of Opalinus Clay in a number of railway and road tunnels, (ii) the findings from more than 20 years of experience in developing and characterising engineered barrier system components within the Swiss programme, as well as by the availability of a strong

international information basis and (iii) the availability of a detailed model waste inventory for SF, HLW and ILW.

- Adequacy of the methodology and the models, codes and databases that are available to assess radiological consequences for a broad spectrum of cases. This is supported by adherence to the assessment principles outlined in chapter 2 (of Nagra, 2002).
- Multiple arguments for safety that include compliance with regulatory safety criteria, the use of complementary safety indicators, the existence of reserve FEPs and the lack of outstanding issues that have the potential to compromise safety.

In an international peer review commissioned by the Swiss Federal Council, this procedure was judged as being in line with the international state of the art (NEA, 2004b). It also underwent a thorough national consultation process, involving expertise from technical authorities and federal commissions, as well as political and other stakeholders. Eventually, in 2006, the Federal Council accepted this (generic) demonstration of disposal feasibility for spent fuel and high-level and intermediate-level waste.

The CCS community is far from such an intricate buildup of scientific evidence and confidence, even though the issue has been recognised and a degree of international harmonisation is under way (IEA Greenhouse Gas R&D Programme) and science has become institutionalised (for example, by creating a dedicated journal, the *International Journal of Greenhouse Gas Control*, Gale, 2007). The needed technical system goes beyond the admittedly well-known injection of CO_2 but must still encompass the aspects described above. This means that CCS technology is not yet 'mature', as maintained earlier (Bachu, 2000, p. 957). Maul et al. (2007) recognises the following research demands in CCS: 'Dealing with the various types of uncertainty, using systematic methodologies to ensure an auditable and transparent assessment process, developing whole system models and gaining confidence to model the long-term system evolution by considering information from natural systems. An important area of data shortage remains the potential impacts on humans and ecosystems' (p. 444). Grünwald (2008) and Jacobson (2009) list the research demands on leakage phenomena (nature, timing, mechanisms, etc.) while Savage, Maul, Benbow, and Walke (2004), add scenario development to the agenda.

Site-Selection Procedure

By definition, disposal facilities also concentrate hazards, which means that the selection of suitable sites is decisive. In the case of radwaste, in 2008 the Swiss government started a stepwise site-selection procedure in order to find a suitable site, both for LLW and SF-HLW-ILW waste types, within 10 years. Although this may not be the case in every country, efforts have been made in a range of national programmes to establish a systematic, transparent and participatory concept for site selection (these include AkEnd for Germany (AkEnd, 2002), NWMO for Canada (NWMO, 2003) and CoRWM for the UK (CoRWM, w/o yr.)).

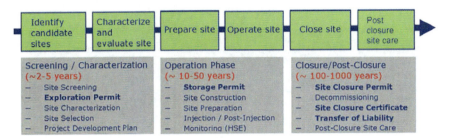

Fig. 10.12 Phases and permits typically associated with CO_2 storage projects
Source: Aarnes, Selmer-Olsen, Carpenter, and Flach (2008, p. 1736)

The institutional situation in the CCS field (involving unclear regulation, many global players, and lacking financial incentives and funds) has meant that no standard site-selection process has been developed. Despite this, many individual and detached projects have been pursued to date. The diverging degree of detail can be seen from Fig. 10.12 for CCS in general or in applications (Ramírez, Hagedoorn, Kramers, Wildenborg, & Hendriks, 2008a; de Visser, de Vos, & Hendriks, 2008; Chalmers, Jakeman, Pearson, & Gibbins, 2009; Jacobs, Cohen, Kostakidis, & Rundell, 2009), as opposed to the site-selection procedure for radioactive waste in general (NEA, 2004a), or as applied in a national programme (see Fig. 10.13).

10.3.3 Criterion #3: Dedicated Internationally Harmonised Regulation and Control

Because CCS is seen as part of the mitigation portfolio addressing climate change, straightforward international regulation is pivotal. In fact, it is a prerequisite before its commercial deployment can take place (Mace, Hendricks, & Coenraads, 2007; Steeneveldt, Berger, & Torp, 2006). Basic research (for example, on modelling, system definitions, side-effects, cost-benefit analyses of the heterogeneous basket of ideas in geo-engineering[14]) requires massive but standardised resources. Applicants of non-commercial activities (such as demonstration projects) need funds, while affected stakeholders and non-governmental organisations need confidence in effective regulators, monitoring and verification depends on international standards, trans-boundary transfer and trading requires the existence of international inventories and transnational treaties, and liability issues need to be regulated. The current IEA Roadmap proposes such comprehensive frameworks by 2020 (IEA, 2009d, pp. 36–37) and demands (Gale & Read, 2005), while recommendations exist in

---

[14] In a systematic assessment (e.g., as proposed by Boyd, 2008). See also Fischedick, Esken, Luhmann, Schüwer, and Supersberger (2007), Wilson and Gerard (2007), Wilson, Friedmann, and Pollak (2007), Wilson et al. (2008), Total (2008), Buesseler et al. (2008), Blackstock and Long (2010).

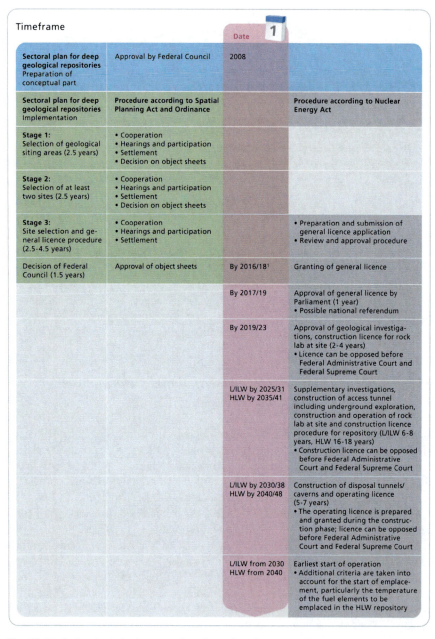

Fig. 10.13 Swiss site selection procedure for radioactive waste disposal in three stages over a decade, followed by site investigation and construction at the chosen sites, for low- and intermediate-level waste (L/ILW, 10–20 years) and for high-level waste and spent fuel (HLW, 20–30 years), respectively. The monitoring phase (not shown) after operation (around 20 years) will take another 50–100 years until the facility is closed
Source: BFE (2008, p. 32)

rather generic forms (IRGC, 2007, 2008; WRI, 2008). Such steps revert to the criteria above; in order to integrate CCS into the Clean Development Mechanism (CDM), for example, technical and organisational aspects must be fully covered, in the form of thorough site-specific risk assessments, good site characterisations and appropriate monitoring during and after operation (Dixon, 2009), including corresponding accounting (Haefeli, Bosi, & Philibert, 2005). Decisions are critical in cases where host country regulatory capacities are insufficient. In such cases, Pollak and Wilson (2009) suggest giving 'preference to sites that present lower seepage risk, such as offshore sites or sites that have well-characterized geology' (p. 84). They rightfully conclude that 'if a geological storage site approved by the CDM were to cause environmental, health or safety problems for the host country, it would damage not just the reputation of the CDM but also the reputation of CCS as a greenhouse gas mitigation technology' (p. 85). The issues appear complicated in terms of attempts to set up multinational repositories for radioactive waste (IAEA, 1998, 2004). For the speedy set-up of a comprehensive yet effective functional framework, early abatement is judged less costly and more effective than postponement (Stern, 2007; Krey & Riahi, 2009).

If CCS is to be a plausible part of the mitigation portfolio addressing climate change, 'control' will also have to include measures for enforcement. CCS is likely to be so costly that it could be tempting to agree to a range of international treaties and then delay their implementation. Before embarking on the road to full-scale CCS, rules and institutions must be put in place that provide reasonable assurance that treaties can be enforced and project targets met. The recent financial crisis in the European Union clearly shows that a functioning mechanism for enforcing regulation must be part of any plausible regime.

10.3.4 Criterion #4: Economic Aspects (Costs and Incentives)

Costs depend on technology and a range of externally set assumptions (including size, inception and promotion strategies in early phases and coverage of carbon penalty). Differences in abatement technology lines range from -100 EUR/t CO_2 avoided (for high-tech lighting) to $+60$ EUR/t CO_2 avoided in the case of a retrofitted gas power plant with CCS, as compiled by McKinsey (2009) (Fig. 10.14).

The abatement potential for CCS is substantial, albeit at a high cost, according to energy carrier, from 30 (coal) to 55–60 EUR/t CO_2 avoided (gas, retrofitted). One asset of CCS is the fact that it can also be used in cement, iron and steel and chemical industries, or in biomass production where it could even amount to negative emissions (because CO_2 would be yielded twice: when taken out of the atmosphere by plants and when buried later on). If developing and emerging countries delay their participation in curbing CO_2 emissions and if the climate target is overshot due to late abatement, CCS is necessary as a mitigation measure to keep the carbon prices from skyrocketing (Krey & Riahi, 2009, p. S96). In any case, costs would still be high for former free riders (and for countries that could not afford to contribute to the internalisation of costs). (Krey & Riahi, 2009, p. S106).

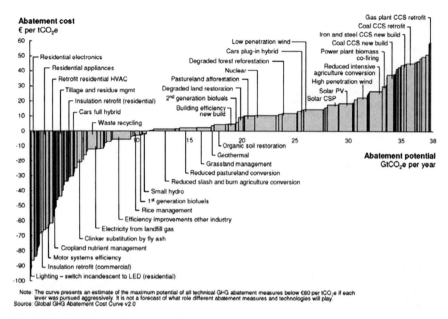

Fig. 10.14 Global greenhouse gas abatement cost curve for 2030 and beyond business as usual
Source: McKinsey (2009)

Usually, the cost breakdown for CCS is as follows: the majority (75 percent) of costs are directed towards capture technology, whereas only some 25 percent go to transportation, injection and monitoring (Plasynski, Litynski, McIlvried, & Srivastava, 2009). The IEA, for example, predicts CCS will cost just USD 35–60 per tonne of emissions reductions by 2030. McKinsey (2009) foresees a price of EUR 30–45 when the technology is mature, sometime after 2030. At any rate, CCS has to be competitive with other technologies; costs must be at least as high as for fossil-fuelled power plants without CO_2 separation. As to storage costs alone, Friedmann, Dooley, Held, and Edenhofer (2006) assume that only a small fraction of costs would go towards accurate geological characterisation. They estimate these one-time costs to be on the order of USD 0.1/t CO_2 as the costs are spread over millions of tons that are likely to be injected into a field over a period of several decades. McKinsey (2008) assumes EUR 4–12/t for early commercial plants. In light of the experience of the nuclear community, this optimistically low share may be questioned because research expenditure has risen with greater knowledge and insight into the complex proof of long-term safety. This is despite the fact that specific costs remain low (since comparatively little waste accrues per kWh produced) and, unlike in the CCS community, financing is secured due to the polluter-pays principle (approximately one EURc/kWh is charged to every electricity customer). Bonds and trust funds have been suggested as means of overcoming financial impasses in CCS (Gerard & Wilson, 2009; Peña & Rubin, 2008, respectively). Monitoring costs will depend on the time frame and specific quality requirements set by regulations. During the stage when CCS technology has not yet matured, R&D expenses accrue as economies of scale have not yet been realised, and the lack of experience may explain the lower quality of cost estimates. Furthermore, social costs also accumulate, such as for licensing procedures, decision uncertainties and delays.

This section has thus far discussed cost exclusively in terms of the specific cost per kWh or per t of CO_2. In view of criterion #3 above, however, it is also necessary to consider the extraordinary cost for a national or regional economy, for example, the cost of a full-fledged operational CCS system as a fraction of GDP (or, for that matter, gross world product). The costs of most environmental measures as a fraction of GDP are usually well below, even orders of magnitude below, one-thousandth of a percent of GDP. This is presumably not the case for CCS, which might cost on the order of one percent of GDP. The order of magnitude can be shown using a rough estimate: fossil fuel resources usually correspond to a small percentage, for example, five percent, of GDP. What if CCS cost 50 percent of this share if a comprehensive surveillance and validation programme had to be implemented over decades or longer?[15] Assuming that 20 percent of the fossil fuel stream ends up in the CCS process means that the CCS would cost 0.5 percent of GDP. Delaying the implementation of CCS for only two years would enable a country to 'save' 1 percent of its GDP. This simple calculation demonstrates the potentially huge economic impact of the deployment or non-deployment of CCS on nations' economies and their global competitiveness. Such considerations, of course, have to be put balanced against potential costs due to damage from unmitigated climate change. Stern estimates that the total cost of global warming could amount to between 5 and 20 percent of the ('one time') gross world product (Stern, 2007).

10.3.5 Criterion #5: Implementation

Since the mid-1990s, CO_2 has been captured in the Sleipner field, which is StatoilHydro's natural gas reserve in the North Sea.[16] The CO_2 is then injected into the saline aquifer of the so-called Utsira sands, one kilometre below the seabed and 3,000 m away from the reserve. The CO_2 needs to be separated in order to lower the natural concentration of nine percent to the safe 2.5 percent of Europe's highest yielding natural gas reserve. The CO_2 would have been blown off had the Norwegian government not introduced a CO_2 tax of 25 EUR/t in 1991. To date, no incidents have occurred in approximately 25 CCS projects worldwide.[17] There

---

[15] Such scenarios are apparently not being considered today, cf.: 'Since the storage process is, in general, based on established oil and gas drilling technologies and practices, learning effects are expected to be relatively limited' and, therefore, assumed storage costs level off at around EUR10/t CO_2 (McKinsey, 2008, p. 24) or even half of that (Bellona, 2011, p. 34/38). See also van der Zwaan and Gerlagh (2009).

[16] http://www.statoil.com (> Technology & Innovation > New energy > CO_2 capture and storage > Sleipner West). See also OED (2007).

[17] For an international database, see: http://www.co2captureandstorage.info/co2db.php, http://www.bellona.org/ccs/Artikler/ccs_web_sites

has been an enormous upturn in the number of projects, experiments and industrial activities being launched.[18] Half a dozen demonstration plants are in operation and dozens more are planned (Beck, 2009; IEA, 2009a). Despite this, large-scale deployment of CCS requires a satisfactory regulatory regime (Mace et al., 2007) and projects by major players such as Statoil, Shell and BP have been shelved due to insecure financing, lacking public subsidies and inadequate tax incentives and carbon trading.[19]

From a broader perspective, to varying degrees carbon 'lock-in' (Unruh, 2000) is a feature of energy systems in many countries today. As coal is the highest CO_2-emitting energy carrier, this source is the focus here. Worldwide use of coal has actually risen (IEA, 2009b), with the USA and China responsible for over 40 percent of the world's CO_2 emissions (IEA, 2009b, p. 11). Nearly all emissions from Chinese power generation stem from coal (IEA, 2009b, p. 21), while more than one-third of the US's CO_2 emissions are due to coal-fired power plants (US EIA, 2009). This means that a large share of the power infrastructure is tied to a conventionally high-emission path, which is systemically determined (centralised, cost-intensive) and associated with vested interests. Furthermore, in both countries coal is one of the major domestic energy sources and is considered key to their energy security. This is a powerful reason for continuing to use coal and to invest in R&D (Dooley, Dahowski, & Davidson, 2009; Pew Center, 2007; CSLF for US; Liang, Reiner, Gibbins, & Li, 2009; Dapeng & Weiwei, 2009; Liu & Sims Gallagher, 2010 for China). Other countries have followed the same reasoning, such as Germany (Praetorius & Schumacher, 2009), the European Union as a whole (EU CO_2 Network, 2004) and India (Shackley & Verma, 2008; Garg & Shukla, 2009). According to the IEA, worldwide one-third of all coal-fired power plants that are not suitable for CCS will need to close before the end of their technical life (IEA, 2008a). There are, apparently, many sources of system inertia.

10.3.6 Criterion #6: Societal Issues

Apart from regulatory aspects, CCS system decisions involve many societal issues. After all, CCS' long-term management reflects fundamental distributional risk-benefit asymmetry, just as in the case of radioactive waste (after Flüeler, 2005, p. ii):

- Local cost and risk vs. general benefit (intra-generational issue)
- Lay persons' vs. experts' perspectives and knowledge (evidentiary issue; that is, related to substantive and technical arguments)
- Current generations' benefits and decision making power vs. that of future generations (intergenerational issue).

---

[18] http://www.ghgt.info, http://www.ieaghg.org

[19] Times, 21 Dec 2007; Economist, 5 Mar 2009.

The crucial point is probably the long-term dimension. Archer addresses this concisely in comparing the two waste systems, focusing on public perspectives: 'One could sensibly argue that public discussion should focus on a time frame within which we live our lives, rather than concern ourselves with climate impacts tens of thousands of years in the future. On the other hand, the 10 kyr lifetime of nuclear waste seems quite relevant to public perception of nuclear energy decisions today' (Archer, 2005, p. 5). Ha-Duong and Loisel (2009) reach the following conclusion from their stakeholder analysis: 'Zero is the only acceptable leakage rate for geologically stored CO_2'. Various studies have analysed the so-called 'acceptance' of CCS.[20] A common denominator of these studies is the rather poor state of knowledge and the need for 'outreach' (Plasynski et al., 2009). This means that specialists only have to inform non-specialists at large to overcome the 'information gap' between them and the public. The strenuous experience in radioactive waste governance shows that this cannot be enough. The decision-making process from a generic approval of CCS to site selection, risk assessment, monitoring, accounting and closure must be transparent, open, fair and accountable, with well-defined decisive criteria in order for CCS to contribute to a sustainable climate policy. The recent German draft of the CCS act, notably 'on the demonstration and application of technologies', reflects the increasing cautiousness governments show in introducing CCS as the German Länder (states) have the power of veto on their territory (BMU, 2011).

10.4 Some Conclusions

Since innovation consists not only of technological progress but is also an integration of technical, conceptual, organisational and societal processes, a one-sided perspective is prone to failure. The following aspects are crucial to the CCS innovation process:

- Integrate the notion of time: needed period of isolation, safety assessment, processes (geological formations, mineralisation), implementation of the CCS system
- Consider adequate integration: integrate scales (geographical, temporal, institutional) into overall energy system(s), markets, and regulation
- Consider CO_2 disposal as a goal in itself, not a 'by-product' of other economic activities

---

[20] For example, Curry (2004), Ramírez, Hoogwijk, Hendriks, and Faaij (2008b), Wuppertal Institut et al. (2008b), Sharp, Jaccard, and Keith (2009), de Coninck et al. (2009), Anderson (2009), Singleton, Herzog, and Ansolabehere (2009), Malone, Bradbury, and Dooley (2009), Wallquist, Visschers, and Siegrist (2009), de Best-Waldhober, Daamen, and Faaij (2009), Ashworth et al. (2009a, 2009b), Pisarski and Thambimuthu (2009), Ha-Duong, Nadai, and Campos (2009), Shackley et al. (2007, 2009), Terwel, Harinck, Ellemers, and Daamen (2009). Respective guidelines exist (e.g., Bellona Europe, 2009).

- Optimise CCS to environmental performance (with regard to effectiveness, efficiency and timing; for example, new 'capture-ready' coal plants to be commissioned on the premise that old ones are shut down)
- Settle major issues (above) and harmonise regulation before large actors (states, companies) deploy CCS on a scale of a *fait accompli* (technically and in terms of generating accountable CO_2 certificates)
- Employ social sciences and humanities, recognising that they can contribute more than just assessing the technology and increasing public acceptance
- Before CCS can be considered a viable and sustainable option, rules and institutions for an international, reliably working CCS regime must be developed. Suitable mechanisms for enforcing regulation are part of any plausible regime; the recent global financial crisis may serve as an instructive case. Perhaps it is necessary to have an 'entrance exam' for countries wanting to use CCS as an emissions reduction measure, similar to the scrutiny subjected to applicants for admission to the eurozone. Funds for CCS may have to be deposited in an internationally monitored fund before CCS can start.
- Take action before political pressures and/or environmental failures (such as spillouts) make it compulsory.

References

Aarnes, J. E., Selmer-Olsen, S., Carpenter, M. E., & Flach, T. A. (2008). Towards guidelines for selection, characterization and qualification of sites and projects for geological storage of CO_2. *Procedia, 1*(1), 1735–1742.

Adams, D., & Davison, J. (2007). *Capturing CO_2*. Cheltenham: IEA Greenhouse Gas R&D Programme.

AkEnd. (2002). *Selection procedure for repository sites. Recommendations of the AkEnd – Committee on a selection procedure for repository sites, AkEnd*. December. Salzgitter: Federal Office for Radiation Protection, BfS. All web links accessed November 16, 2011, http://www.bmu.de

Anderson, J. (2009). Results from the project 'Acceptance of CO_2 capture and storage: Economics, policy and technology (ACCSEPT)'. *Energy Procedia, 1*, 4649–4653.

Archer, D. (2005). Fate of fossil fuel CO_2 in geologic time. *Journal of Geophysical Research, 110*, C09S05, 1–6.

Ashworth, P., Carr-Cornish, S., Boughen, N., & Thambimuthu, K. (2009a). Engaging the public on carbon dioxide capture and storage: Does a large group process work? *Energy Procedia, 1*(1), 4765–4773.

Ashworth, P., Pisarski, A., & Thambimuthu, K. (2009b). Public acceptance of carbon dioxide capture and storage in a proposed demonstration area. *Proceedings of the Institution of Mechanical Engineers, Part A: Journal of Power and Energy, 223*(3), 299–304.

Bachu, S. (2000). Sequestration of CO_2 in geological media: Criteria and approach for site selection in response to climate change. *Energy Conversion Management, 41*, 953–970.

Beck, B. (2009, January). *The technology and status of carbon dioxide capture and storage*. IEA Greenhouse Gas R&D Programme (Presentation, City University, London).

Bellona. (2011). *Insuring energy independence. A CCS roadmap for Poland*. Krakow: The Bellona Foundation.

Bellona Europe. (2009). *Guidelines for public consultation and participation in CCS projects*. Brussels: Bellona.

Benson, S. M. (2004, September). *Monitoring protocols and life-cycle costs for geologic storage of carbon dioxide*. Carbon Sequestration Leadership Forum, Melbourne, Australia.

Benson, S. M. (2005). Lessons learned from industrial and natural analogs for health, safety and environmental risk assessment for geologic storage of carbon dioxide. In S. M. Benson, C. Oldenburg, M. Hoversten, & S. Imbus (Eds.), *Carbon dioxide capture for storage in deep geologic formations – Results from the CO_2 capture project, Vol. 2* (pp. 1133–1141). Amsterdam: Elsevier.

BFE, Bundesamt für Energie (Swiss Federal Office of Energy). (2008). *Sectoral plan for deep geological repositories. Conceptual part*. Bern: Bundesamt für Energie (BFE).

Bijker, W. E., Hughes, T. P., & Pinch, T. (1989). *The social construction of technological systems. New directions in the sociology and history of technology*. Cambridge, MA: MIT Press.

Blackstock, J. J., & Long, J. C. S. (2010). The politics of geoengineering. *Science, 327*, 527.

Blowers, A., Lowry, D., & Solomon, B. D. (1991). *The international politics of nuclear waste*. New York: St. Martin's Press.

BMU, Bundesministerium für Umwelt. (2011). *Gesetz zur Demonstration und Anwendung von Technologien zur Abscheidung, zum Transport und zur dauerhaften Speicherung von Kohlendioxid*. Vom Bundeskabinett am 13. April 2011 beschlossen. Entwurf. Berlin: BMU, The German Federal Ministry for the Environment, Nature Conservation and Nuclear Safety.

Bowden, A. R., & Rigg, A. (2005). Assessing reservoir performance risk in CO_2 storage projects. In E. S. Rubin, D. W. Keith, & C. F. Gilboy (Eds.), *Greenhouse gas control technologies, Vol. I* (pp. 683–691). Amsterdam: Elsevier.

Boyd, P. W. (2008). Ranking geo-engineering schemes. *Nature Geoscience, 1*, 722–724.

Bradshaw, J., Bachu, S., Bonijoly, D., Burruss, R., Holloway, S., Christensen, N. P., et al. (2007). CO_2 storage capacity estimation: Issues and development of standards. *International Journal of Greenhouse Gas Control, 1*, 62–68.

Buesseler, K. O., et al. (2008). Ocean iron fertilization – moving forward in a sea of uncertainty. *Science, 319*, 162.

Carter, L. J. 1987. *Nuclear imperatives and public trust: Dealing with radioactive waste*. Washington, DC: Resources of the Future.

CEC, Commission of the European Communities. (2001): *European governance. A white paper*. COM(2001) 428 final. 2001-7-25. Brussels: CEC.

CEC. (2002). *Amended proposals for council decisions concerning the specific programmes implementing the sixth framework programme of the European ... Atomic Energy Community for research and training activities (2002–2006)*. COM(2002) 43 final. 30.1.2002. Brussels: CEC.

Celia, M. A., Bachu, S., Nordbotten, J. M., Gasda, S. E., & Dahle, H. K. (2005). Quantitative estimation of CO_2 leakage from geological storage: Analytical models, numerical models, and data needs. In: E. S. Rubin, D. W. Keith, & C. F. Gilboy (Eds.), *Greenhouse gas control technologies, Vol. I* (pp. 663–671). Amsterdam: Elsevier.

Chalmers, H., Jakeman, N., Pearson, P., & Gibbins, J. (2009). Carbon capture and storage deployment in the UK: What next after the UK Government's competition? *Proceedings of the Institution of Mechanical Engineers; Part A; Journal of Power and Energy, 223*(3), 305–319.

Clarke, L., Edmonds, J., Krey, V., Richels, R., Rose, S., & Tavoni, M. (2009). International climate policy architectures: overview of the EMF22 International Scenarios. *Energy Economics, 31*, S64–S81.

Colglazier, E. W. (1982). *The politics of nuclear waste*. New York: Pergamon.

Collins, H. M. (1981). Stages in the empirical programme of relativism. *Social Studies of Science, 11*, 3–10.

CoRWM, Committee on Radioactive Waste Management. (w/o yr.). *The options for long-term management of higher active solid radioactive wastes in the United Kingdom*. TS 046, 12123/TR/0001.

CoRWM. (2006). Managing our radioactive waste safely. *CoRWM's recommendations to Government*. CoRWM Doc 700. London: CoRWM.

220 T. Flüeler

Crabbe, M. J. C. (2009). Modelling effects of geoengineering options in response to climate change and global warming: Implications for coral reefs. *Computational Biology and Chemistry, 33,* 415–420.

CSLF, Carbon Sequestration Leadership Forum. (established in June 2003). http://www.cslforum.org/projects/index.html?cid=nav_projects

Curry, T. E. (2004). *Public awareness of carbon capture and storage: A survey of attitudes toward climate change mitigation.* Cambridge, MA: MIT.

Dapeng, L., & Weiwei, W. (2009). Barriers and incentives of CCS deployment in China: Results from semi-structured interviews. *Energy Policy, 37,* 2421–2432.

Davison, J. (2009). Electricity systems with near-zero emissions of CO_2 based on wind energy and coal gasification with CCS and hydrogen storage. *International Journal of Greenhouse Gas Control, 3*(6), 683–692.

de Best-Waldhober, M., Daamen, D., & Faaij, A. (2009). Informed and uninformed public opinions on CO_2 capture and storage technologies in the Netherlands. *International Journal of Greenhouse Gas Control, 3*(3), 322–332.

de Coninck, H. (2008). Trojan horse or horn of plenty? Reflections on allowing CCS in the CDM. *Energy Policy, 36,* 929–936.

de Coninck, H., et al. (2009). The acceptability of CO_2 capture and storage (CCS) in Europe: An assessment of the key determining factors. Part 1. Scientific, technical and economic dimensions. *International Journal of Greenhouse Gas Control, 3,* 333–343.

de Coninck, H., Stephens, J. C., & Metz, B. (2009). Global learning on carbon capture and storage: A call for strong international cooperation on CCS demonstration. *Energy Policy, 37*(6), 2161–2165.

de Visser, E., de Vos, R., & Hendriks, Ch. (Eds.). (2008). *CATO. Catching carbon to clear the skies. Experience and highlights of the Dutch R&D programme on CCS.* The Hague: Dutch Ministry of Economic Affairs.

Dixon, T. (2009). International legal and regulatory developments for carbon dioxide capture and storage: From the London convention to the clean development mechanism. *Proceedings of the Institution of Mechanical Engineers, Part A: Journal of Power and Energy, 223*(3), 293–297.

Dooley, J. J., Dahowski, R. T., & Davidson, C. L. (2009). The potential for increased atmospheric CO_2 emissions and accelerated consumption of deep geologic CO_2 storage resources resulting from the large-scale deployment of a CCS-enabled unconventional fossil fuels industry in the U.S. *International Journal of Greenhouse Gas Control, 3*(6), 720–730.

EC, European Commission, European Parliament. (2009, January–February). Europeans' attitudes towards climate change. *Special Eurobarometer 313/Wave 71.1. Report.* Brussels: Directorate-General for Communication.

EU CO_2 Network, European Carbon Dioxide Network. (2004). *Capturing and storing carbon dioxide: Technical lessons learned.* http://www.co2net.eu

Feroz, E. H., Raab, R. L., Ulleberg, G. T., & Alsharif, K. (2009). Global warming and environmental production efficiency ranking of the Kyoto Protocol nations. *Journal of Environmental Management, 90*(2), 1178–1183.

Fischedick, M., Esken, A., Luhmann, H.-J., Schüwer, D., & Supersberger, N. (2007). *Geologische CO_2-Speicherung als klimapolitische Handlungsoption. Technologien, Konzepte, Perspektiven. Wuppertal Spezial 25.* Wuppertal: Wuppertal Institut.

Flüeler, T. (2005). Long-term knowledge generation and transfer in environmental issues – A challenge to a knowledge-based society. Setting the scene. In J. V. Carrasquero, F. Welsch, A. Oropeza, T. Flüeler, & N. Callaos (Eds.), *Proceedings PISTA 2005. The 3rd International Conference on Politics and Information Systems: Technologies and Applications.* July 14–17, 2005, Orlando, Florida. Orlando, FA: International Institute of Informatics and Systemics, IIIS Copyright Manager:ii.

Flüeler, T. (2006) *Decision making for complex socio-technical systems. Robustness from lessons learned in long-term radioactive waste governance. Vol. 42 Series Environment & Policy.* Dordrecht: Springer.

Freudenburg, W. R. (2004). Can we learn from failure? Examining US experiences with nuclear repository siting. *Journal of Risk Research, 7*(2), 153–169.

Friedmann, S. J., Dooley, J. J., Held, H., & Edenhofer, O. (2006). The low cost of geological assessment for underground CO_2 storage: Policy and economic implications. *Energy Conversion and Management, 47*, 1894–1901.

Gale, J. (2007). To store or not to store? *International Journal of Greenhouse Gas Control, 1*(1), 1.

Gale, J., & Read, T. (2005). Rules and standards for CO_2 capture and storage: Considering the options. In M. Wilson, et al. (Eds.), *Greenhouse gas control technologies, Volume II* (pp. 1461–1466). Amsterdam: Elsevier.

Garg, A., & Shukla, P. R. (2009). Coal and energy security for India: Role of carbon dioxide (CO_2) capture and storage (CCS). *Energy, 34*(8), 1032–1041.

Gerard, D., & Wilson, E. J. (2009). Environmental bonds and the challenge of long-term carbon sequestration. *Journal of Environmental Management, 90*(2), 1097–1105.

Gibbins, J., & Chalmers, H. (2008). Carbon capture and storage. *Energy Policy, 36*(12), 4317–4322.

Goldstein, G., & Tosato, G. C. (2008). Global energy systems and common analyses. Final report of Annex X (2005–2008). *International energy agency implementing agreement for a programme of energy technology systems analysis, ETSAP.* Paris: IEA/OECD/ETSAP.

Gough, C., Shackley, S., Holloway, S., Cockerill, T. T., Bentham, M., Bulatov, I., et al. (2006). *An integrated assessment of carbon dioxide capture and storage in the UK.* Monograph. London: University College.

Gough, C., & Shackley, S. (2006). Towards a multi-criteria methodology for assessment of geological carbon storage options. *Climatic Change, 74*(1–3), 141–174.

Grünwald, R. (2008). *Treibhausgas – ab in die Versenkung? Möglichkeiten und Risiken der Abscheidung und Lagerung von CO_2. Studien des Büros für Technikfolgen-Abschätzung beim Deutschen Bundestag. Global zukunftsfähige Entwicklung, Bd. 25.* Berlin: Edition Sigma.

Hadjipaschalis, I., Kourtis, G., & Poullikkas, A. (2009). Assessment of oxyfuel power generation technologies. *Renewable and Sustainable Energy Reviews, 13*(9), 2637–2644.

Ha-Duong, M., & Loisel, R. (2009). Zero is the only acceptable leakage rate for geologically stored CO_2: An editorial comment. *Climate Change, 93*(3–4), 311–317.

Ha-Duong, M., Nadai, A., & Campos, A. S. (2009). A survey on the public perception of CCS in France. *International Journal of Greenhouse Gas Control, 3*(5), 633–640.

Haefeli, S., Bosi, M., & Philibert, C. (2005). Important accounting issues for carbon dioxide capture and storage projects under the UNFCCC. In E. S. Rubin, D. W. Keith, & C. F. Gilboy (Eds.), *Greenhouse gas control technologies, Vol. I* (pp. 953–960). Amsterdam: Elsevier.

Hawkins, D. G. (2003). Passing gas: Policy implications of leakage from geologic carbon storage sites. In J. Gale & Y. Kaya (Eds.), *Greenhouse gas control technologies.* Proceedings of the 6th international conference, Kyoto, Japan (pp. 249–254). Amsterdam: Pergamon.

IAEA, International Atomic Energy Agency. (1998, June). *Technical, institutional and economic factors important for developing a multinational radioactive waste repository.* TECDOC-1021. Vienna: IAEA.

IAEA. (1999). *Topical issues in nuclear, radiation and radioactive waste safety.* Proceedings of an international conference. Vienna, 31 Aug–4 Sep 1998 (pp. 233–255). Vienna: IAEA.

IAEA. (2003). *Scientific and technical basis for the geological disposal of radioactive wastes.* Technical Reports Series. No. 413. Vienna: IAEA.

IAEA. (2004, October). *Developing multinational radioactive waste repositories: Infrastructural framework and scenarios of cooperation.* TECDOC-1413. Vienna: IAEA.

IAEA. (2006). Safety standards. *Geological disposal of radioactive waste. Safety Requirements.* No. WS-R-4. Vienna: IAEA.

ICRP, International Commission on Radiological Protection. (1993). *Protection from potential exposure: A conceptual framework. ICRP-Publication 64. Annals of the ICRP, 23*(1), Paris: ICRP.

IEA, International Energy Agency. (2007). *Storing CO$_2$ underground. International energy agency greenhouse gas R&D programme 2007*. Cheltenham: IEA.

IEA. (2008a). *Energy technology perspectives 2008. Scenarios and strategies to 2050. Executive summary*. Paris: OECD/IEA.

IEA. (2008b). *CO$_2$ capture and storage: A key abatement option*. Paris: OECD/IEA.

IEA. (2009a, 8 July). *Carbon capture and storage. Full-scale demonstration progress update*. Paris: OECD/IEA.

IEA. (2009b). *IEA statistics. CO$_2$ emissions from fuel combustion. Highlights*. Paris: OECD/IEA.

IEA. (2009c). *World energy outlook 2009. Summary*. Paris: OECD/IEA.

IEA. (2009d). *Technology roadmap. Carbon capture and storage*. Paris: OECD/IEA.

IOP, Institute of Physics, Royal Society of Chemistry, Institute of Biology. (2008). *Carbon capture and storage*. Seminar, December 5, 2007. London: IOP.

IPCC, Intergovernmental Panel on Climate Change. (2005). *IPCC special report on carbon dioxide capture and storage*. Prepared by Working Group III [B. Metz, O. Davidson, H. C. de Coninck, M. Loos & L. A. Meyer (Eds.)]. Cambridge: Cambridge University Press.

IRGC, International Risk Governance Council. (2007). *Workshop on regulation of carbon capture and storage. March 15 and 16*. Geneva: IRGC.

IRGC. (2008). *Regulation of carbon capture and storage*. Policy Brief. Geneva: IRGC.

Jacobs, W., Cohen, L., Kostakidis, L., & Rundell, S. (2009). *Proposed roadmap for overcoming legal and financial obstacles to carbon capture and sequestration*. Discussion paper 2009-04. Cambridge, MA: Belfer Center for Science and International Affairs.

Jacobson, M. Z. (2009). Review of solutions to global warming, air pollution, and energy security. *Energy and Environmental Sciences, 2*, 148–173.

Junker, B., Flüeler, T., Stauffacher, M., & Scholz, R. W. (2008). *Description of the safety case for long-term disposal of radioactive waste – the iterative safety analysis approach as utilized in Switzerland*. Technical paper as part of the 'Long-term dimension of radioactive waste disposal: The role of the time dimension for risk perception' project. Natural and Social Science Interface (NSSI). Zurich: ETH Zurich.

Kasperson, R. E. (Ed.). (1983). *Equity issues in radioactive waste management*. Cambridge: Oelgeschlager, Gunn & Hain.

Kasperson, R. E., Golding, D., & Tuler, S. (1992). Social distrust as a factor in siting hazardous facilities and communicating risks. *Social Issues, 48*(4), 161–187.

Kelemen, P. B., & Matter, J. (2008). In situ carbonation of peridotite for CO$_2$ storage. *PNAS, 105*(45), 17295–17300.

Koornneef, J., van Harmelen, T., van Horssen, A., van Gijlswijk, R., Ramírez, A., Faaij, A., Turkenburg, W. (2008). The impacts of CO$_2$ capture on transboundary air pollution in the Netherlands. GHGT-9. *Energy Procedia, 1*, 3787–3794.

Krey, V., & Riahi, K. (2009). Implications of delayed participation and technology failure for the feasibility, costs, and likelihood of staying below temperature targets – Greenhouse gas mitigation scenarios for the 21st century. *Energy Economics, 31, Suppl. 2, International, U.S. and E.U. climate change control scenarios: Results from EMF 22, December 2009*, S94–S106.

Latour, B. (1987). *Science in action. How to follow scientists and engineers through society*. Cambridge, MA: Harvard University Press.

LDC, London Dumping Convention. (1972). *Convention on the prevention of marine pollution by dumping of wastes and other matter*. http://www.londonprotocol.imo.org

Liang, X., Reiner, D., Gibbins, J., & Li, J. (2009). Assessing the value of CO$_2$ capture ready in new-build pulverised coal-fired power plants in China. *International Journal of Greenhouse Gas Control, 3*(6), 787–792.

Liu, H., & Sims Gallagher, K. (2010). Catalyzing strategic transformation to a low-carbon economy: A CCS roadmap for China. *Energy Policy, 38*(1), 59–74.

Mace, M. J., Hendricks, C., & Coenraads, R. (2007). Regulatory challenges to the implementation of carbon capture and geological storage within the European Union under EU and international law. *International Journal of Greenhouse Gas Control, 1*, 253–260.

Malone, E. L., Bradbury, J. A., & Dooley, J. J. (2009). Keeping CCS stakeholder involvement in perspective. *Energy Procedia, 1*(1), *Greenhouse Gas Control Technologies 9*, Proceedings of the 9th International Conference on Greenhouse Gas Control Technologies (GHGT-9), 4789–4794.

Maul, P. R., Metcalfe, R., Pearce, J., Savage, D., & West, J. M. (2007). Performance assessments for the geological storage of carbon dioxide: Learning from the radioactive waste disposal experience. *International Journal of Greenhouse Gas Control, 1*(4), 444–455.

McKinsey. (2008). *Carbon capture and storage. Assessing the economics.* McKinsey & Company. http://www.mckinsey.com (use search function)

McKinsey. (2009). *Pathways to a low-carbon economy. Version 2 of the global greenhouse has abatement cost curve.* McKinsey & Company. http://www.mckinsey.com (use search function)

Meinshausen, M., Meinshausen, N., Hare, W., Raper, S. C. B., Frieler, K., Knutti, R., et al. (2009). Greenhouse-gas emission targets for limiting global warming to 2°C. *Nature, 458*, 1158–1162.

Mondol, J. D., McIlveen-Wright, D., Rezvani, S., Huang, Y., & Hewitt, N. (2009). Techno-economic evaluation of advanced IGCC lignite coal fuelled power plants with CO_2 capture. *Fuel, 88*(12), 2495–2506.

Morone, J. G., & Woodhouse, E. J. (1986). *Averting catastrophe. Strategies for regulating risky technologies.* Berkeley: University of California Press.

Nagra. (1994). *Bericht zur Langzeitsicherheit des Endlagers SMA am Standort Wellenberg (Gemeinde Wolfenschiessen, NW).* Nagra Technical Report NTB 94-06. Wettingen, Switzerland: Nagra.

Nagra. (2002). *Project Opalinuston – Safety report. Demonstration of disposal feasibility for spent fuel, vitrified high-level waste and long-lived intermediate-level waste (Entsorgungsnachweis).* Nagra Technical Report NTB 02-05. Wettingen, Switzerland: Nagra.

Nagra. (2008). *Modellhaftes Inventar für radioaktive Materialien MIRAM 08.* NTB 08-06. Wettingen: Nagra.

NEA, Nuclear Energy Agency. (1999). *Confidence in the long-term safety of deep geological repositories: Its development and communication.* Paris: OECD.

NEA. (2000). *Nuclear energy in a sustainable development perspective.* Paris: OECD.

NEA. (2004a). *Stepwise approach to decision making for long-term radioactive waste management.* No. 4429. Paris: OECD.

NEA. (2004b). *Safety of disposal of spent fuel, HLW and long-lived ILW in Switzerland. An international peer review of the post-closure radiological safety assessment for disposal in the Opalinus Clay of the Zürcher Weinland.* No. 5568. Paris: OECD.

NEA. (2008a). *Moving forward with geologic disposal of radioactive waste. A collective statement.* Paris: OECD.

NEA. (2008b). *Safety cases for deep geological disposal of radioactive waste: Where do we stand?* Symposium proceedings. Paris, France. January 23–25, 2007. NEA No. 6319. Paris: OECD.

NEA. (2008c). *Moving forward with geological disposal of radioactive waste: An NEA RWMC collective statement.* NEA/RWM(2008)5/Rev2. Paris: OECD.

NEA. (2009). *Considering timescales in the post-closure safety of geological disposal of radioactive waste.* Paris: OECD.

NEA/IAEA/CEC. (1991). *Disposal of radioactive waste: Can long-term safety be evaluated? An international collective opinion.* Paris: OECD.

North, D. W. (1999). A perspective on nuclear waste. *Risk Analysis, 19*(4), 751–758.

NWMO, Nuclear Waste Management Organization. (2003, November). *Asking the right questions? The future management of Canada's used nuclear fuel.* Toronto: NWMO (papers and extensive discussion on http://www.nwmo.ca)

OED, Norwegian Ministry of Petroleum and Energy. (2007). Fact sheet: Carbon capture and geological storage. http://www.regjeringen.no/en/dep/oed.html (> Press centre > Facts and background)

Ormerod, W. G., Freund, P., & Smith, A. (2002). *Ocean storage of CO_2.* Cheltenham: IEA Greenhouse Gas R&D Programme.

OSPAR. (2007). *Decision 2007/1 to prohibit the storage of carbon dioxide streams in the water column or on the sea-bed.* http://www.ospar.org (> Work areas > Offshore Oil & Gas Industry > Decisions ....)

Pacala, S., & Socolow, R. (2004, August 13). Stabilization wedges: Solving the climate problem for the next 50 years with current technologies. *Science, 305*(5686), 968–972.

Pachauri, R. K., & Reisinger, A. (Eds.). (2007). *Climate change 2007: Synthesis report.* Cambridge: Intergovernmental Panel on Climate Change.

Page, S. C., Williamson, A. G., & Mason, I. G. (2009). Carbon capture and storage: Fundamental thermodynamics and current technology. *Energy Policy, 37*(9), 3314–3324.

Peña, N., & Rubin, E. S. (2008). *A trust fund approach to accelerating deployment of CCS: Options and considerations.* Coal Initiative Reports. White Paper Series. Arlington, VA: Pew Center.

Pew Center. (2007). *A program to accelerate the deployment of CO_2 capture and storage (CCS): Rationale, objectives, and costs.* Coal Initiative Reports. White Paper Series. Arlington, VA: Pew Center.

Plasynski, S. I., Litynski, J. T., McIlvried, H. G., & Srivastava, R. D. (2009). Progress and new developments in carbon capture and storage. *Critical Reviews in Plant Sciences, 28*(3), 123–138.

Pollak, M., & Wilson, E. J. (2009). Risk governance for geological storage of CO_2 under the Clean Development Mechanism. *Climate Policy, 9*(1), 71–87.

Pollard, R. T., Salter, I., Sanders, R. J., Lucas, M. I., Moore, C. M., Mills, R. A., et al. (2009). Southern Ocean deep-water carbon export enhanced by natural iron fertilization. *Nature, 457,* 577–580.

Powicki, C. R. (2007). Closing the carbon cycle. *EPRI Journal. Spring 2007.* Palo Alto, CA: Electric Power Research Institute (EPRI).

Praetorius, B., & Schumacher, K. (2009). Greenhouse gas mitigation in a carbon constrained world: The role of carbon capture and storage. *Energy Policy, 37*(12), 5081–5093.

Ramírez, A., Hagedoorn, S., Kramers, L., Wildenborg, T., & Hendriks, C. (2008a). Screening CO_2 storage options in the Netherlands. *Energy Procedia, 1*(1), 2801–2808.

Ramírez, A., Hoogwijk, M., Hendriks, C., & Faaij, A. (2008b). Using a participatory approach to develop a sustainability framework for carbon capture and storage systems in The Netherlands. *International Journal of Greenhouse Gas Control, 2*(1), 136–156.

Ravetz, J. R. (1980). Public perceptions of acceptable risks as evidence for their cognitive, technical and social structure. In M. Dierkes, S. Edward, & R. Coppock (Eds.), *Technological risk. Its perception and handling in the European Community* (pp. 45–54). Königstein: Oelgeschlager, Gunn & Hain.

Richardson, K., Steffen, W., Schellnhuber, H. J., Alcamo, J., Barker, T., Kammen, D. M., et al. (2009). *Synthesis report. Climate change. Global risks, challenges & decisions.* Copenhagen, March 10–12, 2009. International Alliance of Research Universities. Copenhagen: University of Copenhagen.

Rohner, H. (2008, September 8–10). *CO_2 storage – back to where it belongs.* Paper presented at smart energy strategies. Meeting the climate change challenge. ETH Zurich.

Ropohl, G. (1978/1999). *Allgemeine Technologie. Eine Systemtheorie der Technik.* München: Carl Hanser Verlag.

Royal Society. (2009). *Geoengineering the climate: Science, governance and uncertainty.* London: The Royal Society.

Savage, D., Maul, P. R., Benbow, S. J., & Walke, R. C. (2004). *A generic FEP database for the assessment of long-term performance and safety of geological storage of CO2.* Version 1.0. Henley-on-Thames, Oxfordshire: Quintessa.

Schiermeier, Q. (2009). Ocean fertilization: Dead in the water? *Nature, 457,* 520–521.

Schneider, S. H. (2008). Geoengineering: Could we or should we make it work? *Philosophical Transactions of the Royal Society, A.* doi:10.1098/rsta.2008.0145

Seifritz, W. (1990). CO_2 disposal by means of silicates. *Nature, 345,* 486.

10 Technical Fixes Under Surveillance – CCS and Lessons Learned from...

Shackley, S., Waterman, H., Godfroij, P., Reiner, D., Anderson, J., Draxlbauer, K., et al. (2007). *Stakeholder perceptions of CO_2 capture and storage in Europe: Results from the EU-funded ACCSEPT Survey. Deliverable D3.1 from ACCSEPT – Main Report.* Manchester: Manchester Business School, University of Manchester.

Shackley, S., & Verma, P. (2008). Tackling CO_2 reduction in India through use of CO_2 capture and storage (CCS). Prospects and challenges. *Energy Policy, 36*, 3554–3561.

Shackley, S., Reiner, D., Upham, P., de Coninck, H., Sigurthorsson, G., & Anderson, J. (2009). The acceptability of CO_2 capture and storage (CCS) in Europe: An assessment of the key determining factors: Part 2. The social acceptability of CCS and the wider impacts and repercussions of its implementation. *International Journal of Greenhouse Gas Control, 3*(3), 344–356.

Sharp, J. D., Jaccard, M. K., & Keith, D. W. (2009). Anticipating public attitudes toward underground CO_2 storage. *International Journal of Greenhouse Gas Control, 3*, 641–651.

Singleton, G., Herzog, H., & Ansolabehere, S. (2009). Public risk perspectives on the geologic storage of carbon dioxide. *International Journal of Greenhouse Gas Control, 3*(1), 100–107.

Solomon, S., Plattner, G.-K., Knutti, R., & Friedlingstein, P. (2009, February 10). Irreversible climate change due to carbon dioxide emissions. *PNAS, 106*(6), 1704–1709 (incl. supporting information: pnas.0812721106).

Spreng, D., Marland, G., & Weinberg, A. M. (2007). CO_2 capture and storage: Another Faustian Bargain? *Energy Policy, 35*, 850–854.

Steeneveldt, R., Berger, B., & Torp, T. A. (2006). CO_2 capture and storage: Closing the knowing-doing gap. *Chemical Engineering Research and Design, 84*(9), 739–763.

Stenhouse, M. J., Galeb, J., & Zhou, W. (2009). Current status of risk assessment and regulatory frameworks for geological CO_2 storage. GHGT-9. *Energy Procedia, 1*, 2455–2462.

Stern, N. (2007). *The economics of climate change.* Cambridge: Cambridge University Press.

Stucki, S., Schuler, A., & Constantinescu, M. (1995). Coupled CO_2 recovery from the atmosphere and water electrolysis: Feasibility of a new process for hydrogen storage. *International Journal of Hydrogen Energy, 20*, 653–663.

Terwel, B. W., Harinck, F., Ellemers, N., & Daamen, D. D. L. (2009). Competence-based and integrity-based trust as predictors of acceptance of carbon dioxide capture and storage (CCS). *Risk Analysis, 29*(8), 1129–1140.

Tilmes, S., Müller, R., & Salawitch, R. (2008). The sensitivity of Polar ozone depletion to proposed geoengineering schemes. *Science, 320*, 1201–1204.

Total. (2008). *Results of stakeholder consultation. Lacq Basin CO_2 capture and storage pilot project.* http://www.total.com/MEDIAS/MEDIAS_INFOS/2186/EN/results-stakeholder-consultation.pdf

Toth, F. L. (Ed.). (2011). *Geological disposal of carbon dioxide and radioactive waste: A comparative assessment. Series Advances in Global Change Research, 44.* Dordrecht: Springer.

Unruh, G. C. (2000). Understanding carbon lock-in. *Energy Policy, 28*, 817–830.

US EIA, Energy Information Administration. (2009). Emissions of greenhouse gases report. http://www.eia.gov/oiaf/1605/ggrpt/carbon.html#total

Vallentin, D. (2007). *Inducing the international diffusion of carbon capture and storage technologies in the power sector.* Wuppertal Papers n. 162. Wuppertal: Wuppertal Institute.

Van Alphen, K., Hekkert, M. P., & Turkenburg, W. C. (2009). Comparing the development and deployment of Carbon Capture and Storage technologies in Norway, the Netherlands, Australia, Canada and the United States – An innovation system perspective. *Energy Procedia, 1*, 4591–4599.

van der Zwaan, B., & Gerlagh, R. (2009). Economics of geological CO_2 storage and leakage. *Climate Change, 93*(3–4), 285–309.

van Vuuren, D. P., den Elzen, M. G. J., Lucas, P. L., Eickhout, B., Strengers, B. J., van Ruijven, B., et al. (2007). Stabilizing greenhouse gas concentrations at low levels: An assessment of reduction strategies and costs. *Climatic Change, 81*, 119–159.

Victor, D. G., Morgan, M. G., Apt, F., Steinbruner, J., & Ricke, K. (2009). The geoengineering option – A last resort against global warming. *Foreign Affairs, 88*(March/April), 64.

Viebahn, P., Nitsch, J., Fischedick, M., Esken, A., Schüwer, D., Supersberger, N., et al. (2007). Comparison of carbon capture and storage with renewable energy technologies regarding structural, economic, and ecological aspects in Germany. *International Journal of Greenhouse Gas Control, 1*(1), 121–133.

Vlek, C., & Stallén, P.-J. (1980). Rational and personal aspects of risk. *Acta Psychologica, 45*, 273–300.

Wallquist, L., Visschers, V. H. M., & Siegrist, M. (2009). Lay concepts on CCS deployment in Switzerland based on qualitative interviews. *International Journal of Greenhouse Gas Control, 3*(5), 652–657.

Weinberg, A. M. (1972, July 7). Social institutions and nuclear energy. *Science, 177*, 27–34.

Wilson, E., & Gerard, D. (Eds.). (2007). *Carbon capture and sequestration: Integrating technology, monitoring, regulation.* Ames, IA: Blackwell Publishing Professional.

Wilson, E., et al. (2008, April 15). Regulating the geological sequestration of CO_2. *Environmental Science & Technology, 42*, 2718–2722.

Wilson, E. J., Friedmann, S. J., & Pollak, M. F. (2007). Research for deployment: Incorporating risk, regulation, and liability for carbon capture and sequestration. *Environment, Science & Technology, 41*, 5945–5952.

WRI, World Resource Institute. (2008). *CCS guidelines. Guidelines for carbon dioxide capture, transport and storage.* Washington, DC: WRI.

Wuppertal Institute, DLR, ZSW, PIK. (2008a). *RECCS. Ecological, economic and structural comparison of renewable energy technologies (RE) with carbon capture and storage (CCS). An integrated approach.* Published by the Federal Ministry for the Environment, Nature Conservation and Nuclear Safety BMU. Wuppertal: Wuppertal Institute.

Wuppertal Institut, et al. (2008b). *Sozioökonomische Begleitforschung zur gesellschaftlichen Akzeptanz von Carbon Capture and Storage (CCS) auf nationaler und internationaler Ebene.* Wuppertal: Wuppertal Institute.

Wynne, B. (1983). Technologie, Risiko und Partizipation: Zum gesellschaftlichen Umgang mit Unsicherheit. In J. Conrad (Ed.). *Gesellschaft, Technik und Risikopolitik* (pp. 156–187). Berlin: Springer.

Chapter 11
Learning from the Transdisciplinary Case Study Approach: A Functional-Dynamic Approach to Collaboration Among Diverse Actors in Applied Energy Settings

Michael Stauffacher, Pius Krütli, Thomas Flüeler, and Roland W. Scholz

Abstract Participation of a variety of actors has been observed in both energy research and the transition process of energy systems, and more participation is commonly advocated. Despite this, 'participation' seems to be an all-purpose term with an unclear definition. To give it meaning, the following key questions must be addressed: *Why* and *when* should different actors be involved? *Who* should be involved and who should involve them? In *which* specific issues should the participants be involved and *what* are they expected to contribute? *Which techniques* allow for appropriate participation? How can *informal* participation techniques be combined to develop a staged, *formal* process? Finally, what role do research methods play in such processes? This contribution addresses these questions conceptually and then more concretely with illustrations from the authors' own experiences in collaborating with diverse actors in a transdisciplinary research process. The chapter concludes that a functional-dynamic approach to addressing collaboration is necessary, further knowledge integration is crucial and a systematic and analytical framework is thus essential. These elements are presented in the transdisciplinary case study (TdCS) design. Appropriate and tailored participation techniques and research methods were selected and integrated in order to provide the prerequisites for inclusive collaboration, depending on the goals and phase of the research process in question.

Keywords Collaboration · Decision process · Dynamic approach · Knowledge integration · Public participation · Transdisciplinary research

---

M. Stauffacher (✉)
Institute for Environmental Decisions (IED), ETH Zurich, Natural and Social Science
Interface (NSSI), CH-8092 Zurich, Switzerland
e-mail: michael.stauffacher@env.ethz.ch

D. Spreng et al. (eds.), *Tackling Long-Term Global Energy Problems*,
Environment & Policy 52, DOI 10.1007/978-94-007-2333-7_11,
© Springer Science+Business Media B.V. 2012

11.1 Introduction

The process of collaboration among a variety of actors has been the subject of a number of recent publications in energy research and it has featured in the implementation of novel energy technologies and transition processes of energy systems. In energy policy development, stakeholder involvement has become commonplace (Devine-Wright, 2007) and so-called 'community renewables' are being promoted (Walker & Devine-Wright, 2008, p. 497). This implies a 'need for more interactive, deliberative communication' (Owens & Driffill, 2008, p. 4414) among technical experts, stakeholders, the public and decision makers. However, such a process must go beyond the traditional concept of consulting 'communities potentially affected by new proposed energy projects' (Walker et al., 2007, p. 65). New methods in line with so-called 'upstream engagement' are necessary (Flynn & Bellaby, 2007, p. 11). More generally, the need for more intensive public and stakeholder involvement in societal decision-making processes, particularly in those concerning risk issues, as well as in participatory research and similar kinds, has been a topic of discussion for decades in many different academic fields (Arnstein, 1969; Beierle & Cayford, 2002; Rowe & Frewer, 2005; van Asselt & Rijkens-Klomp, 2002; Webler, 1999; Stern & Fineberg, 1996).

Yet 'participation' seems to be an all-purpose term and it is unclear what it actually means, especially in relation to complex issues like the transition of energy systems. In order for collaboration among diverse actors in such processes to be potentially successful, the following key questions must be addressed: *Why* and *when* should different actors be involved? *Who* should be involved and who should involve them? In *which* specific issues should the participants be involved and *what* are they expected to contribute? *Which techniques* allow for appropriate participation? How can *informal* participation techniques be combined to develop a staged, *formal* process? Finally, what role do research methods play in such processes?

These questions will be addressed in turn, first conceptually and then more concretely with illustrations from the authors' own experience in collaborating with diverse actors in a transdisciplinary research process.

11.2 Key Considerations in Planning Public or Stakeholder Involvement

11.2.1 Why Should Different Actors Be Involved?

The literature offers several lines of reasoning that justify the advantages of collaboration among a variety of actors. Fiorino (1990) identified three major arguments in favour of public participation: normative, instrumental and substantive. The normative argument references the democratic ideal that citizens are best able to judge what is in their own interests (Fiorino, 1990, p. 227). It has long been recognised

11 Learning from the Transdisciplinary Case Study Approach

that a concerned and affected public is best suited to identify its own problems and needs. In line with this, Otway (1987) called 'the people whose lives are affected ... the true experts on questions of value regarding the risks of technology' (p. 125). According to instrumental reasoning, public involvement ensures that policy decisions enjoy increased legitimacy, countering the crisis of confidence in policy processes and flagging trust in decision makers (Fiorino, 1990; Chambers, 2003). In line with this argument, the term 'public participation' is restricted in the present work to processes in which the public at large is intentionally involved in political decision making (Beierle & Cayford, 2002). On the substantive level, some risk researchers stress that 'lay judgments about risk are as sound as or more so than those of experts' (Fiorino, 1990, p. 227; cf. e.g., Renn, 2005; Wynne, 1996). Further, Beierle and Cayford (2002) identified the five social goals of public participation: incorporating public values into decisions, increasing the substantive quality of decisions, resolving conflicts among competing interests, building trust in institutions and informing and educating the public. The process character of collaboration and the learning process to which all those participating aspire has received increased attention in recent years (McDaniels & Gregory, 2004; Stauffacher, 2006; Stringer et al., 2006). Collaboration among different actors can serve a variety of purposes and these aims should be defined *before* the process begins in order to be able to accurately determine whom to involve.

11.2.2 Who Should Be Involved?

Making the first decision in this respect requires differentiating between stakeholders and the public. This chapter follows Chilvers (2007) in distinguishing between stakeholders, who represent group interests, and the public, who primarily represent themselves but may potentially embody the interests of different societal groups (cf. Pahl-Wostl, 2002). Further, the term 'actor' is used here, as this is a more general term that reflects the fact that the distinction between a stakeholder and the affected public is often unclear in the early stages of a process. Detailed actor analysis may help signal who could or should be involved. In practice, it is common to involve representatives of well-known stakeholder groups and rare to make explicit reference to the general public. Furthermore, the question of who has the right to be involved is closely linked to the definition of system boundaries, administrative borders and similar issues. In a spatially-oriented conflict such as the siting of hazardous waste or contested, large-scale point-source facilities, this very question is extremely sensitive and deserves attention. It is generally not enough to rely on administrative borders and only involve the inhabitants of the host community because, for example, transport routes and water flow may impact risk factors and thus expand the affected area. In addition, the distinction that is often made between those (objectively) affected and those who feel concerned is again a major one. Whereas the former benchmark might, for example, lead to the exclusion of national stakeholder groups from a local siting conflict, the latter would require that all those potentially or actually concerned be involved and their concerns addressed.

11.2.3 Who Does the Involving?

This question can be considered the mirror image of the former. It is often neglected that the fact that somebody is involved in a decision-making process (DMP) implies that somebody else does the involving and initiating and facilitating the process. In formal participation processes, this might be a consultant contracted by the administration to facilitate the process, the administration itself or a local pressure group trying to increase public involvement from below. In informal processes, rules and criteria for involvement rarely exist and the actor allowing participation often remains obscure, as do the criteria for it. In formal procedures (such as consultations and hearings), clear guidance is generally given by the legal requirements and legitimate actors are assigned the right to choose participants according to predefined rules. This question raises issues relating to the allocation of power, resources and legitimacy among the parties. In general, the body authorised to make involvement decisions also possesses the power and legitimacy. And yet it is also necessary to consider the perspective of those being involved. Why do they choose to participate, what functions do they consider relevant, what are their objectives, what are their needs and, furthermore, what is their mandate related to both process and outcome; i.e., how representative are they? To represent the interests of both sides, it is necessary to have two-way communication resulting in a negotiation process at the very beginning of the DMP.

11.2.4 On Which Specific Issues Should Different Actors Be Involved and What Are They Expected to Contribute?

These two questions are addressed together because they are closely linked. Overall, the answer to this question largely depends on the function of involvement, the step in the DMP and the issue at hand. If, for example, the function is meant to be substantive and the actual project step requires technological risk assessment, those involved should possess the relevant knowledge (Collins & Evans, 2002). Again, with respect to the substantive function, assessment criteria need to be defined. Since this is an important valuation step, in many cases the public at large should have a say, as in the evaluation of the criteria and the risks involved in the – normative – question of 'how safe is safe enough?' (Fischhoff, Slovic, Lichtenstein, Read, & Combs, 1978). This chapter argues that if the function relates more to legitimacy, it is necessary to involve all parties from the start and throughout the process, as this allows for full process transparency and thereby builds trust among all DMP participants. In general, involvement from the very beginning seems to help the participants identify with the outcome. In the course of narrowing down potential sites, as a decision-making process becomes linked to a certain spatial area, more extensive, inclusive and highly targeted forms of involvement may prove necessary and useful throughout the whole process. Regions and groups that 'drop out' lose interest, whereas other people's interest in the remaining areas is aroused. In some situations, involvement may only be considered necessary at the end of a DMP,

11 Learning from the Transdisciplinary Case Study Approach

when the final decision is made, such as by a public vote. Again, the key importance of specifying the objective of involvement becomes apparent.

11.2.5 Which Technique Allows for Appropriate Collaboration?

The literature offers several typologies of involvement for categorising the available techniques (e.g., Arnstein, 1969; Bishop & Davis, 2002; Pahl-Wostl, 2002; Pretty, 1995; Rowe & Frewer, 2005; van Asselt & Rijkens-Klomp, 2002; Webler, 1999). In her 'ladder of citizen participation,' Sherry Arnstein (1969) distinguished eight levels within three groups, classified according to the degree of empowerment, whereby each function may require a tailored technique: non-participation (manipulation, therapy), degrees of tokenism (informing, consultation, placation) and degrees of citizen power (partnership, delegated control, citizen control). The present study, however, questions the idea that 'one technique is adequate for one problem' (Krütli, Stauffacher, Flüeler, & Scholz, 2006). Based on evaluation studies, as well as on the authors' own experiences, this chapter concludes that a one-size-fits-all technique does not exist; instead, different techniques should be carefully combined so as to complement each other (Beierle, 2002; Beierle & Cayford, 2002; Fiorino, 1990; Krütli et al., 2006; Krütli, Stauffacher, Flüeler, & Scholz, 2010b). A large and growing number of different techniques are available (cf. Rowe & Frewer, 2005), which further indicates that collaboration among different actors should take the right form at the right time rather than supporting the notion that more intensive teamwork is always better. Collaboration may require different techniques at the same time and within the same step or a consecutive set of particular techniques. In essence, the techniques need to be adequate for the issue at hand and the respective persons or groups involved.

11.2.6 How Can These Techniques Be Combined to Form a Staged Process?

The combination of different techniques for various aspects of a staged DMP necessitates a dynamic and adapted approach. The importance of a dynamic understanding of collaboration between different actors can be elucidated by referring to two prototypical patterns (Krütli et al., 2006). In the 'expert approach', the problem is perceived as a technical one and participation and problem-solving is largely limited to experts. Involvement of the general public is usually limited to information and consultation. The counterpart of this method is the 'grassroots approach', whereby stakeholders and the public are fully empowered during the entire process. In the context of many current societal decision problems – with the many uncertainties inherent in complex scientific issues and with multiple actors affected at the same time – each of these pure approaches is likely to fail. This chapter argues that each phase of a DMP has its own specific and adequate form of involvement

(Flüeler, Krütli, & Stauffacher, 2007). In fact, there is not a single stage of a complex decision problem that calls for merely one level of involvement; rather, each stage spans different levels at different points in time. The level of participation should (and in the authors' experience, does (see below)) depend on the phase and goals of the process and its context.

11.2.7 What Role Do Research Methods Play in Such Processes?

The discussion thus far has remained within the framework of the classical participation literature. It is necessary to go one step further, however, as *knowledge* and its counterpart, *uncertainty*, play a key role in many decision-making processes. Analytical approaches appear indispensable given the complexity of many present decision problems. With a set of different options for future development, there is a need to include knowledge and values from diverse actors as well as to gauge diverging goals and scrutinise possible trade-offs given high system uncertainties (Funtowicz & Ravetz, 1993). Complex decisional issues require a framework that integrates multiple research methods, allowing for assessment from different angles and by multiple stakeholders (Petts, 2004). Such frameworks for an analytical approach to collaboration have gained a more prominent role recently, at least in sustainable development (see, e.g., Brown et al., 2001; McDaniels & Trousdale, 2005; Loukopoulos & Scholz, 2004; Sheppard & Meitner, 2005). Sheppard and Meitner (2005, p. 184) emphasised that this can 'help bridge the gap between general participatory processes and complex decision-support systems'. Based on our research in large-scale transdisciplinary projects[1] in domains such as transportation, urban and rural development, tourism, radioactive waste management and regional clustering, the present chapter concurs with this concept. The research process identified several distinct phases: goal formation, system analysis, scenario construction, multi-criteria assessment and generation of orientations (Scholz et al., 2006). Throughout these phases, different research methods were selected and combined and various actors were involved accordingly. The next section illustrates this with a concrete example from our own work.

11.3 A Transdisciplinary Case Study on Sustainable Landscape Development

As outlined above, the study proposes a functional-dynamic approach to collaboration among different actors. Unlike most of the referenced literature on participation, the focus of the current study is not placed on involvement of the public in a decision-making process but rather on collaboration among different actors in a research-practice project (transdisciplinary process). The substantive function

---

[1] For an overview, see Scholz, Lang, Walter, Wiek, and Stauffacher (2006).

11 Learning from the Transdisciplinary Case Study Approach

(Fiorino, 1990) is thus essential and knowledge from different actors needs to be integrated. However, other outcomes are possible, such as capacity-building in collaborators, contribution to consensus building guided by the mapping of diversity and even support of the negotiation process when conflicts of interest are detected (Scholz, 2011). Consequently, appropriate research methods should be selected and integrated in order to provide the prerequisites for inclusive collaboration, depending on the goals and phase of the process in question. The approach should be sufficiently diversified and adaptable to cope with a continuously changing context. In the following sections, our approach is illustrated through a case study on sustainable landscape development in the Swiss pre-alpine region of Appenzell Ausserrhoden.

11.3.1 Background

Appenzell Ausserrhoden (AR) is a canton of 20 municipalities spread over 242 km^2 with 53,500 inhabitants in Eastern Switzerland. It lies in the vicinity of St. Gallen and has been historically shaped by traditional industries (Scholz & Stauffacher, 2007). At the time of this study, some municipalities far away from the national traffic infrastructure and the larger cities were suffering from decreasing population and labour opportunities – and thus decreasing wealth. The canton's landscape is its main asset for agriculture and tourism, as well as housing (Scholz & Stauffacher, 2002). For the cantonal government, it was unclear how the region could utilise its landscape without undermining the landscape's natural potential. In order to answer this question, the president of the canton contacted ETH-NSSI (Environmental Sciences – Natural and Social Science Interface chair at ETH Zurich) to conduct a case study in the region.

11.3.2 A Transdisciplinary Approach

For a subject as encompassing and complex as sustainable landscape development, the knowledge and experience of both scientists and people outside academia should be combined. This chapter denotes such an approach as *transdisciplinary*. This term refers to a new form of knowledge production with a change from research *for* society to research *with* society, aiming at a mutual learning process between science and society (Scholz, 2000; Scholz, Mieg, & Oswald, 2000; Hirsch Hardon, Bradley, Pohl, Rist, & Wiesmann, 2006). As such, it is a capacity-building process for society and science: 'Transdisciplinarity is a way of increasing its [society's] unrealized intellectual potential, and, ultimately, its effectiveness ... Transdisciplinarity is a new form of learning and problem solving involving collaboration among different parts of society and academia in order to meet challenges of society. Transdisciplinarity starts from tangible, real-world problems' (Häberli, Bill, Thompson Klein, Scholz, & Welti, 2001, pp. 4, 7).

11.3.3 The Procedure – Step by Step

The core elements of the project were implemented in six steps (see Fig. 11.1): (i) Define a guiding question, (ii) Facet the case, (iii) Perform system analysis and (iv) Construct scenarios, (v) Perform multi-criteria analysis (MCA) by referring both to science-based arguments (MCA I) and obtaining individual preferences from different stakeholder groups (MCA II) and finally (vi) Discuss the results and develop orientations (Scholz & Tietje, 2002; Scholz et al., 2006).

In describing the involvement technique applied, an adapted distinction of involvement was used based on the general idea of increasing degree of empowerment by Arnstein (1969) and direction of communication by Rowe and Frewer (2005): information, consultation, cooperation, collaboration and empowerment. Information and consultation were considered rather weak forms of involvement as they generally have a non-committal character and utilise one-way communication. Cooperation and collaboration, for their part, both use two-way communication. In the former, a hierarchical relationship between those involving and those being involved remains. In the latter, however, all collaborators are equally responsible for the progress of process and output. On the upper end, the public or the stakeholders are empowered[2] by holding decision-making authority over content and process.

Overall, the study aimed at a substantive function of involvement, focusing on knowledge integration and mutual learning as prerequisites for tackling the complex

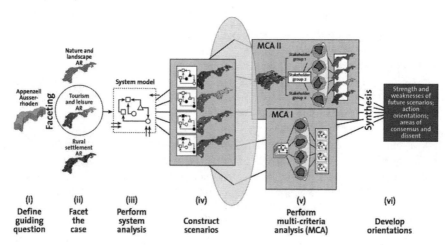

Fig. 11.1 Six steps in the transdisciplinary case study on landscape development in Appenzell Ausserrhoden
Source: Adapted from Scholz et al. (2006)

---

[2] When using a more general definition of the term empowerment as 'to provide with the means or opportunity' (Merriam-Webster's Online Dictionary), one could claim that an involvement process, in general, contributes to the empowerment of those participating.

11 Learning from the Transdisciplinary Case Study Approach

issue at hand. The study attempted to involve those regional actors who had concrete (experiential) knowledge and were able to contribute substantively. In order to strengthen the legitimacy of the case study, co-leadership and several groups composed of different actors from the region and the authors were established. In the main group with decisive power (*steering group*), decisions on whom to involve were jointly made by scientific experts and non-academics.

(i) Define a guiding question (February 2000 to December 2000)

As mentioned above, the initiative for a case study in the region came from the canton. The project was jointly led by the president of the canton and the chair of our group (*co-leadership*). As such, it represents a specific form of participation, as we were seeking a real partnership in which both parties (scientists and people from outside academia) would join on equal footing, bringing their own interests and goals to the table. The problem and the guiding question was defined during a round of intense discussions in the *steering group*. This group was strongly involved during the entire project in defining the project framework and continuously evaluating the quality of the project (Scholz et al., 2006). It is at this stage that *collaboration* on the project was established. The steering group was composed of the following *key players* (see Scholz & Tietje, 2002): the head of the office for the promotion of the economy, two heads of cantonal government agencies (agriculture/forestry, environmental protection), the cantonal historian/archivist, one farmer/mayor of a municipality, a mayor of another municipality, one independent expert for tourism and landscape and three independent experts for regional planning/development. The ETH project team (four senior researchers) also participated. Some preliminary media analysis helped identify how the problem was perceived in the region and how different stakeholders framed the issue of landscape development (Andsager, 2000; Dahinden, 2006). In-depth interviews with key players from the region (*consultation*) and first *experiential case encounters* (Scholz & Tietje, 2002, pp. 241–246) furthered our understanding of the case. Finally, the public were informed about the case study with the help of newspaper articles (*information*).

(ii) Facet the case (January 2001 to September 2001)

The discussion regarding which perspectives to choose again required multiple workshops with the *steering group*. In addition to the steering group, an *advisory board* was established consisting of further actors from the region (such as CEOs of industrial companies, farmers, mayors of municipalities, bank managers) from the region and scholars from ETH and other research institutes. In total, 18 stakeholders were involved in this step at the *consultation* level. As a complement to this step, the research literature was reviewed to assess which disciplines would help address the issue. In transdisciplinary projects, it is not the scientific discipline that defines the problem; it is the problem that suggests the scientific approach and the disciplinary focus to follow (Hirsch Hardon et al., 2006). In addition, the available information was investigated and a basic actor and stakeholder group analysis was executed. With the help of Formative Scenario Analysis (FSA, cf. Scholz & Tietje, 2002; Wiek, Scholz, Deér, Liechtenhan, & Tietje, 2002), frame scenarios were developed for the region. Again, two regional experts were involved in this step (*collaboration*).

(iii) Perform system analysis (October 2001 to November 2001)

A *reference group* was established in each of the subgroups to regularly discuss the project work. In contrast to the steering group and the advisory board, these groups were created in order to enable involvement of broader segments of the public. A range of people participated in these groups, including farmers, teachers, a hotel owner, housewives, a medical practitioner, a bank manager, architects, planners, foresters and a pastor. In these groups, methods such as focus group interviews were used to collect information systematically. In the *system analysis*, these reference groups offered their detailed qualitative insights into the components and functions of the system. Owing to the technicalities of the research method (FSA), which requires an abstracted system understanding, major parts of the analytical work were carried out at ETH. The results were then fed back to and discussed with the reference groups. Hence, the groups were involved for *consultation* and to *cooperatively* develop a better understanding of the problem. Overall, system analysis was used to synthesise results from a wide range of different sources (secondary data analyses, expert interviews, workshops with people from the region and media analyses) and to further our insights into the case. Various research methods (FSA, social research methods) played a prominent role in this step but were supplemented with further *experiential case encounters* and qualitative in-depth interviews (*consultation*) to deepen the level of understanding.

(iv) Construct scenarios (December 2001 to January 2002)

This step was largely driven by FSA, a technique in which intuitive and analytical scenario construction is combined (Wiek, Binder, & Scholz, 2006). Here, input from the region was essential, particularly for the intuitive scenarios. A variety of techniques that stimulate creative thinking, such as brainstorming and mind mapping, were employed in moderated group discussions with the reference groups in order to elicit various perspectives for the future development of the region (*consultation*). These intuitive scenarios were then contrasted, adapted and fine-tuned with the help of a formalised approach that combined different levels of essential impact factors from the system analysis (FSA). The developed scenarios were finally discussed in the reference groups and supported by various visualisation techniques (such as, maps, collages and photo compositions). The scenarios to be evaluated in the subsequent step (*cooperation*) were also selected.

11 Learning from the Transdisciplinary Case Study Approach

(v) Perform Multi-Criteria Analysis (MCA) (January 2002 to February 2002)

A postal survey was used to assess the views of the general public and expert interviews were guided to provide input for a data-based evaluation of the scenarios using multiple criteria (MCA I). In addition, various stakeholder groups from the region provided detailed information in individual assessment sessions (MCA II). These evaluation sessions lasted at least one hour per participant and were conducted in line with the rules governing psychological experiments. Respondents had to provide detailed evaluations in two steps: first, an overall 'holistic' assessment and then another one using the criteria from the MCA I. In addition to detailed quantitative data, qualitative and in-depth information was also collected. Overall, these elements were on the level of *consultation*. The results were then presented and discussed: first independently, in the respective reference groups for each subsystem, subsequently as a comprehensive overview in the steering group, and finally with all involved in a large presentation event (*information*).

(vi) Develop orientations (February 2002 to October 2002)

A round of moderated workshops with the steering group and the advisory board helped us to share interpretations of the results and to jointly develop conclusions and elaborate follow-up activities (*collaboration*). Writing and reviewing a publication (more than 300 pages, see Scholz et al., 2002) was essential in this case.

Following the development of orientations, outcome presentation and discussion events for a diverse range of audiences made it possible to circulate and discuss results and conclusions from the project. The numerous newspaper articles published throughout the process supported the dissemination of knowledge; hence, the level of involvement in this step was *information*.

11.3.4 The Dynamics of Involvement and Research Methods Used

Many different forms of involvement were used throughout the case study in Appenzell Ausserrhoden. Table 11.1 shows the actors and the level of intensity of their involvement. Figure 11.2 shows the project's progress compared to the intensity of involvement and indicates a selection of the applied involvement techniques and research methods.

There is a clear dynamic pattern with varying intensities of involvement throughout the whole process. Collaboration on equal terms between participating actors only became visible after initial preparatory workshops were held to inform people about the project and its organisation and to promote discussion. Involvement in the form of consultation also took place at regular intervals. Here, research methods were applied and knowledge from outside academia was integrated into a classical scientific mode. On the other hand, several workshops were conducted regularly

Table 11.1 Levels of involvement in the case study on landscape development in Appenzell Ausserrhoden. Beyond the involvement depicted here, numerous media articles informed the public about the study and several presentations and poster sessions were held with a total of approximately 150 participants (based on Stauffacher, Flüeler, Krütli, & Scholz, 2008a)

Step	People involved	Level of involvement
(i) Define a guiding question	Steering group (10 'key players' from the region)	Collaboration
(ii) Facet the case	Steering group	Collaboration
	Case study advisory council (eight stakeholders from the region)	Consultation
(iii) Perform system analysis	Personal interviews with approximately 30 members of the public	Consultation
	Discussion of the system analysis in three reference groups with a total of 34 people (from the public)	Cooperation
	Progress review in the steering group	Collaboration
(iv) Construct scenarios using FSA	Elicitation of intuitive scenarios in the reference groups	Consultation
	Discussion and final selection of scenarios in the reference groups	Cooperation
	Progress review in the steering group	Collaboration
(v) Perform MCA	Surveys (approximately 200 questionnaires with members of the public) and approximately 20 expert interviews in MCA I	Consultation
	78 people in MCA II (stakeholders)	Consultation
	Discussion of results in the reference groups	Cooperation
	Progress review in the steering group	Collaboration
(vi) Develop orientations	Discussion of results and elaboration of orientations with steering group and case study advisory council	Collaboration

in a collaborative manner in which decisions on follow-up to the projects were jointly taken based on scientific input on research requirements and preferences or expectations from non-academia. Empowerment in the sense of conferring decision-making capabilities to non-academia only took place in the follow-up phase, when concrete action was taken. Research has little impact here, which exemplifies the symmetric character of the intensity scale: if a particular group's power increases, another group's power is reduced.

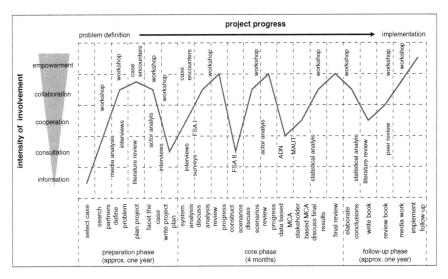

Fig. 11.2 Varying degrees of involvement and selection of involvement techniques and research methods applied in the case study on landscape development in Appenzell Ausserrhoden
Source: Stauffacher et al. (2008a)

11.4 Preliminary Applications to the Energy Field

The transdisciplinary case study design was recently applied to energy issues, adapting the approach while keeping most of the major features intact. The first case study in the energy field focused on siting nuclear waste (Scholz et al., 2007; Stauffacher, Krütli, & Scholz, 2008b; Krütli, Flüeler, Stauffacher, Wiek, & Scholz, 2010a) The aim was to investigate how the opinions of people concerned and affected by a dedicated disposal project impacted the interplay of safety and process, how these opinions changed over time, if at all, and what could be learned for the future design of the Swiss radioactive waste programme. The authors of this chapter, along with seniors and advanced MSc students at ETH, investigated a twice-rejected siting project for low- and intermediate-level waste in Wellenberg, located in the canton of Nidwalden (Central Switzerland). The Wellenberg project was selected because it was the only 'full' societal decision process regarding long-term radioactive waste management in Switzerland. On the basis of actor analysis, an *advisory board* was established with 20 people from the canton of Nidwalden who were intimately familiar with the issue and the case but who had different views of it. The advisory board's function was to share specific local information, experience and interests, to establish contact with additional stakeholders and local experts, to identify conflict areas and to ensure quality control throughout, up to reviewing the final report (Scholz et al., 2007). The students worked in three groups, each with a different perspective and method. The *media group* identified the main actors and their arguments by means of content analysis of newspaper articles and letters to the editors of local newspapers. The *process group* carried out in-depth interviews with 41 respondents representing different stakeholder groups and a wide range of views, both for and against. Participants had to weigh decision-making aspects, such as site-selection criteria, public involvement and monitoring, and subsequently rank the aspects' precise design. The *survey group* polled a representative sample of the Nidwalden population on the interplay of aspects like experience, risk and benefit perception, level of concern, justice, values, emotions, knowledge and trust.

In another case study in the same region as the landscape development study discussed above, we developed and assessed energy strategies for small communities (Trutnevyte, Stauffacher, & Scholz, 2011). Since it was already the third study in the same region, it was much easier and more straightforward to involve local actors because existing personal networks could be used. Furthermore, most of the participants were already experienced in transdisciplinary approaches and, therefore, had a clear and suitable understanding of participants' respective roles in the process. A steering board was established that included the mayor of the core community, an independent energy consultant and head of the district heating system, the head of the cantonal Office for Environment and Energy and the head of a cantonal NGO for the promotion of renewable energy and energy efficiency. Other people, including representatives from the community council, the public at large and other energy producers, were involved as the steering board deemed necessary. With respect to the analytical approach, the three following project groups were chosen: (i) Development and public perception of energy strategies, (ii) development and multi-criteria assessment of energy strategies and (iii) barriers for carbon projects (including energy efficiency) within small and medium enterprises (SMEs). Groups (i) and (ii) first worked together in developing different options for a local energy strategy. Assessment was then done separately with each of the groups being responsible either for the data-based MCA (see MCA I above) or the public perception (similar to MCA II above). Group (iii) focused specifically on the role of SMEs as significant energy users in the region.

11.5 Discussion and Conclusion

For decades, broader involvement of the public in decision-making has been called for (Lengwiler, 2008). However, some approaches have been guided by a rather naïve, 'the-more-the-better' understanding of participation. In addition, participation literature tends to present a static and generic picture that neglects the specific and contextualised underlying decision-making processes. This chapter argues for a functional-dynamic view that considers the intensity of participation, the issue at any specific step, the corresponding technique(s) and the function of participation in the entire decision-making process. Collaboration among different actors necessitates reflection on the *why* question in particular: what is the function and the rationale for collaboration? Fiorino (1990) proposed 'three quite starkly distinguishable types of imperative' (Stirling, 2008, p. 268): meeting democratic ideals, fairness and robustness of decisions (Beierle & Cayford, 2002; Petts, 2004). These objectives may increase acceptability and, from a decision maker's perspective, legitimacy. However, they could be perceived as tokenism, which could jeopardise any participatory efforts that are governed by the perspectives and the normative standpoints of the representatives. Participants are likely to prefer different functions – some might prefer only to be informed, whereas others may want to have decision-making power (cf. Chilvers, 2007). In our view, such a complementary and symmetric view of participation is crucial: the question of who does the involving (of whom) seems to often be overlooked or underestimated.

The three examples we have presented above exemplify the functional-dynamic view of organising a collaborative process among different actors. It is possible to go still one step further, as our projects are research-driven and it is necessary to take a systematic and analytical approach to tackling the research question. In contrast to the less systematic approaches of collaborative planning (Forester, 1999), all input from the region is documented and can be traced in the process; the transparency of the collaboration process is thus ensured (Joubert, Leiman, de Klerk, Katua, & Aggenbach, 1997). At the same time, the systematic integration of multiple methods makes scientists' input traceable and the results more robust. In our original TdCS design, the very process of assessing the actual situation and developing and evaluating future scenarios generates a learning process. This can enable and motivate people from the region to contribute more actively to the subsequent implementation or to formal decision processes, an outcome documented elsewhere in similar studies (Brown et al., 2001; Sheppard & Meitner, 2005).

Nevertheless, our TdCS design cannot always be applied one-to-one and must instead be adapted to the context. In the Wellenberg study, both the analytical procedure and the form of involvement had to be modified, mainly due to the contentious character of the issue. For example, it was impossible to assemble a co-leadership team from the region, as none of the executives from the canton wanted to be officially responsible for a study on this controversial issue. Nevertheless, involvement was still functional-dynamic and we were eventually able to establish an advisory board and involve all the important parties. Because the goal was more about learning from the past than about planning for the future, the analytical approach also had to be adapted. In fact, the chosen groups or facets more closely echoed various disciplinary approaches, and the character of the overall project was therefore more of an interdisciplinary research project. This contrasts with our TdCS design, where knowledge from different sources is integrated from a unifying system-science perspective. Still, it was possible to apply some of the elements of the TdCS design, and the rationale of the Formative Scenario Analysis was used to construct different variants of a site-selection decision process. Again, a slightly adapted approach is evident in the Urnäsch energy case study but with a similar emphasis on the combination of analytical rigour and functional collaboration. As for collaboration, the biggest difference between the energy study and the landscape development study was in the number of groups established: one in the former and a total of five in the latter. This was due to the smaller number of participants from our side and, more importantly, to the great expertise of all participants from both academia and non-academia. In contrast, the analytical approach was more or less identical, as we were again dealing with future developments.

In conclusion, the crucial elements in this TdCS design are the functional-dynamic approach to addressing collaboration between various actors and the systematic-analytical approach to the concrete research issue under scrutiny. We are convinced that collaboration on complex problems is only sensible if it is organised in such a way. Otherwise, knowledge aspects become overshadowed by large involvement exercises with unclear goals. This causes a loss of transparency in the whole process: the input requested from various actors has no clear effect and responsibilities for results and potential outcomes remain unclear. Our TdCS design offers some advantages as a collaborative approach in a decision-analytic framework: input is sought, facilitated, documented and traceable throughout the process, which guarantees transparency and proof of the intensity of collaboration. It is important to recognise, however, that all participants put their autonomy at stake and expect to gain something from the collaborative efforts – additional insights for research on one side and better evidence for sound societal decisions on the other. Insisting on existing boundaries and the division of labour between science and practice makes boundary work essential; in other words, boundaries must be actively discussed and reviewed throughout the project (Midgley, 2003). The TdCS design further enables negotiation and deliberation among a large group of actors. Consequently, the TdCS design offers a means for societal learning in sustainable development (Scholz et al., 2006; Stauffacher, Walter, Lang, Wiek, & Scholz, 2006). For many areas of research on energy and energy system transitions, lessons can be drawn from our experiences in applying analytical frameworks for a functional-dynamic approach to collaboration among different actors.

References

Andsager, J. L. (2000). How interest groups attempt to shape public opinion with competing news frames. *J&MC Quarterly, 77*(3), 577–592.

Arnstein, S. R. (1969). A ladder of citizen participation. *Journal of the American Institute of Planners, 35*, 216–224.

Beierle, T. C. (2002). The quality of stakeholder-based decisions. *Risk Analysis, 22*, 739–749.

Beierle, T. C., & Cayford, J. (2002). *Democracy in practice. Public participation in environmental decisions*. Washington, DC: Resources for the Future Press.

Bishop, P., & Davis, G. (2002). Mapping public participation in policy choices. *Australian Journal of Public Administration, 61*, 14–29.

Brown, K., Adger, W. N., Tompkins, E., Bacon, P., Shim, D., & Young, K. (2001). Trade-off analysis for marine protected area management. *Ecological Economics, 37*, 417–434.

Chambers, S. (2003). Deliberative democratic theory. *Annual Review of Political Science, 6*, 307–326.

Chilvers, J. (2007). Towards analytic-deliberative forms of risk governance in the UK? Reflecting on learning in radioactive waste. *Journal of Risk Research, 10*(2), 197–222.

Collins, H. M., & Evans, R. (2002). The third wave of science studies: Studies of expertise and experience. *Social Studies of Science, 32*, 235–296.

Dahinden, U. (2006). *Framing – Eine integrative Theorie der Massenkommunikation*. Konstanz: UVK.

11 Learning from the Transdisciplinary Case Study Approach 243

Devine-Wright, P. (2007). Energy citizenship: Psychological aspects of evolution in sustainable energy technologies. In J. Murphy (Ed.), *Framing the present, shaping the future: Contemporary governance of sustainable technologies*. London: Earthscan.

Fiorino, D. J. (1990). Citizen participation and environmental risk: A survey of institutional mechanisms. *Science, Technology, & Human Values, 15*, 226–243.

Fischhoff, B., Slovic, P., Lichtenstein, S., Read, S., & Combs, B. (1978). How safe is safe enough? A psychometric study of attitudes towards technological risks and benefits. *Policy Sciences, 8*, 127–152.

Flüeler, T., Krütli, P., & Stauffacher, M. (2007). *Tools for local stakeholders in radioactive waste governance: Challenges and benefits of selected participatory technology assessment techniques*. Final Report, April 2007. Contribution to the EU STREP Community Waste Management COWAM 2, Work Package 1: Implementing Local Democracy and Participatory Assessment Methods. All web links accessed November 16, 2011, http://www.cowam.com/? COWAM-2-Final-Reports

Flynn, R., & Bellaby, P. (Eds.). (2007). *Risk and the public acceptance of new technologies*. Basingstoke: Palgrave-Macmillan.

Forester, J. (1999). *The deliberative practitioner: Encouraging participatory planning processes*. Cambridge, MA: MIT Press.

Funtowicz, S. O., & Ravetz, J. T. (1993). Science for the post-normal age. *Futures, 25*(7), 739–755.

Häberli, R., Bill, A., Thompson Klein, J., Scholz, R. W., & Welti, M. (2001). Summary and synthesis. In J. Thompson Klein, W. Grossenbacher-Mansuy, R. Häberli, A. Bill, R. W. Scholz, & M. Welti (Eds.), *Transdisciplinarity: Joint problem solving among science, technology, and society* (pp. 3–22). Basel: Birkhäuser Verlag AG.

Hirsch Hardon, G., Bradley, D., Pohl, C., Rist, S., & Wiesmann, U. (2006). Implications of transdisciplinarity for sustainability research. *Ecological Economics, 60*, 119–128.

Joubert, A. R., Leiman, A., de Klerk, H. M., Katua, S., & Aggenbach, J. C. (1997). Fynbos (fine bush) vegetation and the supply of water: A comparison of multi-criteria decision analysis and cost-benefit analysis. *Ecological Economics, 22*, 123–140.

Krütli, P., Flüeler, T., Stauffacher, M., Wiek, A., & Scholz, R.W. (2010a) Technical safety vs. public involvement? A case study on the unrealised project for the disposal of nuclear waste at Wellenberg (Switzerland). *Journal of Integrative Environmental Sciences, 7*(3): 229–244.

Krütli, P., Stauffacher, M., Flüeler, T., & Scholz, R. W. (2006). *Public involvement in repository site selection for nuclear waste: Towards a more dynamic view in the decision-making process*. Conference proceedings. VALDOR 2006 – VALues in Decisions On Risk. Stockholm, May 14–18, 2006. SKI, SEPA, SGI, SRCE, OECD/NEA, UK Nirex, 96–105.

Krütli, P., Stauffacher, M., Flüeler, T., & Scholz, R.W. (2010b). Functional-dynamic public participation in technological decision making: Site selection processes of nuclear waste repositories. *Journal of Risk Research, 13*(7): 861–875.

Lengwiler, M. (2008). Participatory approaches in science and technology – Historical origins and current practices in critical perspective. *Science Technology, & Human Values, 33*, 186–200.

Loukopoulos, P., & Scholz, R. W. (2004). Sustainable future urban mobility: Using 'area development' negotiations' for scenario assessment and participatory strategic planning. *Environment and Planning A, 36*, 2203–2226.

McDaniels, T. L., & Gregory, R. (2004). Learning as an objective within a structured risk management decision process. *Environmental Science & Technology, 38*(7), 1921–1926.

McDaniels, T. L., & Trousdale, W. (2005). Resource compensation and negotiation support in an aboriginal context: Using community-based multi-attribute analysis to evaluate non-market losses. *Ecological Economics, 55*, 173–186.

Midgley, G. (2003). Science as systemic intervention: Some implications of systems thinking and complexity for the philosophy of science. *Systemic Practice and Action Research, 16*(2), 77–97.

Otway, H. (1987). Experts, risk communication, and democracy. *Risk Analysis, 7*(2), 125–129.

Owens, S., & Driffill, L. (2008). How to change attitudes and behaviours in the context of energy. *Energy Policy, 36*, 4412–4418.

Pahl-Wostl, C. (2002). Participative and stakeholder-based policy design, evaluation and modeling processes. *Integrated Assessment, 3*(1), 3–14.

Petts, J. (2004). Barriers to participation and deliberation in risk decisions: Evidence from waste management. *Journal of Risk Research, 7*, 115–133.

Pretty, J. N. (1995). Participatory learning for sustainable agriculture. *World Development, 23*(8), 1247–1263.

Renn, O. (2005). Partizipation – ein schillernder Begriff. Reaktion auf drei Beiträge zum Thema 'Partizipation'. *GAIA, 14*(3), 227–228.

Rowe, G., & Frewer, L. L. (2005). A typology of public engagement mechanisms. *Science, Technology, & Human Values, 30*(2), 251–290.

Scholz, R. W. (2000). Mutual learning as a basic principle for transdisciplinarity. In R. W. Scholz, R. Häberli, A. Bill, & M. Welti (Eds.), *Transdisciplinarity. Joint problem-solving among science, technology and society.* Proceedings of the international transdisciplinarity 2000 conference, Zurich. Workbook II: Mutual learning sessions (pp. 13–17). Zürich: Haffmans Sachbuch.

Scholz, R. W. (2011). *Environmental literacy in science and society: From knowledge to decision.* Cambridge: Cambridge University Press.

Scholz, R. W., Lang, D., Walter, A. I., Wiek, A., & Stauffacher, M. (2006). Transdisciplinary case studies as a means of sustainability learning: Historical framework and theory. *International Journal of Sustainability in Higher Education, 7*(3), 226–251.

Scholz, R. W., Mieg, H. A., & Oswald, J. E. (2000). Transdisciplinarity in groundwater management: Towards mutual learning of science and society. *Water, Air, and Soil Pollution, 123*, 477–487.

Scholz, R. W., & Stauffacher, M. (2002). Unsere Landschaft ist unser Kapital: Überblick zur ETH-UNS Fallstudie 'Landschaftsnutzung für die Zukunft: der Fall Appenzell Ausserrhoden'. In R. W. Scholz, M. Stauffacher, S. Bösch, & A. Wiek (Eds.), *Landschaftsnutzung für die Zukunft: der Fall Appenzell Ausserrhoden. ETH-UNS Fallstudie 2001* (pp. 13–47). Zürich: Rüegger und Pabst.

Scholz, R. W., & Stauffacher, M. (2007). Managing transition in clusters: Area development negotiations as a tool for sustaining traditional industries in a Swiss prealpine region. *Environment and Planning A, 39*(10), 2518–2539.

Scholz, R. W., Stauffacher, M., Bösch, S., Krütli, P., & Wiek, A. (2007). *Entscheidungsprozesse Wellenberg – Lagerung radioaktiver Abfälle in der Schweiz. ETH-UNS Fallstudie 2006.* Zürich, Chur: Rüegger.

Scholz, R. W., Stauffacher, M., Bösch, S., & Wiek, A. (2002). *Landschaftsnutzung für die Zukunft: der Fall Appenzell Ausserrhoden. ETH-UNS Fallstudie 2001.* Zürich: Rüegger und Pabst.

Scholz, R. W., & Tietje, O. (2002). *Embedded case study methods: Integrating quantitative and qualitative knowledge.* Thousand Oaks, CA: Sage.

Sheppard, S. R. J., & Meitner, M. (2005). Using multi-criteria analysis and visualisation for sustainable forest management planning with stakeholder groups. *Forest Ecology and Management, 207*, 171–187.

Stauffacher, M. (2006). *Beyond neocorporatism: New practices of collective decision making. Transdisciplinary case studies as a means for societal learning in sustainable development.* Thesis for Doctor of Philosophy. Faculty of Arts, University of Zurich. Available online: http://www.dissertationen.unizh.ch (> Michael Stauffacher)

Stauffacher, M., Flüeler, T., Krütli, P., & Scholz, R. W. (2008a). Analytic and dynamic approach to collaborative planning: A transdisciplinary case study on sustainable landscape development in a Swiss pre-alpine region. *Systemic Practice and Action Research, 21*(6), 409–422.

Stauffacher, M., Krütli, P., & Scholz, R. W. (2008b). *Gesellschaft und radioaktive Abfälle: Ergebnisse einer schweizweiten Befragung.* Zürich, Chur: Rüegger.

Stauffacher, M., Walter, A., Lang, D., Wiek, A., & Scholz, R. W. (2006). Learning to research environmental problems from a functional socio-cultural constructivism perspective: The transdisciplinary case study approach. *International Journal of Sustainability in Higher Education, 7*(3), 252–275.

Stern, P. C., & Fineberg, V. (Eds.). (1996). *Understanding risk: Informing decisions in a democratic society*. Washington, DC: National Academies Press.

Stirling, A. (2008). 'Opening up' and 'closing down' – Power, participation, and pluralism in the social appraisal of technology. *Science Technology, & Human Values, 33*, 262–294.

Stringer, L. C., Dougill, A. J., Fraser, E., Hubacek, K., Prell, C., & Reed, M. S. (2006). Unpacking 'participation' in the adaptive management of social-ecological systems: A critical review. *Ecology and Society, 11*(2), 39. http://www.ecologyandsociety.org/articles/1896.html

Trutnevyte, E., Stauffacher, M., & Scholz, R. W. (2011). Supporting energy initiatives in small communities by linking visions with energy scenarios and multi-criteria assessment. *Energy Policy, 39*, 7884–7895.

van Asselt, M. B. A., & Rijkens-Klomp, N. (2002). A look in the mirror: Reflection on participation in Integrated Assessment from a methodological perspective. *Global Environmental Change, 12*, 167–184.

Walker, G., & Devine-Wright, P. (2008). Community renewable energy: What should it mean? *Energy Policy, 36*, 497–500.

Walker, G., Hunter, S., Devine-Wright, P., Evans, B., Hunter, S., & Fay, H. (2007). Harnessing community energies: Explaining and evaluating community-based localism in renewable energy policy in the UK. *Global Environmental Politics, 7*, 64–82.

Webler, T. (1999). The craft and theory of public participation: A dialectical process. *Journal of Risk Research, 2*(1), 55–71.

Wiek, A., Scholz, R. W., Deér, S., Liechtenhan, W., & Tietje, O. (2002). Rahmenszenarien für die Entwicklung der Landschaftsnutzung im Kanton Appenzell Ausserrhoden – Kurzfassung. In R. W. Scholz, M. Stauffacher, S. Bösch, & A. Wiek (Eds.), *Landschaftsnutzung für die Zukunft: der Fall Appenzell Ausserrhoden. ETH-UNS Fallstudie 2001* (pp. 249–268). Zürich: Rüegger und Pabst.

Wiek, A. H., Binder, C. R., & Scholz, R. W. (2006). Functions of scenarios in transition processes. *Futures, 38*(7), 740–766.

Wynne, B. (1996). May the sheep safely graze? A reflexive view of the expert-lay knowledge divide. In S. Lash, B. Szerszynski, & B. Wynne (Eds.), *Risk, environment and modernity* (pp. 44–83). London: Sage.

Chapter 12
Lessons from the Invited Contributions

Daniel Spreng and David L. Goldblatt

Abstract This chapter selectively highlights and synthesises lessons from the invited contributions Part II in light of the energy-related challenges presented in Chapter 2. Under the lens of each challenge rubric, the contributions are considered in turn, covering a wide variety of issues. Chapter 14 complements the conclusions drawn here with findings from the recent social science literature and offers suggestions for further research.

Keywords Access · Climate change · Economic and social development · Environmental impacts · Security

12.1 Lessons for Challenge I: Access and Security

12.1.1 Pachauri and Spreng[1] (Chapter 5)

Pachauri and Spreng demonstrate that in India, access to clean and readily available energy at the household level is crucial to national development and to the populace's sense of integration into modern society. Their study does not prove a causal relationship between access to modern energy carriers and development, but it shows a strong statistical correlation. Energy access is not the main driver for development, and access is itself influenced by other factors: Microeconomic analysis[2] shows that households that decide to access modern energy carriers tend to have relatively highly educated members and to be located in neighbourhoods with generally more advanced utility infrastructure, including clean water and sanitation. Monetary wealth is not a decisive factor.

---

[1] As a metastudy, Sovacool and colleagues' Chapter 4 does not directly address the energy-related challenges and is therefore not included in the analysis in this chapter.

[2] See in particular Kemmler and Spreng (2007).

D. Spreng (✉)
Energy Science Center (ESC), ETH Zurich, 8032 Zurich, Switzerland
e-mail: dspreng@ethz.ch

D. Spreng et al. (eds.), *Tackling Long-Term Global Energy Problems*,
Environment & Policy 52, DOI 10.1007/978-94-007-2333-7_12,
© Springer Science+Business Media B.V. 2012

Energy access has to be sought after, and sometimes fought for, to be acquired. On the one hand, authorities' reluctance to grant access to a region or community typically stems from a jealous guarding of political and economic power. On the other hand, householders do not always view access to modern fuels and technologies and the new lifestyles they facilitate in a positive light. Pachauri and Spreng review historical and linguistic studies of poverty that highlight possible mechanisms that mutually reinforce and perpetuate rich and poor segments of the population. Despite the prominence of poverty alleviation on political agendas, such mechanisms are easily overlooked in present-day politics. The lack of continuing progress since the early 1990s in alleviating poverty in many areas of the world, particularly in Southern Asia – and even poverty's growth in many African countries – is related to the lack of improvement in energy access. But the precise causes are likely to be much more complicated than design problems and inefficiencies in national and international energy programmes.

12.1.2 Wilhite (Chapter 6)

This study explores the energy demand side in India from an anthropological perspective, focusing on increasing residential electricity demand. Wilhite shows that access to modern energy services and technology is often sought after for the status such energy services confer: For example, many middle class families in urban India acquire household appliances such as refrigerators, freezers and air conditioners more as status symbols than for the direct services they provide, at least initially. Access to inefficient technology along with shifts towards more energy-intensive lifestyles and their normalisation across large population segments often drive aggregate energy consumption. Wilhite's analysis here is more subtle and rigourous than Commoner's (1971) classic criticism of faulty technologies deliberately promoted by conspiring corporations. Wilhite points to the possibility for societies to choose lifestyles more consciously and critically. He analyses how the proliferation of energy-consuming devices and the normalisation of new energy-intensive practices also increase the dependence of Indian communities on imported energy sources, thereby reducing their energy security.

12.1.3 Shankleman (Chapter 7)

Shankleman describes how individuals and small groups often use exclusive control over access to energy resources in resource-rich countries to exert economic and political control over others. Of course, national and international control of access to natural resources is exploited in many other sectors besides energy, but a 'resource curse' occurs when the power this control affords is used to repress a population. In addition to political repression, this can mean denying citizens access to a wide range of other resources and opportunities and thereby retarding overall development. A resource curse has wide-ranging effects on society.

12 Lessons from the Invited Contributions 249

A resource curse can threaten supply security and access for both a country's inhabitants and its trading partners. Although the political situation in countries so afflicted tends to be unstable, the money flow that the resource produces can maintain the status quo for an extended period of time. An example of how one resource curse can breed others is the proliferation of large-scale purchases of territory by rich companies and states (e.g., the Persian Gulf states, South Korea and China) in economically weak countries (in Africa, Asia and Latin America) for agricultural development, often for the cultivation of biofuel crops. This practice has been termed a land-grab.[3]

International organisations such as the World Bank, committed to assisting states caught in this dynamic, encourage more productive use of their energy resources through revenue sharing and long-term investments. Implementing their recommendations has resulted in political and economic improvements for some countries and also brought a few regimes greater respect and international prestige. Whether by the World Bank's traditional means or via the new Chinese approach, 'banishing' the resource curse in weak but resource-rich states has the potential to greatly improve the lot of many, although in the latter case at least, improvements may well come at the cost of long-term economic or political independence.

12.1.4 Muller (Chapter 8)

Muller's study deals with the regulation of access to the atmosphere as a sink for CO_2. This regulation is politically contentious because it involves limiting the power and profitability of numerous and diverse large CO_2 emitters. Market-based regulation must be supported by manageable administrative processes that contain effective compliance incentives and enforcement mechanisms. An emissions permit trading system that is egalitarian in its treatment of emitters of all sizes, even within a given sector, and entirely avoids grandfathering allowances is politically infeasible and perhaps economically unjustified.

It would require an enormous effort to devise a second-best solution that is sufficiently effective, efficient and equitable. The chapter explains how problems with the current European Union Emissions Trading System (ETS) stem less from the shortcomings of the economic theory underlying it than from the limited or faulty application of economics by ETS proponents. (Extra-economic shortcomings of the ETS are taken up in Sections 12.2 and 12.3 below.)

12.1.5 Jørgensen (Chapter 9)

In the sixth study, Jørgensen describes the importance of institutional arrangements to the success of an innovation system for renewable energy technologies. In terms

---

[3] See for example Daniel and Mittal (2009) or http://www.ifpri.org (search for "land grabbing"). Web link accessed November 16, 2011.

of energy access and supply security, his discussion of the development of renewable energy in Denmark raises three distinct questions:

1. Who controls access to energy technology?
2. How much supply security is gained by making use of local resources?
3. How much supply security is afforded by intermittent energy sources?

The first question, addressed in some detail in the study, shows the great advantage of well-functioning local institutions and an innovative environment. The second question points to the significant advantages of local and renewable energy sources, and the third suggests a supply security disadvantage that can be mitigated by setting up sufficient backup capacity. The author advocates combining wind with biomass to build up an intermittent renewable energy source concurrently with an easily storable renewable one. In this way the entire energy system can be based on renewable energy sources and still take into account the requirements of access (local control over technology development) and security of supply.

12.1.6 Flüeler (Chapter 10)

Access, security and power-sharing are also themes in the seventh chapter. In it, Flüeler discusses parallels and differences between the disposal of radioactive waste from nuclear power generation and the capture of CO_2 waste from fossil fuel combustion using carbon capture and storage (CCS).[4] The supply security provided by the energy systems is closely related to their respective safety features, in particular those of their disposal systems.[5] As soon as the safety of one element of the energy system is compromised, the security of the entire supply chain or even the entire energy system is jeopardised.

In addition to many technical aspects, the safety of nuclear waste disposal involves many institutional aspects as well as elements of (public) perception and risk communication. At issue is not only the size of safety margins but also who assesses the safety of a given scheme. In addition to proper checklists and procedures, what is needed is an appropriate safety culture that involves a full range of stakeholders, among them pluralistic experts, and the public.

Institutional flaws in the schemes proposed for using CCS as part of an international emissions reduction framework – particularly the lack of a plausible means of effectively enforcing sanctions in the absence of a world government – may make it impossible for CCS to function on the political level in time to make the technical concept a viable option for emissions reduction.

---

[4] This is large-scale CCS for greenhouse gas reductions in addition to the more highly developed but much smaller-scale CCS technology for enhanced oil recovery.

[5] Security in the nuclear industry has the additional crucial dimension of protecting against proliferation of nuclear material for weapons purposes.

CCS may still prove to be viable technology. But the feasibility of deploying CCS as a CO_2 mitigation option on a large scale – like that of any new technology in a developing energy system – rests as much with the strength of the political, legal and social framework built up around it as it does with its technical soundness.

12.1.7 Stauffacher et al. (Chapter 11)

The eighth study discusses transdisciplinary approaches to involving the public in planning and implementing energy facilities. Although it does not explicitly consider questions of access to or security of energy systems, the notion of 'access to knowledge', in particular to expert knowledge, is very much the focus of the chapter. Naturally, like physical access, access to knowledge is closely linked to questions of power, and asymmetries in access to knowledge can be misused as easily as asymmetries in access to energy resources. The chapter recommends involving the various stakeholders to the extent that symmetry in access to knowledge and, to some degree, participation in decision making can be achieved. The best time to involve stakeholders is at the point when this symmetry can be most easily achieved.

12.2 Lessons for Challenge II: Climate Change and Other Environmental Impacts

12.2.1 Pachauri and Spreng (Chapter 5)

National and international programmes and policies aimed at facilitating energy transitions in developing countries have followed diverse strategies, but many have been expensive failures. Instead of advocating narrow performance criteria such as local electrification or modern stove penetration rates, the study argues for broadening perspectives and invoking multiple assessment criteria, such as those embodied in the United Nations Millennium Development Goals.

Cleaner and decarbonised energy production and consumption in developing countries are rapidly becoming overriding climate priorities as China overtakes the United States in its rate of CO_2 output. But the authors show that although the perspectives of individuals are multidimensional, they tend to be short-term and rarely encompass broader ecological criteria. Local overuse of biomass is eroding the natural capital base of communities and destroying their future livelihoods: in this context, 'renewable resource' is losing its meaning and becoming obsolete. Environmental education here will have little effect: ecological criteria must be applied top-down through policy, integrated into market instruments where markets exist or built into the socioeconomic fabric of societies.

However, education and awareness-raising may well be the most effective means of ameliorating the direct environment health effects – as well as the potent climate-forcing particulate products – of indoor air pollution caused by incomplete combustion of biomass; here, such measures will yield more benefits than programmes that blindly promote new stove technologies.

The chapter promotes a per capita measure of energy consumption for its utility as an ecological criterion when its upper limit is subject to a cap derived in some fashion from national emissions targets. This inevitably introduces issues of both international and intranational equity; in particular, statistical analyses show that the energy consumption of the richest deciles in most countries – even the poorest countries as measured by per capita income – greatly and disproportionately exceeds the recent world average and ecological threshold of 2,000 watts per capita.

12.2.2 Wilhite (Chapter 6)

The explosive growth of aggregate energy consumption and greenhouse gas (GHG) emissions from developing countries, particularly India and China, has exceeded the upper bounds of most projections and, if left unchecked, promises to be the overriding factor in future climate developments. Increasing electricity demand in both countries is largely being met by coal-fired power plants. Coal is one of the least desirable energy sources because of the local environmental consequences of mining it and the regional and global pollution its combustion generates. Additional hydropower plants face considerable ecological problems of their own in addition to generating difficult social and political resistance. The proliferation of household electrical appliances such as washing machines is also causing increased water pollution and scarcity of drinking water because they vastly increase the use of water, detergents and soaps in towns and cities that lack adequate infrastructure for wastewater treatment.

The scale of the threat to climate stabilisation is not adequately addressed by focusing on carbon emissions trading in developed countries (cf. Chapter 8 Muller), even with increased investment in Clean Development Mechanisms (CDM) projects in the developing world. Nor does a gradual, organic, bottom-up evolution of renewable power sources (cf. Chapter 9 Jørgensen) seem to be the answer for development that dwarfs the North's own historical industrialisation in its potential scope and pace.

Wilhite describes the penetration of air conditioning in building construction design in Kerala, India and many other areas around the world and its displacement of traditional low-energy, passive-design methods. He identifies this trend as one of the single most significant factors driving increases in urban electricity demand. The direct results include a loss of culturally and aesthetically rooted traditional buildings that require little external energy input to cool them; degradation of naturally cooling microclimates and generation of urban heat islands; blackouts from unmanageable strains on local power infrastructure; local environmental damage from increased power production; and rapid increases in GHG emissions from both residential and commercial construction sectors in developing countries. Wilhite's ethnographic approach, with its emphasis on a contextualisation of consumption in social relations and cultural practices, looks to the cultural drivers of change in such energy practices and provides a fresh look at barriers to and incentives for energy saving and consumption. For example, appliances not only consume electricity but also predispose householders to use them in certain ways within their specific socioeconomic and cultural contexts.

12.2.3 Shankleman (Chapter 7)

Discussion of the resource curse tends to focus on the socioeconomic effects of large resource endowments in weak states, rather than the direct environmental effects of resource extraction. Yet the effects of oil development on the local environment can be severe. The Social and Environmental Performance Standards set out in the Equator Principles for assessing, avoiding, minimising or compensating for the local effects of natural resource exploration and mining are an essential element of the London-Washington consensus on how to avoid or ameliorate the resource curse. Shankleman notes the difficulty of applying these standards under the prevailing political conditions in many oil-rich countries.

The Chinese approach largely addresses environmental issues indirectly by investing in national infrastructure development, which has increased access to clean water and other resources. More recently, in response to the Chinese government's new emphasis on the environment, some Chinese oil companies and banks are beginning to actively apply their own environmental standards, to consider embracing the Equator Principles, or both.

12.2.4 Muller (Chapter 8)

This study offers a detailed discussion of the economics of carbon trading; here we highlight only certain issues touching on the potential and actual effectiveness of such systems as a politico-economic answer to climate change and the implications for alternative policy measures.

Muller's comprehensive cataloguing of the state of the art and his methodical treatment demonstrate that, in approaching a social science discipline such as economics on its own terms, it is possible to identify its limits through critical analysis and reflection. This will enable (1) identification of weak points or contradictions within the discipline (e.g., the practical impossibility of gathering sufficient information to accurately monetise the costs of damage from climate change), and (2) the pinpointing of system boundaries, the points at which other disciplines must be invoked (to continue the example, problems of intrinsic value and justice in monetisation). This sort of reflective analysis is helpful for building interdisciplinary and transdisciplinary teams of researchers to investigate cross-cutting environmental issues like climate change.

But Muller also concedes that even when emissions reduction is the primary objective, the realisation of an ecologically effective emissions trading system, i.e., one that adequately mitigates global climate change, is not guaranteed even under favourable conditions of system design and implementation. The record of the largest and oldest international carbon trading system to date, the EU ETS, is not encouraging in this regard: It has produced fairly negligible emissions reductions from the baseline. Muller would attribute this mostly to design flaws involving the cap, coverage, allocation, power asymmetries and international harmonisation – flaws that proper attention to economics could remedy. Given the regular upward revision of assessments of the scope of emissions from developing countries and the acceleration of climate change already under way, ecological effectiveness will be critical in other carbon trading systems or any future phase of the EU ETS. Combining Muller's insights with Jørgensen's analysis of the factors and history behind the successful emergence of renewable energy technologies (Chapter 9) is therefore also necessary.

Experience with the early phase of the EU ETS has underscored the tensions between the environmental effectiveness and political feasibility of environmental regimes (also cf. recent US federal legislative experiences). Is it better to water down proposals (for example, limiting sectoral and/or national coverage, grandfathering or providing free allocation of permits to industry) in order to pass a bill establishing a basic permit-trading system that can be improved and strengthened at a later point, or does this pose too great a risk that the new regime and accompanying bureaucracy will just act as a cover for environmentally meaningless steps and engender a false sense of complacency?

12.2.5 Jørgensen (Chapter 9)

Wind and biomass are now well-established renewable technologies, although biomass in particular faces concerns over competitive land use and other environmental impacts. The scale of their potential contribution to overall reductions in CO_2 from the energy system is debatable.

In this chapter, Jørgensen uses a science and technology studies (STS) approach to describing technologies in sociotechnical regimes, asserting that 'basic qualities assigned to technologies are not intrinsic properties but rather depend on the network of relations to actors specifying these qualities and reflect the perspectives and context in which a technology operates' (pp. 168–169). Accordingly, there is no 'history' as such, but a variety of 'stories' (versions of history) that vary according to the perspectives of the actors involved. Such a constructivist view of technology systems clashes with the traditional technocratic 'environmental' view according to which components, effectiveness, yield factors and other system features can be definitively determined. Here again, the conclusion is that a successful energy system relies not just on hardware but also on software as in actor networks and public discussion of objectives. This conception of dynamic, multipolar sociotechnical systems has bearing on adaptation strategies that societies will need to develop to contend with environmental threats, particularly climate change.

12 Lessons from the Invited Contributions

Jørgensen asserts that the short-term and incremental focus of existing international climate programmes such as the Kyoto Protocol and the European ETS has 'helped to sustain existing power regimes and fossil-fuel-based energy technologies' (p. 184) and has worked against the swift conversion to sustainable energy systems. A strong emphasis on market mechanisms, especially CO_2 permit trading and European power market liberalisation, has also encouraged short-term optimisation by financial markets and large investors over the interests of local investors and entrepreneurs. In Denmark, as Jørgensen points out, these local actors were crucial to the successful emergence of wind and biomass.

The success of biogas in Denmark was ensured by presenting the technology not as part of an alternative energy vision but as a means of ameliorating water pollution from large-scale Danish pig farming – a multiple-benefit approach. In this way biogas is helping to 'sanitise' a part of the Danish food sector that is arguably unsustainable from a broader systems perspective, a useful role but one unworthy of its potential contribution to broader long-term energy sustainability.

12.2.6 Flüeler (Chapter 10)

In this chapter, Flüeler uses the hard lessons learned over decades in the contentious field of radioactive waste to reflect on current pilot implementation of carbon capture and storage (CCS) and its potential for large-scale deployment as part of a global CO_2-mitigation strategy. He finds quite a few parallel developments in the two sectors, such as technological constraints, path dependencies, and top-down and isolated planning. CCS engineering and associated international institutional development have a long way to go to match the state of knowledge, risk and safety concepts, institutional safeguards and the degree of public confidence and trust (inconsistent as it is) that have developed in the field of radioactive waste management. In many areas where the system characteristics of the two technologies converge, the more highly developed governance of nuclear waste management may serve as a good example for CCS, and the CCS community has much to learn from the nuclear waste field. For example, the single most important technical design element for CCS is system longevity; yet as Flüeler notes, it is telling that the term the CCS community uses is CO_2 storage, not disposal.

Well-designed rules and institutions are a prerequisite for an effective and expedient international CCS regime. The immature state of CCS technological development does not permit an adequate assessment of its long-term performance and consequently its environmental effectiveness in mitigating CO_2 emissions. CO_2 isolation cannot be its only performance indicator; environmental effectiveness, efficiency and timing must extend over the entire chain of CO_2 production and disposal. To promote system-wide assessments, the use of a global warming and environmental production efficiency ranking is recommended.

A main policy challenge is to ensure that the role of CCS is limited to a bridging function on the path to a renewables-based energy system, rather than perpetuating existing fossil fuel-based systems subject to political carbon constraints. This requires that CCS be embedded in a suitably diversified and far-reaching international carbon regime, via life cycle analysis or integrative technology assessment, for example.

12.2.7 Stauffacher et al. (Chapter 11)

Stauffacher and colleagues note that increased public and stakeholder involvement and participatory research have been called for in energy research and policy for many years. Disenchantment with and distrust of governmental decision making are on the rise in civil society, so the need for legitimacy and public trust in energy decisions would be well served by increased participation and improved participatory methods of decision making. To increase the legitimacy and regional embedding of a power plant, for example, local residents must have avenues to voice their interests, contribute their knowledge and express their risk valuations. Frameworks for an analytical approach to collaboration such as the one presented in this chapter can 'help bridge the gap between general participatory processes and complex decision-support systems' (Sheppard and Meitner, 2005, p. 184) that have evolved to manage the complexity of decisions in contemporary energy infrastructure planning.

But whether effective participatory processes are more favourable from an environmental perspective is uncertain even where science allows for a relatively clear environmental ranking of various options. Environmental considerations are just one of a mass of competing claims from citizen and consumer groups. Especially under stringent economic constraints, more often than not, short-term economic and political considerations trump longer-term energy and environmental goals – as some West European countries in the throes of recession are demonstrating in their retreat from support for alternative energy development.

12.3 Lessons for Challenge III: Economic and Social Development

12.3.1 Pachauri and Spreng (Chapter 5)

Pachauri and Spreng's multidisciplinary study examines how energy systems can constructively support economic and social development in developing countries, particularly among the poorest segments of the population. They highlight transitions to new energy carriers and arrangements in Indian households.

In developing countries the poorest very often barely eke out a living by their traditional means, and they have little extra time and human energy to explore new ways of coping. As suggested by the linguistic and historical analyses described in this chapter, both longstanding practices in household provisioning, including fuel and food gathering, as well as traditional, ingrained social relationships between the rich and poor, must be taken into account. Ignorance or disregard of these realities

12 Lessons from the Invited Contributions

in ill-conceived development programmes and misguided subsidies has wasted huge amounts of money and thousands of unused imported cooking stoves.

Economic and social development is occurring rapidly in many regions, with increased access to better education, food and improved infrastructure (including access to modern energy carriers) often going hand in hand. Here, indirect energy is as important as energy per se, as most other aspects of development require some energy inputs to be realised.

Sustainable development should not neglect heretofore hidden transitions among the populace, in particular the transformation of users of traditional, non-commercial energy sources into consumers of modern, centralised commercial energy, as well as the emergence of 'new consumers' joining the ranks of the formal market economy. Accounting for these transitions is also potentially important for improving global energy use prognoses and planning.

It is well recognised that indicators of economic performance need to be expanded beyond national economic welfare to include broader measures of human well-being (e.g., Stiglitz, Sen, & Fitoussi, 2009). Per capita energy use is a promising candidate for addition to the roster of indicators, but the focus should not be restricted to minimum levels for subsistence. Two new indicators of per capita energy use should be put into wider use: upper-limit thresholds imposed by aggregate global and national targets driven by climate exigencies; and the standard deviation of national and international energy consumption, which can be a useful supplement to conventional monetary measures of equity.

12.3.2 Wilhite (Chapter 6)

In this chapter Wilhite convincingly demonstrates how 'energy [use] is bound up in complex social and material interactions' in the Indian state of Kerala. He shows the 'grounding of [domestic] energy consumption in household practices' (p. 111): As these practices change significantly within the context of Kerala residents' changing relationships to work, gender, communications media and globalising capitalism, energy consumption is also rapidly evolving, and in an upward trajectory.

Wilhite casts serious doubt on the conventional policy assumption that promoting the sale of more highly energy-efficient electrical appliances will lead to a reduction in the use of electricity for a given practice such as household food preparation or air conditioning. As Kerala modernises its means of food preparation and storage, cooling residences and other practices against a backdrop of changing gender roles and globalising work practices and communication, aggregate energy use is greatly increasing *in tandem with* the rapid proliferation of electrical appliances; in this context, the efficiency of each individual appliance is relatively less important. Wilhite concludes that energy policy should shift to focusing on reducing the energy intensity of energy practice clusters such as home cooling, cooking, entertainment and transport.

As developing countries like India modernise their societies, family life, communications, entertainment and transport; as they replace traditional, locally-based, climate-aware building knowledge with typical Western standardised concrete construction methods that require the use of air conditioning; and as Indians who have worked abroad return with new tastes and expectations for products and services, some of them highly energy-intensive, their standard of living and various public welfare indicators improve. But generating the power and manufacturing and distributing the products required to meet these new demands create an entirely new constellation of problems, including energy access, power grid reliability and the regional and global ecological consequences of massive increases in coal and oil combustion.

The explication of these factors, which are so important as drivers of global energy demand but so little accounted for by conventional energy metrics, is a prerequisite for any attempts to shape them.

12.3.3 Shankleman (Chapter 7)

The 'curse of oil' is a manifestation of a deep dysfunction in economic and social development, according to Sen (1999), especially the repeated operation of two societal deficits that can retard or completely stymie development: a lack of broad economic and political participation and a lack of transparency in the political economy.

As Shankleman describes it, the Washington-London consensus championed by Western oil companies and international organisations has promoted spreading the oil wealth and reducing revenue volatility; increasing transparency and political accountability; and tightening environmental and social standards – with modest but limited success. Over the last few years, the emergence of the Chinese as a player in the global oil industry has 'forced a new perspective onto conventional thinking that redirects the focus away from governance and corporate responsibility issues and towards roads, railways, light, and power' (p. 129). Partly government-owned Chinese oil companies invest in the oil and minerals sectors of developing countries while making loans for large-scale infrastructure development, which implicitly addresses the shortcomings in infrastructure development of the Washington-London policy prescriptions.

In the short time since Shankleman wrote her article, the tectonic shifts in economic power from West to East due to the financial crises and subsequent great recession in many developed economies tend to reinforce her suggestion that the Chinese revenue management approach will continue to increase in importance. There are also signs that amid increasing environmental exigencies in both resource-exporting countries and China, the Chinese are beginning to embrace environmental and social elements of the consensus. As Shankleman describes, a 'development-centred' combined approach could avoid the pitfalls of each approach applied separately. But in each case, applying such an approach must take the specific economic, political, social and cultural context into consideration. Participation, for example, may not always be expressed in forms typical of European democratic

12 Lessons from the Invited Contributions

constitutional states. However, as suggested above, development on Chinese terms may come at significant cost to nations' long-term self-determination.

12.3.4 Muller (Chapter 8)

The prospects for new GHG emissions-trading systems have dimmed considerably since this book was first conceived. The 2009 Copenhagen climate conference failed to achieve international consensus on meaningful climate policy or concrete post-Kyoto global accords. The 2010 Cancún climate conference was more encouraging in terms of diplomatic process, but no successor to Kyoto emerged from it; instead, as this book goes to press, all of the six largest global GHG-emitting nations reject legally binding UN limits on their emissions now or when the Kyoto Protocol expires. In the United States, political and popular support for 'cap and trade' declined precipitously, and the matter was definitively put to rest in this political cycle by the Republicans' majority win in the House of Representatives in November 2010.

Section 12.2 discussed the paper's analysis of the potential environmental impact of emissions-trading systems. From this section's perspective of development, ethics and equity, Muller shows how problematic it is to include developing countries in an international emissions-trading system: power asymmetries, among other factors, can lead to highly unjust outcomes. Yet to stabilise greenhouse gas emissions and arrest climate change, full participation of developing countries in mitigating emissions will eventually be necessary by one means or another (see, for example, Clarke et al., 2009). The other principal market-based mechanism capable of setting a positive price for CO_2 emissions is a carbon tax. But a host of academic arguments can be raised against this too, apart from the potentially serious political obstacles it faces, depending on the country and context.

The chapter describes several other critiques of emissions-trading systems based on questions of equity and justice. Muller concludes that the only non-justice-based critique of emissions trading – the allegation that emissions trading is akin to the Catholic Church's medieval trade in indulgences – is valid only if one prioritises a non-environmental policy goal, here individual moral development, over emissions reduction.

The chapter considers only anthropocentric systems of ethics, and not, for example, biocentric or theocentric ethics. It also provides no indications how economic arguments would hold up against – or how academic ethical critiques would mesh with – ethics as reflected in contemporary religious practice, which is increasingly relevant given the demographic shifts under way in West European societies.

12.3.5 Jørgensen (Chapter 9)

Historical analysis of wind and biogas development in Denmark underscores the importance of bottom-up involvement of local actors for the success of innovations in renewable energy. The impetus for Danish wind turbine development did not come from either large players in the energy industry or the markets: big investors were too risk-averse. Small wind energy entrepreneurs and environmentalists – pursuing their vision of local, independent power generation – were the crucial early drivers. Similarly, technological progress did not primarily develop out of formal research, but rather was born of the experience and knowledge generated through the interaction of users, engineers and entrepreneurs.

In terms of policy lessons, extrapolating from Denmark's experience suggests that renewable energy policies that employ single stimulative measures and focus only on a few large economic actors may lock out key innovators and entrepreneurs, reduce needed learning and innovation processes and thereby unwittingly help to entrench existing power-generation regimes and fossil-fuel technologies (so-called lock-ins). The bottom-up involvement of smaller actors is likely to remain an important impetus for future energy innovation and system transition in some West European countries. This could help limit the dampening effect of the current contraction in public investment in renewable energies, especially the swing from large governmental stimulus to governmental austerity and deficit reduction programmes in the wake of the ongoing European sovereign debt crisis.

12.3.6 Flüeler (Chapter 10)

As described in Sections 12.1 and 12.2, this chapter derives lessons for the emerging sociotechnical framework for carbon capture and storage (CCS) from the institutional governance framework built up around the nuclear waste disposal industry. Many examples and parallels can be identified in safety, effectiveness, oversight, liability, regulatory compliance and enforcement, as well as risk communication, public trust and social acceptance. But one of the most pressing factors that may determine the success or failure of CCS is its economic feasibility. Flüeler shows that in order to be effective, CCS needs the right combination of technologies and implementation timing, energy carriers and international carbon regimes. Since CO_2 has limited economic value, the chances of successful development and widespread implementation of CCS are distinctly lower in the absence of an international regime committed to maintaining a positive carbon price over the long term. Yet even with robust, long-lived carbon markets, CCS may not be cost-competitive without massive government investments in the technology. The history of the nuclear industry provides some lessons here too – on government support, incentives for industry participation and learning curves – and they are not uniformly encouraging for CCS (cf. Rai, Victor, & Thurber, 2010).

12.3.7 Stauffacher et al. (Chapter 11)

Broad participation in the development and application of technology is a hallmark of societal development. In general, empowerment of the public progressively

12 Lessons from the Invited Contributions

increases with access to information and consultation, cooperation and collaborative forms of participation, in that order. This chapter characterises participation in energy systems and develops criteria and guidelines for 'successful' participation on the basis of a review of conceptual methodology and applied transdisciplinary experiments. It identifies actors and their rationales for participating; their timing, sequence and level of involvement; expected inputs; and research and practice in informal collaborative techniques and formal integration.

The chapter recommends the transdisciplinary case study (TdCS) as the preferred functional-dynamic structure and process to ensure broad participation in complex energy issues. The TdCS takes into consideration the 'intensity of participation, the issue at any specific step, the corresponding technique(s), and the function of participation in the entire decision-making process' (p. 240), and it provides clearer goals for the outcome and transparency (and legitimacy) of the participation process. The chapter sketches preliminary applications for nuclear waste repository siting in Switzerland. With appropriate adaptations of the framework, TdCS presumably would be useful for public involvement in CCS development, planning and siting.

Does the method described in this chapter adequately answer the key methodological challenge of knowledge management and integration identified in Chapter 3? From a classical transdisciplinary perspective, it makes progress, especially in increasing the transparency of scientific input, integrating a richer societal discourse and the robustness of results. However, since this type of study is not focused on integration of academic and specialised knowledge, it does not address all the needs of Max-Neef transdisciplinarity (see Chapter 13). The necessity of adapting the approach to each specific case requires developing the methodology towards even more integration.

References

Clarke, L., Edmonds, J., Krey, V., Richels, R., Rose, S., & Tavoni, M. (2009). International climate policy architectures: Overview of the EMF 22 international scenarios. *Energy Economics, 31,* supplement 2, S64–S81.

Commoner, B. (1971). *The closing circle: Nature, man, and technology.* New York: Knopf.

Daniel, S., & Mittal, A. (2009). *The Great Land Grab. Rush for world's farmland threatens food security for the poor.* Oakland, CA: The Oakland Institute.

Kemmler, A., & Spreng, D. (2007). Energy indicators for tracking sustainability in developing countries. *Energy Policy, 35,* 2466–2480.

Rai, V., Victor, D. G., & Thurber, M. C. (2010). Carbon capture and storage at scale: Lessons from the growth of analogous energy technologies. *Energy Policy, 38,* 4089–4098.

Sen, A. (1999). *Development as freedom.* New York: Alfred A. Knopf.

Sheppard, S. R. J., & Meitner, M. (2005). Using multi-criteria analysis and visualisation for sustainable forest management planning with stakeholder groups. *Forest Ecology and Management, 207,* 171–187.

Stiglitz, J., Sen, A., & Fitoussi, J. (2009). *Report by the commission on the measurement of economic performance and social progress.* Paris: Commission on the Measurement of Economic Performance and Social Progress.

Chapter 13
Synthesis: Research Perspectives

David L. Goldblatt, Daniel Spreng, Thomas Flüeler, and Jürg Minsch

Abstract This synthesis chapter of Part II characterises the various research perspectives represented in the studies in Chapters 5–11 and arranges them in a graphical adaptation of the transdisciplinary framework first introduced in Chapter 3. The studies occupy various positions and ranges along the axes of the perspectives cube depending on the number of disciplines involved, the type and scope of participating actors and the nature of the research question. Energy policy implications of multiple disciplinary perspectives are suggested.

Keywords Collaboration · Energy research · Research perspectives · Stakeholder participation · Transdisciplinarity

13.1 'Perspectives Cube' for Characterising Research Approaches

There are many ways of characterising the diverse perspectives represented in the varied scientific studies presented in Chapters 5–11.[1] These include basic vs. applied research, analytic vs. synthetic approach, degree of self-reflection, degree of disciplinary collaboration, degree of practitioner involvement, and the level of the transdisciplinary approach (see, in particular, the weak Max-Neef transdisciplinarity pyramid, Fig. 3.2, in Chapter 3). Treatment of criteria relating to the quality of the scientific research or the studies' practical or policy usefulness is reserved for the discussion in Chapter 14.

Figure 13.1 below presents one way of capturing three of these parameters in a single graph with the following axes:

---

[1] Again, since Chapter 4 does not directly address an energy-related challenge, it is not included in the following analysis.

D.L. Goldblatt (✉)
Department of Management, Technology, and Economics (D-MTEC), Centre for Energy Policy and Economics (CEPE), ETH Zurich, 8032 Zurich, Switzerland
e-mail: dgoldblatt@mtec.ethz.ch

D. Spreng et al. (eds.), *Tackling Long-Term Global Energy Problems*,
Environment & Policy 52, DOI 10.1007/978-94-007-2333-7_13,
© Springer Science+Business Media B.V. 2012

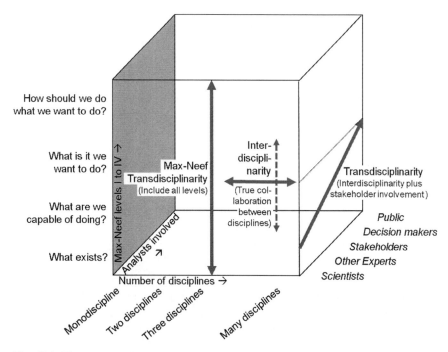

Fig. 13.1 The 'perspectives cube' – Proposed scheme for characterising social-science energy research approaches

- x: Number of disciplines involved
- y: Type and scope of actors involved in the scientific endeavour
- z: Nature of the research question(s) explored.

For the sake of discussion, this chapter makes the simplifying assumption that these three dimensions are not strongly dependent on one another.

13.1.1 X-Axis: Number of Disciplines Involved

The number of disciplines involved in the research is a relatively straightforward piece of information, notwithstanding potential overlap, fuzzy disciplinary borders, controversial definitions,[2] the existence of subdisciplines and emerging disciplines.[3] Where multiple disciplines are involved, they may operate analytically side by side (sometimes called multidisciplinarity) or scientists may collaborate in an interdisciplinary or transdisciplinary fashion, which gives the study a synthetic character.

---

[2] It is arguable, for instance, whether regional planning is its own scientific discipline.

[3] For example, industrial ecology.

13.1.2 Y-Axis: Type and Scope of Actors Involved in the Scientific Endeavour

The axis that identifies the researchers and actors participating in the study is also relatively unambiguous. This is an important aspect of transdisciplinarity (TD) in the conventional sense; transdisciplinary research is not only interdisciplinary but also involves the collaboration of stakeholders from the 'real world' (cf. Section 3.7 in Chapter 3). The form this collaboration takes is less straightforward. Only in rare cases will the stakeholders be willing to act as authors of the reports, which means it requires a special effort to include them authentically.

13.1.3 Z-Axis: Nature of the Research Question(s) Explored – Normativeness and Implementation

The z-axis depicts the nature of the research questions addressed, as captured by the questions in Max-Neef's weak transdisciplinarity pyramid (cf. Max-Neef, 2005 and Fig. 3.2 in Section 3.7, Chapter 3). Research projects encompassing all four levels represent the essence of Max-Neef-transdisciplinarity, the importance of which was underscored in Chapter 3. The social sciences have a special role to play when the research questions extend beyond both the archetypal question of natural and physical scientists – 'what exists?' – and the engineers' question, which is 'what are we capable of doing?' Does a particular engineer base his answer to the question on specialised knowledge or does he apply a broad systems knowledge encompassing a fuller range of sciences? Furthermore, the engineers' question also depends on knowledge of technical and institutional feasibility and means of implementation. The third and fourth levels are described by Max-Neef's questions 'what is it we want to do?' and 'how should we do what we want to do?' (for example, how fast, according to which priority, for whose benefit, that is, how equitably). These questions relate to societal goals and ethics and are therefore areas in which the social sciences and the humanities are at the fore.

Together, the three axes characterise the research perspectives of the studies in Part II better than when they are viewed individually. Imagine that each axis ranges in value from 0 to 1, so that the point in the exact centre of the cube has the coordinate $(x, y, z) = (0.5, 0.5, 0.5)$. A project involving the participation of scientists from multiple disciplines would be located on the plane 1, y, z. In its usual sense, transdisciplinarity occurs if, in addition, $y = 1$. However, conventional transdisciplinarity also usually implies that the collaborative effort goes beyond the questions of 'what exists?' and 'what are we capable of doing?' to include 'what is it we want to do?' Here, therefore, z also tends towards at least 0.75. Conventional TD is situated along the upper vertical line segment on the back right edge of the cube.

As noted, the cube does not explicitly depict the character of the collaboration in cases where more than one discipline is involved, that is, when the work is multidisciplinary, interdisciplinary or transdisciplinary. True collaboration may be more likely above the plane z = 0 (illustrated by the dashed arrow in Fig. 13.1). On the other hand, Max-Neff transdisciplinarity, by definition, takes place in research that is concerned with all four questions on the z-axis.

13.2 Positioning of the Chapter Studies in the Research-Perspectives Cube

This section discusses the perspectives of the seven studies, Chapters 5–11, and, where appropriate and feasible, places them in their proper position in the cube (compare the following discussions with Fig. 13.2 below).

- Chapter 5: In their contribution entitled *Towards an integrative framework for energy transitions of households in developing countries*, Pachauri and Spreng attempt to include multiple disciplines in engineering, social science and even the humanities. Their inclusion of humanities, such as ethics and linguistics, reflects a conscious effort to touch on issues at the upper transdisciplinary normative levels on the z-axis. The study includes statistical analysis of large consumption survey data sets, results from microeconomics and analysis of programmes and initiatives aimed at improving households' access to quality energy in developing countries, as well as historical and linguistic analysis of the nature of poverty. While this small-scale study deliberately takes a broad, multidisciplinary approach, it falls short of being a true transdisciplinary endeavour, or even an

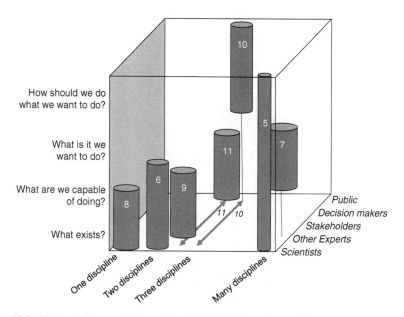

Fig. 13.2 Studies in Chapters 5–11 positioned in the 'perspectives cube'

interdisciplinary one. The disciplines are treated sequentially rather than in an integrated manner and involved only two scientists working together, rather than an intensively collaborating team. The authors do not offer direct policy advice.

- Chapter 6: At both the macro- and micro-level, analysis of energy consumption is usually conducted from an economic or engineering perspective. In *A socio-cultural analysis of changing household electricity consumption in India*, Wilhite uses anthropology to take an unusual and fresh look at factors that influence energy consumption. Although analytical, the study is intended to inspire policy makers to consider expanding their repertoire of measures beyond, for example, appliance standards, infrastructure improvements, and economic instruments, to influence energy consumption by shaping factors that impact people's lifestyle choices.
- Chapter 7: Shankleman's *The changing context for efforts to avoid the 'curse of oil'* is both descriptive and prescriptive. It is a classic piece of policy advice, based not so much on a particular scientific discipline as inspired by Shankleman's personal observations and wide-ranging experience as a consultant. The analysis grows from careful observation of past events and efforts to counteract the resource curse in developing countries with large energy resource endowments. The debate over the extent of the resource curse as an economic phenomenon (see the introduction to Chapter 7) is of limited relevance to Shankleman's practitioner-oriented perspective.
- Chapter 8: In some sense, the fourth study takes the opposite perspective. Muller explores the interface between *Economics- and ethics-based assessments of emissions trading* and thereby reflects on the potentials, limitations and boundaries of theory-based perspectives. Muller's comprehensive cataloguing of the state of the art and his methodical treatment demonstrate that by approaching a social science discipline such as economics on its own terms, it is possible to identify its limits through critical analysis and reflection, specifically: (1) to identify weak points or contradictions within the discipline (such as the practical impossibility of gathering sufficient information to accurately monetise the costs of damage from climate change) and (2) to pinpoint the system boundaries, the points at which other disciplines such as ethics must be invoked. This sort of reflective analysis is helpful for building interdisciplinary and transdisciplinary teams of researchers to investigate cross-cutting environmental issues such as climate change. Muller's study also points to the need for a theory-based, partially empirically validated discipline like economics to be conscious of its own limitations when offering policy advice on complex questions.
- Chapter 9: In *No smooth pathway to sustainable energy systems – Politics, materiality, and visions for wind turbine and biogas technology*, Jørgensen presents a techno-institutional analysis of the development of energy systems. He demonstrates that without accounting for institutional factors, it is not possible to offer a satisfactory explanation of the evolution of renewable energy in Denmark or to offer credible predictions of future developments. This underscores the policy importance of integrated technological and institutional development.

Jørgensen's focus on the relationship between energy technologies and their institutional/regulatory environment has been the subject of extensive research and writing in recent years, particularly in the Netherlands in the field of 'strategic niche management' (compare with Section 12.2). The view constitutes a counterpoint to the general policy prescriptions of engineering economics and simplified microeconomics, which tend to single-mindedly stress the virtue of competition in energy markets.

- Chapter 10: In *Technical fixes under surveillance – CCS and lessons learned from the governance of long-term radioactive waste management*, Flüeler compares and contrasts management approaches for CO_2 and radioactive waste, with particular focus on the interface between technology, politics and business. The author makes a strong argument that the social sciences should not let themselves be instrumentalised to further public acceptance without actively helping to shape the technological regime. He shows how the field of radioactive waste management – marked for years by top-down technology dictates, with social science playing the traditional acceptance role – has been transformed by both political pressure and institutional learning in the radioactive waste community. As an emerging technology with an uncertain future, CCS is only just starting along this developmental management path.

  The contributions in Chapters 10 and 11 discuss the role of researchers and other experts, stakeholders and the public, and they more or less systematically incorporate the perspectives of the full range of these actors. Figure 13.2 depicts this through arrows on the xy-plane that extend across most of the y-axis. However, only the scientists themselves participated in authoring the final articles.

- Chapter 11: In *Functional-dynamic approach to collaboration among different actors: learning from the transdisciplinary case study approach*, Stauffacher et al. contend that 'the public' are not a monolithic block but change with the context and over time and that the trendy research notion of 'public involvement' must be adjusted accordingly. The authors dispute the simplistic notion that the more people involved and the deeper the involvement the better. The authors represent various social sciences and are experts on transdisciplinary case studies (TdCS), having refined their methodology over the course of multiple applications in numerous settings. They regularly include classes of 40–50 students of environmental science in their case studies. Here – as opposed to Max-Neef's use of the term – transdisciplinarity is meant in the more conventional sense of involving a fuller range of actors and stakeholders.

13.3 Preliminary Policy Implications

Science has been likened to a bright spotlight that illuminates a small object in a large dark room; the brighter the spotlight, the darker the backdrop of the object studied.[4]

---

[4] Oral communication by Markus Fierz, physics lecturer at ETH Zurich, in the early 1960s.

The review of the various perspectives of these studies shows how enriching it can be to view and research a problem from various angles. In some cases, multiple perspectives lead to a deeper and clearer picture that can lend greater confidence and certainty to derivative policy advice. Yet multiple perspectives often at least seem to be mutually contradictory. Several possibilities present themselves here.

The first variant is that one perspective dominates the issue, because this one perspective captures the essence of the issue much better than any other perspective. It also has more support within science and leads to findings that translate more readily into policy advice than other findings or that translate into policy advice that is more acceptable to established political circles.

The second variant is that there are two or more perspectives on the issue at hand resulting in findings that lead to contradictory policy advice. This may give rise to various schools of thought and competing research circles. Alternatively, the conflicting results may stimulate the development of further scientific perspectives and of a deeper understanding in order to better capture the issue and resolve the contradictions.

The third variant is that there are two or more perspectives on the issue resulting in findings that lead to better and more subtle policy advice. In this case, the advice may be accepted or rejected. If it is rejected, it is important for scientists to communicate better in order to ensure that the relevant decision makers understand the bases of the advice and the opportunities it presents.

This is not a plea to research all energy issues from multiple disciplinary perspectives. However, policy advice based directly on research with a single perspective should be taken with a grain of salt. Science has more to offer than one-dimensional policy prescriptions.[5]

Reference

Max-Neef, M. A. (2005). Foundations of transdisciplinarity. *Ecological Economics, 53*, 5–16.

---

[5] This notion is explored further in Chapter 14.

Part III

Chapter 14
Lessons for Problem-Solving Energy Research in the Social Sciences

Jürg Minsch, Thomas Flüeler, David L. Goldblatt, and Daniel Spreng

Abstract The prominence of societal, economic and institutional aspects of contemporary global energy challenges makes energy research in the social sciences indispensable. Intensified collaboration on an equal footing among the natural, technical and social sciences is necessary. This concluding chapter offers a synthesis of the editors' insights from the invited contributions in Part II, the overall ASRELEO project as well as recent findings from the social science literature. Research themes highlighted include reflexivity and futurity; energy, poverty and inequality; critical infrastructure; governance; institutions and regimes; markets; systemic analytical approaches; policy strategies, forecasting and path dependencies; change, social learning and scientific communities; value systems, cultures and actors; social acceptability of energy technologies; and risk and energy communication.

Keywords Cross-disciplinary cooperation · Futurity · Governance · Institutions · Policy strategies · Reflexivity · Social science energy research · Social learning

> *[The planner's] would-be solutions are confounded by a still further set of dilemmas posed by the growing pluralism of the contemporary publics, whose valuations of his proposals are judged against an array of different and contradicting scales.*
>   *It should be clear that the expert is also the player in a political game, seeking to promote his private vision of goodness over others.*
>
> *Rittel and Webber (1973, pp. 167, 169).*

This concluding chapter offers a synthesis of the editors' insights drawn from both the invited contributions in Part II and the entire ASRELEO project[1]. In this chapter,

---

[1] ASRELEO: Agenda for Social-Science Research on Long-term Energy Options, the research project (2005–2007) out of which this book was developed (see Chapter 1).

J. Minsch (✉)
minsch sustainability affairs, 8057 Zurich, Switzerland
e-mail: juerg.minsch@bluewin.ch

D. Spreng et al. (eds.), *Tackling Long-Term Global Energy Problems*, Environment & Policy 52, DOI 10.1007/978-94-007-2333-7_14, © Springer Science+Business Media B.V. 2012

we compare our conclusions with findings from the recent social science literature[2] and propose areas for energy research in the social sciences.

The point of departure for this book (ASRELEO) was the assertion that the role of the social sciences in the field of energy should not be reduced to merely assisting and serving technology or special economic and political interests (e.g., supporting acceptance or helping in the introduction of new energy-related technologies on the market). This holds true when an effort is made to seriously confront global energy-related issues and challenges (Chapter 2), including

- energy access and security;
- climate change and other environmental impacts; and
- economic and social development.

This list is not comprehensive; there are other legitimate issues as well. But these were chosen as examples of the tasks facing science and society if they are to undertake forward-looking transformations in the energy sector. In these transformations, human rights must be taken into account as seriously as ecological and global economic issues (see, for example, EC, 2009; AGECC, UN Advisory Group on Energy and Climate Change, 2010; IEA, UNDP, & UNIDO, 2010; WBGU, 2011a, 2011b; UNDP & GEF, 2011).

Even a cursory glance at each of these issues makes it clear that durable solutions require including the gamut of social sciences in the process of formulating and implementing them. A second aspect is also immediately clear: All of these issues are interconnected. In addition to the concrete issues themselves, in a narrow sense, two other levels come into play, each involving at least one additional challenge:

At the social-institutional level an important question is how well societies are currently configured to deal constructively with these issues. The transformation of energy systems is, in this respect, an important question for the social sciences (provided that these transformations are conducted within the framework of human rights, democracy and the rule of law). Following this thread, Chapter 3 explores the concept of the learning society. It describes institutional progress in democracy, civil society and market economies as influenced by the principles of reflexivity, self-organisation, conflict prevention, creativity and innovation as well as constructive approaches to resource scarcities. Chapter 3 identifies the following central functions of socioscientific energy research:

- to contribute to 'reflection and target knowledge';
- to contribute to 'the analysis of social transformation processes';
- to contribute to 'transformation knowledge' (the study of the formation of transformation processes).

---

[2] A number of these sections greatly benefited from contributions by Benjamin K. Sovacool (this volume, Chapter 4).

14 Lessons for Problem-Solving Energy Research in the Social Sciences

At the scientific-institutional level, there is an additional need: Considering the complexity of the issues, how can one move from simply managing knowledge to actually integrating it? Chapter 3 describes the necessity of inter- and transdisciplinary collaboration and suggests some basic means of achieving it. It is clear that this kind of collaboration demands special abilities of researchers (see Box 14.1, de Haan et al., 2008; Rychen and Salganik, 2003; Parkin, Johnston, Buckland, Brookes, & White, 2004), and thus inter- and transdisciplinary methods need to be systematically incorporated in the training of junior research staff. At the same time, inter- and transdisciplinary research needs institutional and financial security within the scientific system.

Box 14.1 Competencies for Sustainable Development

Interactive use of media and resources

- Adopting perspectives: broad-based, new perspectives to support the generation of integrated knowledge
- Anticipation: ability to analyse and assess forward-looking developments
- Cross-disciplinary approach: generating and using interdisciplinary insights
- Dealing with incomplete or overly complex information: ability to recognise and weigh risks, dangers and uncertainties

Interaction in heterogeneous groups

- Cooperation: ability to plan and implement actions with others
- Coping with individual decision-making dilemmas: ability to deal with conflicting objectives by reflecting on strategies for action
- Participation: ability to take part in collective decision-making processes
- Motivation: ability to motivate oneself and others to become active

Independent activities

- Reflection on general principles: ability to reflect on one's own and others' general principles
- Moral actions: ability to incorporate ethical principles such as justice in decisions and actions
- Independent actions: ability to plan and act independently
- Supporting others: ability to show empathy for others

(Source: adapted from de Haan et al., 2008, p. 188 and appendix)

The 2011 flagship report of the German Advisory Council on Global Change (WBGU) *World in Transition – A Social Contract for Sustainability* (WBGU, 2011a, 2011b) demonstrates how far the social sciences are from an adequate level of influence in the energy field. The meagre analysis of social transformation processes (described by the WBGU as transformation research and education) and the elaboration of transformation knowledge (transformative research and education) are pinpointed as the main shortcomings. The report places three social transformation fields at the forefront of sustainability: energy, urbanisation and land use. It almost goes without saying that urbanisation and land use are closely connected to transformations in the field of energy.

The stagnation of energy research in the social sciences can be illustrated by the issue of climate change. As early as 1998, Rayner and Malone's state-of-the-art compendium of social sciences relevant to climate change negatively assessed the analytical and policy tools for climate change available at that time. These tools were 'incapable of providing a reliable basis for rational goal setting and policy implementation' largely because of an over-reliance on the narrow rational actor paradigm (Rayner and Malone, 1998a, p. xxxix). Nevertheless, their suggestion that policymakers 'employ the full range of analytic perspectives and decision aids from the natural and social sciences and the humanities' (Rayner and Malone, 1998c) has not been generally adopted.

Similarly, several other recent reviews lament the relatively modest contributions of the social sciences to climate change research, as well as their undistinguished record of integration and collaboration on the issue with the hard sciences; see, for example, Lohmann (2010) and Härtel and Pearman (2010), who advocate a 'whole-of-science' agenda for research on climate change. These opinions echo the oft-cited critical article by Wilhite and colleagues entitled 'The Legacy of Twenty Years of Energy Demand Management: We Know More about Individual Behaviour but Next to Nothing about Demand' (Wilhite, Shove, Lutzenhiser, & Kempton, 2000). Ten years later, the story is still very much the same: Global aggregate energy demand and CO_2 emissions are higher than ever and are on a skyrocketing trajectory. At the same time, general research on energy demand remains predominantly focused on efficiency, and research in the social sciences is mostly concerned with end-user behaviour.

In many respects, the climate problem is an important aspect of the much broader issue of energy. Energy, broadly construed, is the subject of the ASRELEO project as well as of this volume. As described in Chapter 1, ASRELEO was initiated in response to a request by the European Fusion Development Agreement (EFDA), to design a research programme for the social sciences that would complement the high level of funding supporting energy research in the natural sciences and technology. The agency also realised that their social science research programme had been too heavily focused on short-term issues and wanted a compensatory long-term emphasis. Canvassing the broader research landscape, we found that this was a generic problem. Although huge institutional and societal problems are intricately involved in most energy issues, offering a large potential for rich and fascinating research in the social sciences (Anderies et al., 2004), these sciences had not played their part in shaping emerging sociotechnical energy systems.

14 Lessons for Problem-Solving Energy Research in the Social Sciences   277

The ASRELEO project included a series of workshops with invited presentations followed by discussions. The insights that were gained over the three-year process were worked into the final versions of a number of contributions to this volume (Chapter 4). The discussion rounds, consciously organised as both reflective and prospective, brought additional insights and ideas that were incorporated separately into the final ASRELEO report (Flüeler, Goldblatt, Minsch, & Spreng, 2007, pp. 53). These are the first two main sources informing this chapter's organisation of themes. In addition, the summary of issues covered in social science energy research as described in the ASRELEO final report (Flüeler et al., 2007, 26pp.)[3] as well as findings from recent literature and activities in the social sciences[4] have also been used.

In the following, the research fields we identified as being crucially involved in social science energy research are listed and briefly described. A broad range of subjects are presented. The only deliberate systematic ordering imposed is a progression from general and institutional topics to more specific ones. Indeed, based on the volume's lines of argument, other arrangements may be just as valid, such as, for example, basing the presentation of issues on the type of knowledge involved, the chosen research design (cf. 'perspectives cube', Chapter 13) or the energy-related challenge. The participants at our workshops initially favoured the last form of systematising. However, there it quickly became clear that because of the interdependence of the challenges, such an organisation might well have led to confusion, suggesting system limits that do not exist.

Comparable reports have applied explicit selection criteria when recommending directions that might be taken in further research, including the potential for advancing the relevant disciplines, the impact on policy making, environmental effectiveness or the likelihood that the recommendation would be put into practice.[5] The ASRELEO project investigated the status of R&D (Flüeler et al., 2007, 26pp.) and found that the suitability of selection criteria largely depends on the (national or local) research setting and policy framework. Selection criteria often resist general applicability, which is why we have not used such criteria when choosing the research questions in this chapter.

For these reasons, we decided to present the following in the form of a broad range of research themes and fields:

Perspective and embedding

- Reflexivity and futurity
- Energy, poverty and (in-)equality

---

[3] To update our knowledge and to obtain a richer picture, we asked the participants of the ASRELEO project to share their knowledge of recent research (in summer 2010).

[4] Benjamin K. Sovacool and colleagues (this volume, Chapter 4).

[5] E.g., Diamond and Moezzi (2002), Ekins (2003), Brewer and Stern (2005), Jackson (2006), Webler and Tuler (2010); also related efforts, such as Janda (2009), Driessen, Leroy, and Van Vierssen (2010), Research Council of Norway (2010), Chabay et al. (2011), Berlinbrandenburgische Akademie der Wissenschaften et al. (2011), Helmholtz Gemeinschaft (2011).

'Hardware'

- Critical infrastructure

'Software' (Institutions)

- Governance
- Institutions and regimes
- Markets

Methodology (for analysis and transformation)

- Systemic approaches to analysis
- The long term: policy strategies, forecasting and path dependencies

Dynamic interaction of technology and society

- Change, social learning and scientific communities
- 'Software': Value systems, cultures and actors
- Social acceptability of energy technologies
- Risk and energy communication studies

> Boxes offer non-exclusive illustrations of how we conceive research questions in the respective research areas. The reader is invited to reflect upon the formulated questions and to add some of his or her own.

14.1 Research Themes and Fields

14.1.1 Reflexivity and Futurity

Although ethics and philosophy are branches of the humanities rather than social sciences, ethical and philosophical issues are necessarily interwoven into many of the socioscientific themes that this book has explored. (Muller's study in Chapter 8 explicitly sets out to explore this disciplinary interface.) In this respect, Georg Christoph Lichtenberg's aphorism 'Anyone who only understands chemistry doesn't even understand that correctly' is generally valid for the sciences and scientists (Lichtenberg, 1789).

As emphasised particularly in Chapter 3, ethical and philosophical questions occupy the highest transdisciplinary level of enquiry and thus in transdisciplinary research they must be pursued concurrently with socioscientific investigation in (Max-Neef) transdisciplinary research. Questions about good life, development,

14 Lessons for Problem-Solving Energy Research in the Social Sciences

equity, futurity and distribution are predominately ethical, ontological and epistemological in nature. More than 30 years ago, Caldwell noted the intrinsically social nature of energy extraction, processing and use, remarking that 'if there is a comprehensive energy problem, it is a problem of choice and value in a world of finite capabilities. It is therefore also a moral and political problem, and for this reason will not yield to a purely technical solution' (Caldwell, 1976, p. 32). Many of the central questions regarding energy use – including how to distribute energy resources, how to allocate government funds, whom to burden with environmental pollution, how (much) to discount future impacts, how safe is safe enough – are philosophical and ethical (Davison, 2001; Dobson, 1999; De-Shalit, 1995; Fischhoff, Slovic, Lichtenstein, Read, & Combs, 1978; Möller, Hansson, & Peterson, 2006).

Two promising areas in modern philosophy that appear most applicable to energy are reflexivity and futurity. The philosophical postulate of *reflexivity*, growing out of the philosophy of science, calls on analysts and policymakers to be more aware of their own hidden values (see Bloor, 1991; Chubin and Restivo, 1983; Woolgar, 1988 for introductions to this area of enquiry). This also requires revealing and reflecting upon those phenomena and questions that are ignored – and thus made invisible – due to the habits of research or politics, or due to economic, political and scholarly interests (cf. Gyges and the ring of invisibility, Chapter 1). Understanding reflexivity in this way is, in our opinion, an essential part of 'good science' and cannot be delegated to philosophy or ethics. It is clear that values and interests enter into reflexivity, that reflexivity explores analytical possibilities and limits, recognises the phenomenon of concealment and attempts to take this concealment as an object of enquiry in itself. Reflective science takes 'the whole' into consideration; it is necessarily integrative. This is true to an exceptional degree for energy research. The discussions in this volume (especially in Chapters 2 and 4–11) show that the energy-related challenges can only be successfully met if they are approached in an integrated manner.

These challenges also made it clear that today a central aspect of reflexivity is the ability to plan for the future. The notion of *futurity* asserts that truly prudent energy research and thus strategies for transforming energy systems must take future generations as well as environmental protection adequately into account (Maclean, 1980).[6] Achieving futurity points to energy choices that borrow from nature (by using renewable resources or energy efficiency) rather than those that steal from nature or borrow from the future (by depleting natural resources, locking in options that are highly polluting or presenting high hazard potentials).

In our view it is not an exaggeration to suggest that humanity is being confronted with the challenge of another 'dual' (political-technical) revolution[7] or,

---

[6] The most common translation of the English term 'sustainable development' in Germany is 'Zukunftsfähigkeit' (futurity) (BUND & Misereor, 1996).

[7] The combination of the (political) French Revolution of 1789 and the (technical) English Industrial Revolution was the last 'dual revolution', called by Hobsbawm 'probably the most important event in world history' (Hobsbawm, 1962, p. 62). The German film director C. A. Fechner titled his 2010 movie *The 4th Revolution* – following the agrarian, industrial and digital revolutions comes the energy revolution. (http://www.energyautonomy.org > English, all web links accessed November 16, 2011).

rather, transformation; to steer the globe onto a socially and environmentally sustainable path – with energy (and communication) technologies as the 'industrial' dimension of this revolution.

> What assumptions concerning risk, modernity, culture, time, materialism, human rights, freedom, equity, development, cost and end-use are associated with particular forms of energy research and technologies?
>
> Which corresponding assumptions are the dominant energy policies (national, international) based on? What kinds of energy research do they promote and support?
>
> Which concrete phenomena and questions are ignored or come up short in energy research or policies? Why? For example: the idea of the good life and prosperity based on something other than cheap energy and economic growth.
>
> How do particular energy systems harm the environment, degrade the social structure of local communities or damage traditional cultures?
>
> How do practices and mindsets connected to particular technologies impact the well being of future generations? How should the costs and benefits of energy production and use be distributed?
>
> What kinds of related interests (money, power, reputation) vie for influence within energy research, energy economics and policy? Are these interests transparent?
>
> What is the significance of reflexivity and futurity in the energy debate (politics, sciences, economics)? Do these notions have a systematic (institutionally and financially secure) place in this debate?

14.1.2 Energy, Poverty and (In-)Equality

Energy is not the main driver for development but a necessary input. This is clear from the long-standing practice of exploiting 'inexpensive natural resources', which has focussed on supplying energy as cheaply as possible (cf. the mercantilism syndrome described in Chapter 3). It is thus somewhat hard to understand why energy was not explicitly named as one of the UN Millennium Development Goals (MDG). Later, it was noted that 'A common finding of the ten Task Forces of the UN Millennium Project has been the urgent need to improve access to energy services as essential inputs for meeting each MDG. Without increased investment in the energy sector, the MDGs will not be achieved in the poorest countries' (Sachs in Modi, McDade, Lallement, & Saghir, 2005, p. iii).[8] At the presentation of the IEA World Energy Outlook 2010, Fatih Birol of the International Energy Agency said that 'Meeting the goal [of eradicating extreme poverty by 2015] would not only end extreme poverty but decrease political instability, link the world socially

---

[8] http://www.undp.org/energy

14 Lessons for Problem-Solving Energy Research in the Social Sciences

through mediums such as the Internet, and boost gender equality' (Christian Science Monitor, 21 Sep 2010).

Recently, the definition of energy security has been broadened to encompass wider sociotechnical criteria. While we have treated energy access and security as distinct but related concepts (Chapter 2), for some analysts 'energy security' is an umbrella term for a still broader group of challenges, including availability of energy, affordability, economic efficiency and environmental stewardship (cf. Sovacool & Brown, 2010). From an analytical perspective, the precise content is inconsequential as long as the concept of energy security incorporates complexity and multiple interrelationships. For example, energy efficiency may reduce dependence on cheap imported oil since higher-priced sources are more affordable if energy is used efficiently. But this fact is rarely appreciated because energy efficiency is traditionally the concern of a policy field far removed from energy security. By dealing with efficiency together with availability, Sovacool and colleagues (ibid.) break down traditional conceptual and analytical boundaries.

A much less commonly explored problem is energy supply security for consumers and single households in developing countries. Aspiring electric grid customers in poor areas of many developing countries have two problems to contend with: physical access, i.e., connection to the grid or to a local source of electricity, and duration and reliability of supply. In rural India, for example, households commonly have no more than two hours of electricity per day. This is often adequate for irrigation purposes and thus, no more is necessary to ensure the food supply for the wider region. Other uses of electricity are often not considered essential enough to be provided for. Here, aspects of supply security, physical access and duration of supply are closely related to both economic and political power.

Amartya Sen's research on poverty is fundamental to this issue. According to Sen (1993), poor populations experience poverty not only in their lack of financial resources, knowledge or goods and services, but also in their slim prospects of bettering themselves and, in particular, of shaping their personal and communal destinies. Cast in terms of energy, this implies that energy poverty should be identified not only by a level of energy consumption below that which is necessary to meet basic needs, but also by the absence of opportunity to choose one's consumption level, particularly if access to energy is lacking and unlikely to be available in the future.

A basic energy need in developing countries is energy for cooking. A forthcoming special issue of *Energy Policy* on 'Clean Cooking Fuels and Technologies in Developing Economies' presents research on programmes and initiatives to promote cleaner cooking fuels and stoves in poor areas of developing countries. Nearly one-third of humanity still lacks access to clean-burning cooking fuels and stoves (UN AGECC, 2010). Knowledge of how to remedy this in specific locales is shockingly sparse, and effective programmes are few.

A final poverty-related dimension of energy production and use is gender (IEA et al., 2010, p. 15). The socially defined gender roles of women affect their access to opportunities for income generation and to technology, which presumably influences systems of energy production and use. It is women and children who spend hours each day collecting fuelwood and dung around the world. And it is women who spend even more hours cooking meals and thereby exposing themselves to hazardous pollutants.

Energy is clearly an important instrument for development. However, the politics of cheap energy, which is widely accepted across the world, follows a development ideal that is largely focussed on economic growth. It seems that we can only imagine prosperity and peace by means of cheap energy and economic growth. Yet the impact of climate change on energy systems and the increasing global energy demand, with its associated ecological, economic and political side effects, pose great challenges to the process of peaceful adaptation and transformation.

The reorganisation of world energy systems on the basis of renewable energy also creates losers: products, technologies and branches that are not fit for the future, as well as entire regions that are dependent upon the same. This also includes the owners of reserves of fossil fuels and nations that export crude oil or natural gas. This is a factor that must be taken seriously into account. At the same time, shifting energy systems to renewable energy brings opportunities to decentralise economic structures and thus to stimulate economic development on a broader scale while potentially decentralising decision-making processes. In this respect, systems based on renewable energy can work against the economic and political concentration of power, and against the resource curse as well.

How exactly are energy and development connected? Are the mechanisms at work universal or only situation-dependent?

How important are institutional inertia, political repression, asymmetry of information and lack of human, natural and financial resources? In each case, which policies hold the greatest promise for bettering the lives of people in developing countries?

How do asymmetries in knowledge and power (between experts and laypersons, or rich and poor societies/countries) influence decision making and choices? How should the costs and benefits of energy production and use be generally distributed?

What are some of the gendered aspects of energy production and use, and what constitutes energy justice? Which types of energy technologies and/or systems reflect and entrench masculine or patriarchal values, and which are more egalitarian? Can technology designed to improve efficiency (especially for women) backfire when unaccompanied by broader social and cultural change?

What consequences would a model of prosperity that is neither connected to growth nor dependent on cheap energy have for current or transformed energy systems?

What are the crucial barriers (political, institutional, economic, technical) to fundamental innovations in the energy system? Which instruments or strategies might be effective in overcoming these barriers? Which role could civil societies play in this process?

14 Lessons for Problem-Solving Energy Research in the Social Sciences 283

> Global peace is directly dependent on finding creative and peaceful answers to the question of energy (in all its multifaceted aspects). What might be necessary elements in a world energy constitution?

14.1.3 Critical Infrastructure

The issue of energy security is well recognised and often discussed in the relevant literature. Energy security is a multifaceted and multidimensional issue. In addition to safety (the protection of human beings and the environment), security sometimes also refers to the functioning of individual plants or components of the physical infrastructure (infra-technology, Tassey, 2007). In this regard, statistical data on technical failure rates are maintained, but these do not reflect cases of deliberate attacks, an increasingly relevant aspect of security.[9] Careful attention to the design of energy infrastructure can provide a degree of resilience against the failure of one or more parts of the system, and so-called critical infrastructure is beginning to be designed with this in mind (cf. Farrell, Zerriffi, & Dowlatabadi, 2004).[10] Sound and stable infrastructures are important factors in place marketing and choice (for both companies and individuals).

As the degree of interconnection between energy infrastructures around the world increases, so does their vulnerability.[11] The security and safety of energy infrastructure can only be improved over the long term through large investments in knowledge and institutions (i.e., regimes). This includes the creation of relevant organisations, the implementation of rules and regulations as well as funding for the personnel to enforce them.[12]

The vulnerability of a system is characterised by 'properties . . . that may weaken its ability to survive and perform its mission in the presence of [external or internal] threats' (Einarsson & Rausand, 1998). The converse of this is the system's resilience (referred to sometimes as 'potential'), which enables it to cope with destabilising influences. Systems are designed to make them resilient enough to cope with their vulnerability so that they can maintain their structure and functioning. 'Resilience'

---

[9] In the case of nuclear systems, so-called safeguards add to the complexity of safety concepts. Safeguards commonly refer to measures against the misuse of nuclear material, or dual-use goods, for building atomic weapons.

[10] Critical infrastructure or 'lifelines' include the operation and distribution of fuel and energy, telecommunications, railways, air traffic, water and disposal systems. Support measures have been implemented in several countries, e.g., UK: http://www.cabinetoffice.gov.uk/ukresilience; Canada: http://www.publicsafety.gc.ca/prg/em/cbrne-rap-eng.aspx; USA: http://www.dhs.gov/xabout/gc_1296160752600.shtm

[11] Iran, for instance, has discovered this through its experience with the 2010 Stuxnet computer virus attack on its nuclear facilities (Farwell & Rohozinski, 2011).

[12] For a conceptual treatment see Perrow (1984), applied for example by Le Coze and Dupré (2006) or de Carvalho, dos Santos, Gomes, da Silva Borges, and Huber (2006) in industrial environments or by Westrum (2006) to a natural disaster.

is not a static concept but includes adaptability as well as the dynamic interplay between 'hard' and 'soft' factors, i.e., between infrastructures and social structures. A 'Potential and Vulnerability Tool' (Blumer, forthcoming) attempts to combine existing analyses of energy systems (from life cycle assessment) with system properties, both existent and to be developed, so that the overall energy system can be sustained in the 'long run' (50–100 years). Both vulnerability and resilience are characterised by a host of influences and interactions in all partial capital stocks (environment, technology, economy, society) as well as in systems as a whole. They may be typified by corresponding criteria (e.g., energy intensity) and associated indictors (e.g., relative energy consumption in megajoules per capita or sector).

> How should durable regimes be established and what sort of oversight is appropriate? How can a culture of safety be advanced that successfully provides incentives to willingly and carefully follow rules and procedures?
>
> How can the concepts of critical infrastructure and vulnerability used in the context of energy borrow from other concepts, such as (proactive) resilience and (reactive) adaptive capacity developed in other communities (e.g., Holling, 1973, 1996; Gunderson & Holling, 2002; Folke et al., 2002; Walker, Holling, Carpenter, & Kinzig, 2004; Mileti, 1999; Seville et al., 2009)?
>
> Since it is unavoidable that access to energy and supply security be conflictual in a world of finite resources and high competition, how can actors and states responsibly address their – legitimate – quests for these, and not involve them in political power plays? How can technical options be sensitively embedded as technical components in individual societies rather than becoming mere technical fixes?[13]

14.1.4 Governance

Global energy governance[14] refers to international collective activities that manage and distribute energy resources and services in order to overcome public goods problems and externalities, many of which are increasingly cross-border phenomena

---

[13] As a concrete example: It is not sustainable, in the true sense of the word, to substitute fossil-fuel based colonisation in oil-rich countries with extensive solar farming, as is potentially envisaged in a large-scale European solar energy scheme in Northern Africa (http://www.desertec.com), even though the pressure to change existing energy systems has increased after the recent Deepwater Horizon and Fukushima catastrophes. Desertec faces serious obstacles due to instability in the region, a factor that is also deterring potential investors.

[14] Governance has been defined as 'the set of traditions and institutions by which authority in a country is exercised. The political, economic, and institutional dimensions of governance are captured by six aggregate indicators'. These are in the fields of voice and accountability, political stability and absence of violence, and government effectiveness (IBRD/WB, 2006, p. 3).

(Florini & Sovacool, 2009). The concept of global energy governance reorients the discussion of energy problems around four topics: who sets the rules governing energy and how they are enforced; how current sociopolitical configurations and transformations affect decisions about energy; the adequacy (or inadequacy) of current national approaches to addressing key energy problems; and the broadening of roles and responsibilities for non-state actors to become involved in energy decision making (Biermann & Pattberg, 2008).

Polycentric forms of governance – those that simultaneously mix scales, mechanisms and actors – imply that interaction and sharing of power among numerous stakeholders can provide a more effective response to energy-related problems than actions taken in isolation. A number of studies have argued that polycentrism potentiates certain factors that increase the likelihood of cooperation and resolution. These factors include the availability of reliable information that can be gathered with minimal transaction costs; open and frequent communication among stakeholders regarding the pros and cons of different energy systems; effective rule enforcement, including provisions for enforcing compliance (such as sanctions); and predictable and gradual transitions to new rules and enforcement procedures (see especially Ostrom, 2009, 2010; see also Dietz, Ostrom, & Stern, 2003; Poteete, Janssen, & Ostrom, 2010; Brown & Sovacool, 2011). Environmental sociology has also recently developed a polycentric governance framework for environmental flows that considers, in the context of globalisation, hybrid material (physical) and human (social-institutional) systems, as well as hybrid state-transnational political arrangements (see Spaargaren, Mol, & Bruyninckx, 2006).

If sustainable development implies improving people's quality of life while making less intensive use of resources in order to maintain sufficient natural and social capital for future generations, related factors such as poverty, lifestyles, equality and climate change – which necessarily includes climate policy – entail more than merely reducing emissions. They call for pluralistic decision making, that is, governance, and, consequently, pluralistic science (as is powerfully demonstrated in Rayner, 1998). The earth's carrying capacity must be maintained, but people's response capacity to mitigate and adapt to climate change also must be maintained and developed (Travis, 2010). Nations and regions, together with their energy systems, are nested in their sociotechnical landscapes (cf. Chapter 10 on carbon capture and storage, CCS). It is not enough to 'change technology'; rather, for any viable changes to occur, the relationship between technology and society (societies) must be understood. This requires a comprehensive understanding of governance.

Beck has observed that 'social, political, economic and individual risks increasingly tend to escape the institutions for monitoring and protection in industrial society' (Beck, 1994, pp. 5 f.). He has proposed 'reflexive modernization' as a counterweight to 'the effects of risk society that cannot be dealt with and assimilated in the system of industrial society' (ibid.). In research, technology assessment provides an instrument for coping with unintended consequences. In this vein, Voss and colleagues have developed the notion of 'reflexive governance' and have taken the interconnected issues of complexity, uncertainty, path dependence, ambivalence and distributed control as a starting point to characterise the governance problem surrounding sustainable development. They have derived six strategies to prevent societal development from being undermined by the unintended detrimental effects of steering activities: (1) integrated knowledge production, (2) experiments and adaptivity of strategies and institutions, (3) iterative, participatory goal formulation, (4) anticipation of long-term systemic effects of measures (developments), (5) interactive strategy development and (6) creating congruence between problem space and governance (Voss, Bauknecht, & Kemp, 2006). All of these strategies are in line with what has been observed and propounded in this book.

In particular, policy areas (e.g., climate policy) and energy systems (e.g., nuclear systems) with long-term and wide-ranging ramifications need adequate supervision. New supranational institutions are needed to help cope with such systems. Neither the IPCC nor the International Atomic Energy Agency (IAEA) is fit for this purpose today. Socially robust frameworks that integrate adequate technical, political, legal and social criteria are needed. According to Rip, a system is 'socially robust' if most of its arguments, evidence, social alignments, interests and cultural values lead to consistent options (Rip, 1987, p. 359). This is a formidable challenge for governance, and it needs the support of the social sciences (Weingart, 2008).

Who are the key international governors (that is, the major formative institutions) involved in global decisions about energy? Are there underlying universal principles of global energy governance that transcend individual countries, or are governance principles always rooted in particular locations and contexts? How do best practices in governance (e.g., transparency, accountability, legitimacy and participation) affect energy policy and regulation?

Which governance models (or combinations thereof) are best suited to address climate change and other global energy-related environmental problems? In particular, in an era of globalisation, when traditional models of environmental governance centred on the nation-state are outdated and old monolithic international institutions such as the United Nations and the World Bank have lost much of their traction, what further development of alternative concepts (e.g., environmental flows as found in sociology) is required to provide the necessary analytical rigour? How can fruitful syntheses be created between separate conceptual streams in environmental sociology, political science and integrative environmental studies (e.g., industrial ecology, or environmental systems and flows analysis) (Spaargaren et al., 2006)?

How can the institutional design of an international technological and legal regime for CCS allow for 'adaptive governance of complex systems of human-environment relationships' (Brewer & Stern, 2005)? How can the need for effective monitoring and enforcement be fulfilled?

14.1.5 Institutions and Regimes

Two institutional topics of intense discussion currently are how global climate policy should be organised and what the role of the Intergovernmental Panel on Climate Change (IPCC) should be. This is understandable, since these are fundamental issues for the future. In addition, a huge experiment is underway with important institutional ramifications. Observing and quantifying the impacts of this experiment is important not only for future climate policy but for energy policy and the policies of global development and peace.

To a greater extent than most environmental problems, climate change is a highly complex, ill-defined, ill-structured and 'wicked' problem (Rittel & Webber, 1973). As is typical of wicked problems, the framing of climate policy since the Kyoto Protocol has suffered from epistemological limits in defining and delimiting systems knowledge (cf. Chapter 3, Box 3.1: Social systems knowledge in the climate knowledge infrastructure) and to using systems knowledge as an input to generate transformation knowledge. Part of the intractability of wicked policy problems, and one of their defining characteristics, is their lack of stopping rules. In the case of climate change, such stopping rules would indicate when enough information on related physical and sociotechnical systems had been gathered to realise a deeper understanding. In fact, Prins et al. (2010) suggest that one of the most useful types of knowledge is an understanding of the *gaps* in systems knowledge.

Furthermore, there is no clear linear, unidirectional path from systems knowledge to target knowledge and on to transformation knowledge. Rather, as is typical of wicked problems, there are multiple directional pathways and interactions, innumerable potential problem formulations and any number of solutions – but no decisive tests for these solutions (Rittel & Webber, 1973; CASS/ProClim, 1997).

Devising effective targets in climate policy requires good target knowledge. It is generally understood that adopting specific emissions reduction goals, even if non-binding, has a positive effect on multilateral climate deliberations and perhaps also indirectly on individual behaviour. The 2°C target has emerged as the most prominent interpretation of the controversial objective in the 1992 UN Framework Convention on Climate Change to 'prevent dangerous anthropogenic interference with the climate system' (UNFCCC, 1992, §2). Scientists originally suggested a climate policy target of 2°C as the maximum allowable global average warming. However, politicians have since interpreted the target knowledge of 2°C as scientific systems knowledge, while scientists now treat it as politically determined target knowledge (Jaeger & Jaeger, 2010). Yet 2°C can still be a basis for agreement focal points, though they may shift in the future in light of new information or understanding (Jaeger & Jaeger, 2010).

The choice of the increasingly ambitious level of 2°C as the target knowledge threshold implies the need for highly ambitious transformation knowledge and its aggressive implementation ('the 2°C limit is a strong call for action, and it is understood as such' [Jaeger & Jaeger, 2010]). Of course the function of the IPCC is not to dictate the form of transformation knowledge; rather, it should emerge organically from the interactions of a plurality of actors. Yet winning public acceptance for the IPCC's system knowledge about the state of the climate has been difficult, in part because at every step of its production, this type of knowledge includes socially negotiable – and thus politically assailable – knowledge (see again Chapter 3, Box 3.1). Gaining broad public support for transformation knowledge in the form of effective mitigation measures that are also implementable is all the more difficult.

Proponents of international environmental institutions as being an answer to global environmental problems, and climate change in particular, contend that the IPCC's endeavours are crucial to meeting the challenge of climate change and other environmental impacts (as discussed in Chapter 2). Critical reflection on the IPCC and its institutional deficiencies has already led the way to first reform steps (Committee to Review the IPCC, 2010). The IPCC's loss of public credibility could have negative spill-over effects on other international institutions should it continue to exist in its current form and could even be symptomatic of a more general crisis in legitimacy and public acceptance of global governance institutions.

The need for reform, however, goes far beyond the IPCC. The need for developing new regimes, frameworks and international energy and climate policies has become ever clearer. Prins and colleagues (2010) and others suggest that the failure of the 2009 United Nations climate change conference in Copenhagen called into question the usefulness of multilateral climate diplomacy through large conferences. Despite modest progress at Cancún in December 2010, many insist that the impetus for future emissions reductions lies outside the UN framework. The approach favoured by Prins and colleagues would pursue multiple agendas that have the potential for contingent benefits, including climate benefits: This means they would make the climate an indirect beneficiary of other agendas and would turn climate benefits into contingent benefits of other policies including energy access, environmentally sustainable development and adaptation. Jaeger and Jaeger (2010) would similarly lower expectations for climate policy in isolation and instead concentrate on parallel problems. This was also one of Rayner and Malone's (1998c) core recommendations for policymakers.

Several recent trends are noteworthy. The first trend clearly points towards *integrated solutions*. Climate policy can no longer be understood and conceived as isolated environmental policy. Climate issues are inextricably linked with energy issues and in the final analysis with a just global economic order as well. This is because significantly limiting or reducing GHG emissions, especially CO_2, implies a material expropriation in countries with corresponding natural resources and a devaluation of existing business models and production lines across the world. To the greatest possible extent, meaningful long-term climate policy requires future-oriented planning for a carbon-free world economy – without just shifting the problems onto others. Recent 'global governance' work advocates a comprehensive, skilful integration of various relevant policy areas. Two ground-breaking examples are Stern (2009) and Edenhofer, Wallacher, Reder, and Lotze-Campen (2010).

A second trend is that of *polycentric solutions*. The lack of political desire for change reflects the limits of a 'greater policy' (and research) that is accustomed

14 Lessons for Problem-Solving Energy Research in the Social Sciences 289

to thinking in terms of large-scale, global political blueprints. This neglects the fact that innovation and development take place on the basis of cooperative actors at the micro-level (businesses) and meso-level (regions, innovative environments, networks). It is obvious that global agreements are indispensable for dealing with global goods (e.g., the climate). But it is just as important to identify and support the motors of development and innovation.

> What kind of climate policy targets emerge from the interplay between climate science and politics? To what extent should such targets be science-based? What types of targets are most effective? Is an integrated global energy and climate policy desirable? How can policies be fashioned that effectively and coherently combine multiple fields, even those that are potentially conflicting, including energy, climate, agriculture, trade, finance and monetary policy?
>
> How are the interests of state actors in climate negotiations affected by geographical and regional differences, including fossil fuel resources, population density, energy transport possibilities and the proximity of accessible renewable energy sources? What is the influence of non-state actors, including industry and other interest groups? How would international climate treaties affect the value of natural resource commodities, especially oil and coal, and what would be the corresponding effect on global money flows?
>
> What groups are likely to emerge as dominant players from an increasingly multifaceted, polycentric environment of interests in climate governance? How will the role of state actors change vis-à-vis non-state actors in such an environment? What conditions would support the success of groups following enlightened self-interest based on longer time horizons and more all-encompassing notions of social costs?
>
> What can be learned from the credibility-damaging events connected to the operations of the IPCC or the assessment reports it has produced? What recommendations in the various national and international IPCC reviews can be practically applied in order to reform its structure and strengthen its institutional design as well as its systems for internal review and evaluation?
>
> Which additional strategies for global policies on development, energy and climate protection can be derived from a polycentric understanding of development (Edenhofer et al., 2010, p. 160 ff.)?
>
> What is the potential international or geopolitical risk of violent conflicts being generated if aggressive international climate policies are implemented (Jaeger & Jaeger, 2010)?
>
> In the area of education and public understanding of science, how can scientific findings on complex phenomena related to climate change be integrated into general science education without either overtaxing students or over-trivialising the subject matter?

14.1.6 Markets

Der Wettbewerb als Entdeckungsverfahren (*Competition as a Discovery Procedure*) is the graphic title of a major work by Friedrich A. von Hayek (1968). In it, innovative forces are characterised as one of the major advantages of a market economy. Von Hayek, a contributor to the Austrian School of political economy, emphasised competition, i.e., the process, the freest possible interplay between economic actors. In contrast, the Freiburg School – with Wilhelm Röpke, Alexander Rüstow, Walter Eucken and Ludwig Erhard as important proponents – placed the market in the foreground, that is, the institution and its structures. The tension between process and structure marks the entire history as well as the direction of thought of these two schools of market economy. This was expressed in their conflicting fundamental positions: on one hand, trust in the ability of the market to repair itself and the notion of minimal statism (Austrian School), and on the other, the concept of markets as a politically formed order with a clear framework and (in keeping with the market) intervention in case of market failure, such as externalities and monopolies (Freiburg School). With the awarding of the Nobel Prize to von Hayek (1974) and later to Milton Friedman (1976), the radical market variant became dominant while the Freiburg School declined in importance, at least in English-speaking countries. This had a variety of consequences, some of them less than favourable. The formal principles of the market economy sacrificed their obligations: e.g., responsibility for the consequences of one's actions, true-cost pricing, transparency and accountability and intervention in the case of market failure (cf., for example, Eucken, 1968). The results of this political (regulatory) negligence can be seen most spectacularly in the recent financial market crash.

A first critical point is that politics has relieved the market far too quickly of the requirement of *innovation performance*. While this has seemed to offer short-term solutions, it has actually hindered more in-depth innovations or made them superfluous. This can be seen for example in the operation of international markets for fossil fuels and the national quests for energy independence.

Energy (in)dependence is a central component of the conventional notion of energy security. Dependence is particularly strong where a supplier, nation or firm has monopoly power, especially where that power is not or cannot be regulated. Much of Western Europe is dependent on Russian natural gas; alternative supplies cannot be secured from one day to the next. The European Commission has established a Gas Coordination Group to study ways of making the gas supply more secure. Chile's costly dependence on Argentina for its natural gas recently spurred it to invest in a large liquefied natural gas (LNG) terminal in order to free itself from this dependence.

Parallel to consumers' calls for security of their oil supply, national oil companies now often claim that they need 'demand security' in order to protect their large investment in exploration (Barkindo, 2006). There has also been some research done on the related topic of the importance of secure energy supplies for the preparedness and capabilities of national armed forces (cf. Singer, 2008; Le Billon, 2005).

14 Lessons for Problem-Solving Energy Research in the Social Sciences

Common to these strategies is the fact that while the dependence on solitary suppliers has been reduced, at the same time *the general dependence of the economy on these energy sources and their infrastructures has been reinforced.* More far-reaching innovations would reduce the unstable dependence on these energy systems.

A second critical point is the extent of *energy subsidies and non-internalisation of external effects* (cf. the mercantilism syndrome as described in Chapter 3). This is a central pillar of traditional energy policies. This includes the lack or only partial internalisation of greenhouse gas emissions from fossil fuels as well as the support of nuclear power by means of (a) subsidies, (b) liability limits in the case of nuclear accidents, and (c) inadequate provisions by operators for the decommissioning of nuclear plants or the permanent disposal of nuclear waste (Lenzen, 2008; Costanza, Cleveland, Cooperstein, & Kubiszewski, 2011; Kopolow, 2011). Costanza and colleagues conclude their reflections on the question of whether nuclear energy can be part of the solution with the following recommendations: 'Let's remove the subsidies, require nuclear power plants to be fully insured, and put aside adequate funds for decommissioning and long-term radioactive waste disposal. Let's do the same for all energy sources Then we can use the market mechanism to find out whether nuclear power plants should be part of the energy solution' (Costanza et al., 2011).

This leads to a third critical point: All energy providers should have the opportunity to discover and develop their position in future energy systems. This presupposes that the respective economic and regulatory *framework is fair*, that is, that it *applies equally to all* (see above). At the same time it is clear that initially new energy systems must be given additional support (for a limited period of time) if they are to manage the jump from pilot projects to market maturity. For this, the means of promotion used until now (research and development, education and information, subsidies, feed-in tariffs, bonus models, quotas, facilitated access to loans, etc.) need to be internationally appraised, improved and optimised. Well-intended but inappropriate support must be prevented from disrupting fair market innovations or leading to expensive and undesirable development. A particular challenge will be to find a balance between the need to make technologies marketable that are still expensive today and the obligation to use scarce funds most efficiently (Banfi Frost et al., 2011, pp. 16 ff.).

Fourth, *new markets* will be created, such as markets for emissions certificates (Chapter 8), markets for alternative energies within the framework of liberalising European energy markets and markets for energy services and energy contracting. These must be socially legitimate (pre-economically) and (economically) robust. For this reason, market barriers and obstacles must be recognised and reduced. This is especially important in the energy sector, as it is characterised by a high degree of regulation, information gaps and asymmetry, difficult or limited access to technologies and money markets, high initial investment costs, long pay-back periods, missing cost-effectiveness and small market share, lacking institutions for information dissemination as well as uncertainty with regard to state intervention (Painuly, 2001).

How do national oil companies and their modes of operation vary with different political, economic and geographical conditions in oil-rich areas? How are their huge revenue streams utilised? How can social science compensate for restrictions on access to information in many economically important but politically authoritarian oil-rich countries?

What is the cost-benefit ratio (short-, middle- and long-term) of strategies that pursue supply security for conventional energy carriers in comparison with strategies for developing decentralised systems on the basis of renewable energies? Do traditional principles of supply security undercut innovations in new energy systems?

What effect does increased diversification of the energy mix or the compulsory conversion to renewable energies in industrialised countries have on economic and social development in oil-exporting countries (resource curse, democratisation, economic development)?

Across Europe, there has been a move towards increased liberalisation of energy markets. What are the main motives behind this move, how are the markets constructed and how are the efforts being evaluated? To what extent are suppliers and consumers of energy acting according to expectations?

What can be learned from international experience with strategies to promote renewable energies? How successful have promotion instruments been at improving efficiency (modification of the energy system at the lowest economic cost), innovation, competition, employment and equity? What lessons can be learned for trying to bring renewable energy out the econiche and into the mainstream? How significant in this regard are taxes or levies on fossil fuels and nuclear energy (with lump-sum refunds)?

What could be a polycentric European strategy for 100% renewable energy supply, climate protection and economic development? Which current policies could be used as a basis, and how could this basis be further developed?

14.1.7 Systemic Approaches to Analysis

Integration is sorely needed both at the level of analysis as well as in governance. Analysing the problems at hand, their interactions, the options for coping with them and the systems they are embedded in should proceed from various perspectives. Systemic change of existing energy systems is dealt with below (see 'Change, social learning and scientific communities'). Science and technology studies (STS) offers a promising area of historically-oriented investigation. The sociotechnical system concept is the best-known (and most relevant) concept related to energy.[15] The notion of large-scale, sociotechnical systems that weave together technical artefacts,

---

[15] It originated with the work of Emery and Trist (1960) and Hughes (1969, 1983, 1987) and was extended in the edited compilations of Mayntz and Hughes (1988), La Porte (1991), Summerton (1994) and Coutard (1999).

14 Lessons for Problem-Solving Energy Research in the Social Sciences   293

organisations, cultural values as well as institutional rule systems and structures has featured prominently in this book. For example, viewed as an artefact, a pipeline is just a physical conduit for oil or gas, but viewed as a system, it includes pumping stations, operators, financing institutions, investors, land, import and export terminals, oil refineries and natural gas sweetening facilities as well as energy traders.

The various catastrophes that have occurred in large technical systems have provided particularly drastic examples for scientific examination. Some of the most infamous include the core meltdown of the Three Mile Island nuclear power plant in 1979 (Perrow, 1982), the Challenger space shuttle accident in 1986 (Esser & Lindoerfer, 1988), the nuclear reactor catastrophe at Chernobyl, also in 1986 (Reason, 1987), and recently the Deepwater Horizon drilling rig explosion and oil spill in the Gulf of Mexico in 2010 (BP, 2010; Graham et al., 2011) and the earthquake, tsunami and Fukushima reactor catastrophe in Japan in 2011. Even though these are all unrelated systems, their common denominator is that these events were triggered not only by 'technical' failures but by a combination of factors and a sequence of events that sometimes had developed slowly and imperceptibly. These include component failures, mistakes and errors of actors, inadequate attention to precursors, non-compliance with regulations as well as a lack of safety culture at different levels (from individuals to management and authorities). The Chernobyl disaster led to a committed development of safety culture in the nuclear community (IAEA, 1991).

Energy systems with a high hazard potential require comprehensive oversight for the entire chain of safety culture, from internal quality assurance to the technical and political supervision of these systems. This includes perceptive and observant operators with a questioning attitude, top management that takes a living 'safety first' approach (and that is willing to publish critical reports by experts), board members who approve adequate budgets, tenacious regulators, resolute independent committees as well as informed voters and critical consumers. Devising such structures needs the input of a range of disciplines, including the social sciences.

Other approaches (such as theories of epistemic culture and the social construction of technology and actor networks) hold that technology and scientific practices emerge in society as a 'seamless web'.[16] They are shaped by (and help shape) an array of social, political, cultural and technical elements. The study in Chapter 9 has shown the implications of such approaches by presenting the examples of wind turbine and biogas technology evolution in Denmark. The complex social processes that shape energy systems evolve in a mutually constitutive and dynamic process based on the involved actors and their interests. Technologies possess interpretive flexibility and are the product of negotiation. In this way, energy systems become 'heterogeneous' because their meaning, rather than being fixed, is interpreted and negotiated by the social groups connected to them. Table 14.1 provides an overview of the range of associated concepts.

---

[16] There is not necessarily a contradiction between a truly systemic – integrative, non-mechanistic – approach and Hughes's 'seamless web' (as is suspected by Joerges, 1996, p. 57).

Table 14.1 Summary of science and technology studies (STS) methods and concepts

Approach	Primary authors	Central thesis	Key concepts	Contribution
Social construction of technology (SCOT)	W. Bijker et al. (e.g., 1987), D. MacKenzie (e.g., 1993), T. Pinch, T. Hughes	Technological artefacts are socially constructed	Interpretive flexibility, heterogeneous engineering	Reveals that both social and technical factors concurrently shape technological artefacts
Technological frame	W. Bijker (e.g., 1995)	A single technological artefact is seldom worked on by only one group of people	Relevant social groups	Helps reveal otherwise concealed actors connected to technological systems
Epistemic culture	K. Knorr-Cetina (e.g., 1999)	The sciences produce knowledge differently and are bound by disparate epistemic communities and practices	Knowledge machinery	Reveals that the way practitioners think about problems simultaneously enables and constrains their work
Actor network theory (ANT)	S. Woolgar, B. Latour (e.g., 1979, 1987), M. Callon (e.g., 1986)	Technical Objects are nodes in a network of people and devices in interlocking roles	Enrolment, sociotechnical networks	Reveals that knowledge and power can be equally important in explaining why technologies succeed or fail

Source: Sovacool (2006)

> Which social groups are involved in the production (or benefit from the use) of a particular energy system? How do a nation's historical behaviour and attitudes towards energy influence its contemporary energy consumption? What is the social context in which energy technologies emerge and are used? What are the broader factors inhibiting or fostering the safe and reliable maintenance of existing energy technologies and the adoption of new ones?

Are biofuels really always environmentally and socially friendlier than fossil fuels? When should I replace my refrigerator? Should a German prefer organically grown asparagus from California to asparagus that is conventionally grown in the Upper Rhine Valley? Such issues are topics for flow analysis tools like life cycle assessments (LCA). The growing complexity of the issues, the quest for substantiated valuations and the increased need for policy support have led to

14 Lessons for Problem-Solving Energy Research in the Social Sciences   295

ever more interdisciplinary approaches, including full-cost accounting, agent-based modelling, other industrial ecology tools, sustainability assessments and studies of the metabolism of societies.

Sustainability analysis, which examines systems' capacity to assess the longevity or durability of the systems, has been developed using concepts from LCA combined with sustainability indicators. In contrast to these, it focusses on the sustainability of the activity. The tool of technology assessment (TA) has also evolved over the last decades, moving from the traditional expert TA to participatory TA involving stakeholders and the public (Bellucci & Joss, 2002). A related effort is the 'cultural assessment of technoscience', which has been developed by historians (Hård & Jamison, 2005).

> A large variety of methods and procedures have been developed for assessing the sustainability of systems, processes and services. What are the basic assumptions behind the methods and procedures used to assess the sustainability of systems, processes and services (entities, limitations)? What is the field of application of the various types of assessments (substances, products, space and time)? What are the system boundaries? What is the explanatory power of the indicators that are being used? How solid are the underlying data, both qualitative and quantitative?

14.1.8 The Long Term: Policy Strategies, Forecasting and Path Dependencies

Long-Term (Expert) Approaches to Energy Models and Systems

Modelling and scenario techniques have a long tradition in energy research (e.g., Sweeney & Weyant, 1979; WEC, 2007; Shell, 2008). They have also been used in climate research (e.g., IPCC, 2007a). The focus on climate change has recently shifted energy modelling from rather stable and long-term assumptions concerning energy supply, carriers and demand to switching technologies, changing carriers and efficiencies of system sectors. The use of pure economic models is giving way to adopting more complex system perspectives or combining the models with other types of research on the efficiency and choice of technologies (e.g., Craig, Gadgil, & Koomey, 2002; EMF, 2011).

> How can foresight instruments satisfy the concurrent needs for scientific rigidity and adaptability? (cf. EEA, 2009)
>
> How can soft criteria – such as human behaviour and regimes – be incorporated into long-term models?
>
> How can survey techniques such as those using point source data provide a reliable and sustained base for long-term forecasts?

Municipal and regional energy systems are expected to fulfil a variety of goals, some converging, others mutually conflicting: climate change mitigation, reductions in primary resource consumption, security of energy supply as well as moderation of energy price fluctuations. Bottom-up strategies should be generated on a local level by combining discourse among stakeholders about aims and measures with expert analytical methods for decision-making. Trutnevyte and colleagues have developed a methodology to support decision-making on future energy systems in small communities. The methodology links (i) stakeholders' visions concerning the future energy system, (ii) technology portfolios that reflect engineering expertise, and (iii) stakeholder-based multi-criteria assessment for appraising the potential consequences of implementing the visions (Trutnevyte, Stauffacher, & Scholz, 2010a, 2010b). The transdisciplinary character of this methodology is crucial, since by involving actors who are outside academia, it includes forms of knowledge production that complement traditional disciplinary and interdisciplinary scientific endeavours.

Controls and lock-ins are vital for determining the sustainability of energy systems. A balanced mixture of centralised/decentralised energy production tends to make for more robust systems and more reliable energy security. Path dependencies, whether longstanding (e.g., fossil fuels, nuclear power) or new (CCS) must be carefully scrutinised. Sufficient backup and innovative environments are also crucial.

Naturally, the fundamental economic mechanisms and dynamics of an economy – and through their operation, the economy's innovative and competitive potentials – are contingent on public policies. As a form of societal organisation, the economy depends on fundamental structural conditions and rules. Negligence here can prove very costly to society. Insufficient or poor regulation has led to dangerous economic overheating (e.g., financial crises) and severe ecological and social consequences (e.g., Deepwater Horizon, Chernobyl, Fukushima). Overregulation, however, can dampen economic innovation and generally undermine the power of the market in a liberal economic system. This is also true of strategies for the long-term restructuring of energy systems based on renewable energy. These strategies must carefully straddle the line between providing sufficient impetus and stifling innovation.[17]

> What mechanisms might be helpful in integrating energy issues, particularly energy conservation, into broader policies, such as financial, industrial, science and economic policy? How to cope with "double crises" such as the financial and climate crises (Cohen, 2011)?
>
> What type of energy goal is most appropriate for a given level of aggregation, and what time frame and level of detail does it require? Which goals are most appropriate? How can contradictory political goal-setting best be managed or avoided altogether in crafting energy policy?

---

[17] Examples of recent national energy strategies that attempt this include German Federal Environment Agency (2010), Fachausschuss 'Nachhaltiges Energiesystem 2050' (2010), and Schreurs (2010).

14 Lessons for Problem-Solving Energy Research in the Social Sciences 297

Participatory Strategic Energy Scenarios

The complexity of the possible social and technical choices and the difficulty of predicting future developments due to different types of uncertainty (Stirling, 2010) have created a need for instruments that outline alternative futures and allow for more adaptive and interactive planning. Such instruments should also respond to the need for a more enlightened public dialogue, since expert regimes and political choices are, in themselves, insufficient for providing satisfactory alternatives or dealing with their implications. These fields of research are based on insights from different ways to combine or even produce new interdisciplinary approaches bridging technology, policy and social theory. The need to identify the challenges and choices facing societies today requires receptiveness to alternative futures, as well as identifying the possible actions necessary for preparing and building the capacity to handle them. Awareness of the role of expertise and existing interests[18] opens the door to studies on the processes and institutions involved in providing foresight.

As described above, the impacts of climate change on energy systems and the increasing global demand for energy pose serious challenges to the process of peaceful adaptation and transformation (IPCC, 2007a, 2007b). Challenges for the saliency, credibility, and legitimacy of future IPCC scenarios can be revealed by analysing shortcomings in past scenarios (Girod et al., 2009; Flüeler, 2007; see also Alcamo, 2001). Analytical constructions, such as formative scenario building (Scholz & Tietje, 2002, pp. 79–116), may help improve the basis and transparency of scenarios. A 'Business-as-Past' concept may also allow a comparison between past and present consumption patterns. In emphasising the assumed 'change' in the different scenarios, this concept leaves it up to the scenario users to subjectively rate how likely such changes are. The storylines and their translation into quantitative inputs for the GHG emission models benefit in the traceability, integrity and consistency of formative scenario analysis. Analysing the actors provides a means of measuring consistency, as consistency is increased if future developments follow paths with lower conflict levels.

14.1.9 Change, Social Learning and Scientific Communities

The contribution of the social sciences to energy research can be framed partly in terms of the 'need to understand and manage systemic change' (Jackson, 2006). For example, older industrialised countries are facing the severe challenge of transforming their existing energy systems to fit the needs of the future. (Newly industrialising countries face different challenges, as we have seen above.)

The European Commission recently sponsored a large-scale research project to develop models of sociotechnical change in energy demand-side management programmes (Mourik et al., 2009). Very much in line with the conclusions in Goldblatt

---

[18] Cf., for example, Brand and Karvonen (2007), Hessels and van Lente (2008), Hansson and Bryngelsson (2009), Möller (2009).

(2005b), this work is based on the idea that *'efforts to change end-user behaviour should not only focus on individual change but also include the other stakeholders influencing energy-related social practices and the social level of change'* (Mourik et al., 2009, p. 16, emphasis in original). Programmes need to facilitate *both individual and societal learning* through social networks, movements and intermediary organisations, so that new individual practices will become rooted in supportive social frameworks. This project adds to a growing body of literature that emphasises the need to move from an individual end-user focus to a larger sociotechnical systems perspective.[19]

The Dutch have a well-established school of thought on sociotechnical system innovation as featured in the writings of Raven, Rip, Schot and others. According to their theories, as explained in Hofman and Elzen (2010), new technology niches in electricity systems, for example, may emerge from problems in existing regimes such as power failures or sociotechnical pressures stemming from concern over GHG emissions. With a proper understanding of the requisite conditions for successfully introducing a niche technology into the mainstream, it is possible to tailor and target sociotechnical policy interventions to specific favourable (e.g., renewable) energy technologies.[20]

Technological niches are part of the conceptual framework for a recently developed planning technique called 'sociotechnical scenarios', which explores the interaction between technological and social innovation. Niches interact with sociotechnical landscapes and sociotechnical regimes to shape system innovations. The sociotechnical scenario technique can complement traditional scenario methods and vision-building techniques such as backcasting. For example, Hofman and Elzen (2010) have recently applied sociotechnical scenarios to innovation in electricity systems.

> What are the societal preconditions for long-term energy conservation? For decarbonisation of the economy?
>
> Can evolutionary technological approaches like strategic niche management be used effectively to advance sustainable energy systems?
>
> Can the critical mass for systemic change in energy systems be quantified or qualified in terms of social network analysis?
>
> How can the partly contradictory goals of economic efficiency and the long-term viability of energy systems, as well as of consumer sovereignty and sustainable lifestyles, be reconciled in research and practice?

---

[19] As a practical matter for research in an era of increasing fiscal austerity, investments in networking existing social science knowledge about energy systems may be more cost-effective than new projects that seek to add to it (Ekins, 2003).

[20] For other examples of recent studies in this area, see Kemp (1997), Elzen, Geels, and Green (2004), Schot and Geels (2007), Verbong, Geels, and Raven (2008), Verbong, Christiaens, Raven, and Balkema (2010), and Grin, Rotmans, Schot, Geels, and Loorbach (2010).

14 Lessons for Problem-Solving Energy Research in the Social Sciences

> With regard to adaptive capacity in catastrophic situations, how can communities/societies achieve a new and stable level of 'normality' in a disturbed environment (Lochard & Prêtre, 1995)?
>
> What can we learn from the distinction between single- and double-loop learning (Argyris & Schön, 1978) regarding adaptation to new environments, i.e., instrumental reversible modification vs. deep-rooted change?

This book began with an ancient classical illustration of the political importance of rhetoric, 'the art of persuading assemblies' (cf. Chapter 1, Box 1.1). It suggested that conceptually framing and communicating specialised knowledge is of great strategic political importance for energy issues.

Hence, the fields of rhetoric, argumentation and communication hold a great deal of promise for scholarship in energy. For example, the now classic texts of Merton (1973), Kuhn (1962, 1977) and Price and Beaver (1966) have shown how scientists and experts set rhetorical boundaries that demarcate their disciplines from those of others. Later works focussed intensely on how scientists and engineers have used such boundary setting to enhance their credibility and attract funding for research projects (Ceccarelli, Doyle, & Selzer, 1996; Taylor, 1996; Ceccarelli, 2004, 2005). Porter (1995), Beck (1992) and Möller and Hansson (2008) rigorously analysed the rhetorical tropes and lines of argument, such as quantification and risk assessment, that are commonly used in technical discourse, and Gross (2006) has documented the many logical and stylistic choices utilised by different scientific communities. Others have focussed on the common lines of argument made by key stakeholders in debates on alternative energy technologies (Barry, Geraint, & Robinson, 2008), and Latour (1987) and Collins (1985) have shown the complex ways that scientific research depends on the social networking and persuasion skills of its practitioners.

In any case, it must be recognised that no overarching disciplinary principle is universally valid for society, whether from economics (Vogl, 2010), the natural sciences, or the technical sciences. Proponents of 'sustainability science' emphasise that the crucial point is to be able to connect science with society: The goals (of a transition to conditions of sustainability) should not be 'defined by scientists alone but rather through a dialogue between scientists and the people engaged in [the] practice of "meeting human needs while conserving the earth's life support systems and reducing hunger and poverty"' (Clark & Dickson, 2003, p. 8059, cf. also NAS-BSD, 1999).

The political nature of goal setting notwithstanding, energy is a central input to contemporary societies and is instrumental in their realisation of sustainable development (see Table 14.2). As a hinge (and instrument), energy used sustainably is crucial for achieving these goals, including eradicating poverty, increasing equity within and among societies, maintaining the technological stock and achieving climate stability.

Table 14.2 A recent agenda of priority areas or central challenges of sustainable development constructed from major documents demonstrates the relevance of energy as a persistent and pivotal research area in moving 'knowledge into action'. The core documents are: *Our Common Future* by the Brundtland Commission (WCED, 1987), *Our Common Journey* by the US National Academy of Science (NAS-BSD, 1999), the *Achievable Agenda* of the Secretary-General of the UN (Annan, 2002b),[21] and the *Readings in Sustainability Science and Technology* (Kates, 2010)

Our common future (WCED, 1987)	Our common journey (NAS-BSD, 1999)	Achievable agenda (Annan, 2002)	Sustainability science reader (Kates, 2010)
Population and human resources	Human population		Population
		Health	Health and well-being
Food security	Agriculture	Agriculture	Agriculture and food security
Species/ecosystems	Living resources	Biodiversity	Biodiversity, ecosystem services
Energy	Energy, industry	Energy	Energy, materials
Industry			
Urban	Cities		Urban growth
		Water	Water and sanitation
			Poverty
			Climate change
			Peace/security

Source: Kates (2011)

> Who are the actors in the social discourse on future energy policy? How are they organised and what influence do they have?
>
> How is this discourse being conducted with regard to systems knowledge, target knowledge and transformation knowledge? What role does science play in this? Which other knowledge is brought to bear (or not)? (networking of individuals and institutions as 'know-who'[22] and Polanyi's tacit knowledge as 'know-how-undercover', Polanyi, 1966).[23]

14.1.10 Value Systems, Cultures and Actors

Energy consumption and production are the two sides of the 'energy system' coin. In approaching consumption from the bottom-up, it can be productive to analyse lifestyles and the 'extra-energetic' reasons (e.g., status) for choosing or using certain appliances within various cultural contexts and over longer periods of time.

---

[21] In an earlier report by the Secretary-General, the list was more comprehensive. Additional items included peace, security, disarmament, development and poverty eradication, protecting the common environment, human rights, democracy and good governance, protecting the vulnerable as well as 'strengthening the United Nations' (Annan, 2002a).

[22] See e.g., http://ukerc.rl.ac.uk/index.html, or http://www.eptanetwork.org

[23] See also Cash (2003) or Hessels and van Lente (2008).

Since no 'overarching model for consistently predicting behaviour' (Egan, 2001) has been developed, it makes sense to search for cross-disciplinary, less highly segmented perspectives. Concepts in one research field (like the resource curse) may be usefully applied in others (land-grabbing). An example of how one resource curse can breed others is the increasingly common large-scale purchases of agricultural territory by rich companies and states (e.g., the Persian Gulf states, South Korea and China) in economically weak countries (e.g., Africa, Asia and Latin America), often for the cultivation of biofuel crops.

The importance of systematically analysing the connection between value systems, cultures and behaviour is described in the book *Society, Behaviour, and Climate Change Mitigation* (Jochem, Sathaye, & Bouille, 2000). In the preface, Robert T. Watson, then Chair of the IPCC, writes:

> The patterns of behaviour and underlying value systems and cultures differ substantially among world regions and societal groups. The lead authors of several chapters of WG III [Working Group III: Mitigation of Climate Change] recognized that the social, behavioural and cultural changes involved in mitigating global climate change are poorly understood. They acknowledged the need to broaden the existing conceptual framework and decided to invite psychologists, anthropologists and other social scientists to share their perspectives on the issue of climate change mitigation.

The more radical the necessary changes of behaviour, the more important it is to understand the value systems and cultures underlying such behaviour. Here, the environment in general and energy in particular are in themselves cultural factors.

The natural conditions that form the backdrop for people's daily life and commerce are formative of their culture. They influence their perceptions and valuation of nature, attitudes (carefulness, thriftiness), economics, lifestyle, institutions, legal systems (including property law), and even art. The key resource of energy is of particular importance. Cultures based on renewable energy sources (for example, agricultural and alpine economies, forest economies, irrigation cultures, fishing communities), with their multitudinous mechanisms for collective restraint and conflict resolution, are well known and relatively well researched. With modern, conventional energy carriers (fossil energies, nuclear energy) the culture and general population's connection to the natural realm is much weaker or even non-existent.

The question of power always comes into play here. While in theory every society has the opportunity to build decentralised economic and political decision-making structures on the basis of renewable energies, this is much more difficult where centralised fossil fuel and nuclear energy systems predominate. Centralised energy supply allows an economic and political concentration of power that has a profound impact on economic and political development as well as on culture.

> The provision of energy is highly centralised. What does this mean for the political and economic culture in various countries? Does it threaten a renaissance of the centralised irrigation dictator (cf. Mesopotamia) in a new form as a new large energy-system dictator?

The link between natural conditions and culture – concretely, the culture of cooperation – is aptly illustrated by the following example: Security is an individual and societal political value and rarely has independent meaning. Political science is instrumental for understanding the mindsets, motivations, interests and influence of the diverse actors in the energy landscape and for more accurately judging the risk of politically motivated supply interruptions. Nonetheless, much more highly detailed information is needed to understand the relationships between energy producers and energy consumers and how they are likely to evolve. Recent studies on the interrelations between energy supply and demand have built on decades of research and generations of experience in water management situations in which downstream users depend on upstream users. Recent research has shown that this kind of dependence need not lead only to conflicts or war but can also be a starting point for cooperation (Orttung & Perovic, 2009).

> How are energy security, energy access and international politics interrelated? How can nations' energy interdependence for supply and demand be capitalised upon to improve international relations and reduce the long-term risk of conflict?

Value systems and cultures do not just express themselves through concrete behaviour or specific types of conduct (e.g., cooperation). They are also characterised by the dominant modes of perception of reality. Viewing energy consumption from the perspective of energy services is an important example of this.

Since the mid-1980s, social science behaviouralists such as Paul Stern have advanced the view that the locus of agency in household energy consumption lies with the idiosyncratic user, not the technology. To lower energy use, attitudes, values and lifestyles should be the target of interventions, and thus behavioural (and social) sciences have an important role to play in designing more effective CO_2 reduction measures (Stern, 2006). As has been noted, in the social sciences, research on attitudes, values and lifestyles is quite common, but with regard to energy research and policy, behavioural and social research has as a whole remained very much in the shadow of the technologists' camp. Sociologists and anthropologists active in energy consumption research insist on the importance of focusing on energy service consumption – especially escalating consumption in developing countries – rather than energy consumption per se.[24] They emphasise that technoeconomic approaches tend to presuppose demand and thus take it 'out of the equation' (Shove, 2004). In addition, the people operating the energy systems tend to be neglected in the models. A partial system analogy can illustrate this: Only by investing in drivers (the users), and not just cars (the technology) is traffic safety (the system) lastingly increased,

---

[24] 'Buildings don't use energy: people do' (Janda, 2011). And after all, consumers want power, light and heat, not kilowatts or kilowatt hours.

14 Lessons for Problem-Solving Energy Research in the Social Sciences 303

because it is only then that the user feels useful (and not merely a polluter and taxpayer). Only then is true 'ownership of the problem' achieved.

Recognising that levels and patterns of energy consumption are jointly determined in the interdependent context of technical systems and social practices increases the potential number of policy intervention points: These can be found anywhere in production, provision, access and use. 'Latent potentials in the technology acting together with the changing socio-cultural contexts of everyday life constitute a distributed agency for change in [energy-relevant social] practices' (Wilhite, 2008, p. 125).

Analysis of households' use of energy typically employs quantitative methods from economics, psychology and building science. But to account for the socio-cultural influences on household behaviour, these methods should be supplemented with qualitative ethnographic research methods from sociology and anthropology, e.g., in-depth interviews, focus groups, monitoring of energy consumption, self-report surveys, administered surveys/structured interviews and secondary analysis of data (see Crosbie, 2006; Rayner & Malone, 1998b). This requires changing the negative view of qualitative data analysis held by policymakers and funding bodies and fostering a willingness to supplement universal prescriptions with ones that take local contexts into account more readily.

There are a number of social science disciplines that should be engaged in research along these lines, including the sociology of science and technology (Rammert & Schulz-Schaeffer, 2002); the sociology of consumption (especially the 'systems of provision' approach [Southernton, Cappells, & Van Vliet, 2004]); history of energy and of human use of the environment (Tvedt, Chapman, & Hagen, 2010); and anthropology, especially with regard to cross-cultural differences in energy use and the malleability of use conventions (e.g., Wilhite, 1996).

There is inadequate scientific understanding of household energy use and related environmentally significant behaviour as well as the mutual influence of these two factors and their intervention points. How can this understanding be increased?

How does the evolution of cultural expectations and norms regarding 'cleanliness, comfort and convenience' affect energy consumption? How are these norms, as well as routines and habits, co-evolving with electrical appliances and energy infrastructures?

How are lifestyles that are ever more energy-intensive being accommodated, appropriated and normalised in developing countries?

To what extent are efficiency improvements in 'comfort' technologies like air conditioning and refrigeration, which are rapidly proliferating in developing countries, helping to create and maintain lifestyles that depend on these technologies – and thereby fuel a subtle but pervasive form of rebound? Under what conditions can societies preserve or resurrect less energy-intensive standards of comfort?

> How can research streams from disparate camps of household energy researchers (technology, behaviouralism, economics) be fruitfully combined to expand the research focus beyond efficiency and develop additional effective points for policy intervention?

To conclude, value systems and cultures are also fertile soil for new analytical perspectives and innovative takes on end-user behaviour. They influence how effective new core concepts and behavioural practices are when applied. Sustainable consumption and sufficiency offer further examples of this.

Efficiency seems overrated as an answer to ever-increasing global energy consumption and emissions. It may in fact be a part of the problem, as we have suggested in various places in this book (cf. Wilhite & Norgard, 2004). In economics, the phenomena of rebound and backfire from efficiency improvements are well recognised: Increased efficiency of electrical appliances may be offset by higher utilisation or shifts in usage patterns (cf. Madlener & Alcott, 2009). Social science literature on sustainable consumption suggests that the nature and scope of the problem have been underestimated (cf. Polimeni, Mayumi, Giampietro, & Alcott, 2008; Herring & Sorrell, 2009), and proposes a number of alternative concepts that better contend with this challenge, at least on a conceptual level, such as restraint and sufficiency.

Princen (2005) has shown how unit efficiency ratios that seem to indicate progress in advancing energy-environmental agendas may actually hide increasing external costs. In the globalised economy, the environmental impacts of consuming (direct and particularly grey, i.e., embodied) energy are typically concealed and broadly diffused throughout lengthy chains of production and consumption. Restoring restraint and sufficiency requires reordering social institutions and business as well as financial structures. This will enable resource managers to receive and act on negative feedback related to the integrity of the natural resource base (Princen, 2005; Goldblatt, 2007). Recently, there have been some encouraging examples of heightened attention to changes in consumption behaviour. For example, attempts have been made to incorporate such changes into demand assumptions in life cycle assessment methodology (cf. Girod et al., 2011). In addition, it might also be fruitful to propose per capita goals that directly involve end-users in the societal dialogue (see Box 14.2).

Box 14.2 The 2000 Watt Society – A Research- and Action-Guiding Vision of Prosperity Without Energy Consumption Growth

The term '2000 Watt Society' was first coined in 1994 in the article *A Two-Kilowatt Society – plausible future or illusion?* (Kesselring and Winter, 1994). 2000 watts refers to per capita energy consumption (Jochem, 2004; Novatlantis, 2011). Three points were at the centre of the argumentation:

First, the historical fact that in the year 1960, Switzerland was a 2000 watt society (today Swiss energy consumption is approximately 6,500 per capita, excluding embodied energy).

Second, the assessment that it is possible to become a 2000 watt society again, without a reduction in the quality of life. This could be accomplished using existing technical capabilities together with additional technological advances. However, behavioural changes would also be necessary.

Third, the awareness that a global '2000 Watt Society' is necessary (Spreng, 2005). 2000 watts represents the approximate current mean global per capita energy consumption (but variation around the mean is considerable and highly uneven across and within nations). This level is considered ecologically and globally tolerable provided that (1) the energy mix ensures that in the long run, of the 2000 watts consumed per capita, on average only a quarter is generated from fossil fuels (which corresponds to about one ton of CO_2 per capita), and (2) world population is stabilised at about 10 billion.

The idea of a 2000 Watt Society was taken up in 1998 by the Board of ETH Zurich and promoted – its target later modified to one-ton CO_2 per capita[25] – as one of the central ideas guiding research at ETH. In 2002, the Swiss federal government enshrined the idea as the core of its official sustainability strategy.

How can mechanisms that deliver negative economic-ecologic feedback about the impacts of energy consumption on the resource stock be constructed or restored in order to encourage restraint? How can the emerging concept of sufficiency, which increasingly is theoretically well elaborated, be put into practice? In particular, how can small, local examples of sufficiency be adequately scaled up so that they might constitute a critical mass for social change? Can ecological rationales of sufficiency be generalised so that they apply to various cultures as well as to the global economy? And when attempting to expand sufficiency in developed countries where it seems urgently needed, how can potential consumption rebound effects be avoided with lower resource prices resulting from reductions in consumption in developed countries perversely stimulating increased usage in developing countries (see Alcott, 2008)?

---

[25] In the mid-term there is no effective difference between the two strategies, since efficiency must be drastically improved either way. However, the 1 ton CO_2 Society does not call into question growth in energy consumption per se and envisions a fossil-free (possibly including nuclear-based) electrification of the society (ETH, 2008).

If we accept that freedom is the most important precondition for successful societal development (Sen, 1993), including the development of energy systems, how can the social sciences identify and characterise institutions that are obstacles to freedom? How should such institutions be modified so that they facilitate freedom in society rather than stifle it?

In evaluating responses to catastrophic events, how is it possible to distinguish between 'technology-induced' and 'cultural' rationality?

14.1.11 Social Acceptability of Energy Technologies

Traditionally, research in the social sciences on energy systems approached technoeconomic topics through a top-down and managerial perspective and was closely related to particular developments in technology. Notions of risk, public awareness, attitude and acceptance prevalent in social psychology, the media and communication studies as well as other social science disciplines used to closely reflect the goal of policymakers and technology implementers to advance energy options that were technologically favoured. However, the growing plurality of different actors and interests has opened up the discourse: Today, in some countries decisions are not only negotiated but society has become an active participant, and thus modern decision making processes for energy projects have become very complicated and politicised. Gone are the days of deterministic top-down direction of technology development and implementation.

Especially over the past ten years, a more advanced and reflexive notion of public acceptability has emerged in the social sciences. This is because public resistance has been recognised as a major obstacle to implementing many renewable energy projects (especially wind energy) as well as building conventional fossil fuel and nuclear power-generating facilities and waste disposal plants. A broader definition of participation, not just social acceptance, that involves many social groups and crosses multiple social strata has grown more common. Accordingly, a new technology only achieves societal acceptance for both general and concrete applications when (1) it has the support of experts, national and local policymakers, as well as residents, (2) laypersons are willing to adopt and use it in their own specific contexts, and (3) they have some say in shaping the technology. When the expectations of all these groups are aligned in a common vision, societal acceptance of an energy project can be considered 'modulated' (Raven, Mourik, Feenstra, & Heiskanen, 2009) or 'socially robust' (Rip, 1987). Bi-directional risk communication is part of this modulation process (see 'Rhetoric and communication studies' below). The concept of NIMBY (Not In My Back Yard) has been enlarged to include discourses on value systems (cf. Mazmanian & Morell, 1990; Lidskog & Elander, 1992; Kemp, 1992).

> How can the 'national-local divide' be bridged in (renewable) energy projects, i.e., how can national renewable energy policies and local interests be harmonised in a fashion that allows a flow of initiatives in both directions (national-local and local-national)? How much international learning takes place between countries concerning the proponents and opponents of energy projects at the community level?[26] How relevant and adaptable are new notions of acceptance to the rapidly developing energy infrastructure in transitional countries? How applicable are they given the intensive and far-reaching quest for global resources that has been triggered by the huge energy infrastructure projects that are under construction in developing countries, often involving massive foreign participation (cf. Chapter 7)?

14.1.12 Risk and Energy Communication Studies

As described above in 'Social acceptability of energy technologies', modern concepts of risk communication are crucial to realising progressive forms of societal acceptance of energy technologies. In addition, various levels of trust can be involved in communicating 'creeping' risks such as climate change, electromagnetic fields and radioactive waste. These levels of trust have corresponding routes of persuasion and levels of debate. Researchers in risk communication and psychology find that many energy and environmental issues are too complex for a risk manager to employ central routes of persuasion that use technical arguments and appeal to the audience's experience, rationality and knowledge of the issue; rather, they use peripheral routes of communication that rely on non-substantive cues such as their reputation for establishing the credibility of their messages (Renn & Levine, 1991). Highly contested, ambiguous risks like climate change or radioactive waste disposal also require a very high level of discourse that often involves discussions about fundamental tradeoffs, values and lifestyles (cf. Renn & Levine, 1991; Goldblatt, 2005b; Flüeler, 2001), a discourse that politically is enormously challenging in culturally diverse societies marked by declining trust in political institutions, political fragmentation and acerbic and partisan media. Just as 'Buildings don't use energy: people do' (Janda, 2011), energy systems are not just fuels but social and institutional systems mediated by materials and devices (Rayner, 2010). Trust, fairness and risk are institutional characteristics: 'The system requires incorporation of the minimum level of diversity of engineering technologies and social actors to be sustainable' (ibid., p. 2623). Otherwise, cultural appropriation, i.e., 'processes in which human needs have been met and the human condition [has] been improved' will not take place (Hård & Jamison, 2005, p. 307).

---

[26] Based on proposed research questions in the introductory article for the 2007 special issue in *Energy Policy* on this topic (Wüstenhagen, Wolsink, & Bürer, 2007).

What types of information and feedback have the most influence on energy producers and users? What types of technological or policy failures can occur due to breakdowns in communication? More broadly, how do particular energy technologies alter or enhance communication practices?

In climate communication, how can central and peripheral routes of communication be used to convey the deep complexity, uncertainty and ambiguity of climate change to a public largely ignorant of scientific reasoning and statistical uncertainty? How can high-level discourse on climate issues be conducted under conditions of intense political and ideological divisions?

In energy communication, how can lower-level measures of environmental impact be linked to important individual, municipal and national choices? Specifically, how can conventional household and corporate energy indicators be improved to address issues of scale and aggregate energy consumption (cf. Goldblatt, 2005a)?

In highly controversial risk debates (such as on nuclear energy), can an enhanced notion of rationality (beyond absolute and bounded, towards social rationality) and a dedicated emphasis on process increase the chance of a more enlightened societal discourse on technology and progress (cf. Flüeler, 2006)?

14.2 Promoting Problem-Solving Energy Research in the Social Sciences

To the best of our knowledge, this book represents the first basic consensus to have been developed, as well as the first comprehensive catalogue of key agenda topics to have been compiled, on energy research in the social sciences. Both were specific, strategic goals of ASRELEO. Although further refinements are desirable and necessary, the results of this book nonetheless form an important basis for repositioning the social sciences in energy research and for formulating concrete research agendas.

This consensus must be competently communicated to the appropriate audience if the social sciences are to become a secure part of energy research in the future, especially, as envisaged here, energy research oriented towards the long-term.

In addition to communicating and convincing others of the potential of the social sciences for energy research, it is necessary to firmly embed them in long-term research programmes. Energy policies and energy economics are related to a wide range of activities. For this reason, this embedding should be conducted in research programmes at many different levels:

- at a supranational level (e.g., UN Sustainability Council; International Human Dimensions Programme on Global Environmental Change, IHDP[27]),

---

[27] http://www.ihdp.unu.edu

14 Lessons for Problem-Solving Energy Research in the Social Sciences

- at the European level (e.g., via the European Science Foundation, European Research Council, EU research programmes),
- at the national level (e.g., national energy research programmes in the framework of basic research as well as disciplinary research, for example national science foundations, academies, and energy policy research bodies),
- at the regional and local level (states or cantons, counties; energy and sustainable cities alliances[28]),
- at the level of civil society (e.g., NGOs),
- at the sectoral level (e.g., social science programmes for long-term-oriented energy options launched by key players in the energy sector) and
- at mixed levels (e.g., Sustainable Energy Europe Campaign,[29] Climate Alliance[30]).

In developing an agenda, the following considerations may serve as operational guidelines:

- *Clearly specify the target audience*
  It is insufficient to couch issues in general or overarching terms. Statements or propositions are more productive when they are addressed to defined audiences, such as specific research policy bodies or certain research communities.
- *Stick to challenges, but address them at several different levels*
  The energy-related challenges must be characterised at various levels, scales and time frames (regional, national, global; short-term, mid-term, long-term).
- *Make the link between the challenges and social science explicit*
  This pertains to specifying the audience. The researchers, including research communities, must be able to relate to the challenges in terms of their paradigms, ongoing research and state of the art.
- *Show potential users what the social sciences can contribute ('added value')*
  Users, such as politicians or government researchers and officials, must recognise the value that the research can provide in terms of their own needs.
- *Start with examples ('success/failure stories')*
  Most users are not scientists and therefore are unfamiliar with scientific approaches, frameworks and terminology. Offering them concrete examples set in day-to-day applications in their 'world' will make it easier for them to understand and accept research findings and, for that matter, budget proposals.
- *Emphasise 'learning by doing'*
  Working from externally induced challenges rather than topics defined by scientific disciplines shows how the issues evolve over time, how they are subject to different framing and contexts and how they are treated differently by different players. In contrast to a one-size-fits-all approach, this approach capitalises on the learning abilities of all involved.

---

[28] http://www.energy-cities.eu, http://www.eumayors.eu, http://www.iclei.org

[29] http://www.sustenergy.org

[30] http://www.climatealliance.org

In addition, academic socialisation should be borne in mind: Every researcher, irrespective of scientific field, is socialised in his or her research community and school of thought. In multi-disciplinary collaboration, it is challenging but fruitful to try to transcend conceptual boundaries and reach out to researchers and users with different perspectives.

Normalisation Through Embedding in Research and Teaching

Finally, energy research in the social sciences must become a fixed part of the research and teaching going on at universities, polytechnics, and research institutions. In addition to securing research programmes, this will provide young scientists with needed long-term perspectives. This is the only way to ensure that energy research in the social sciences is more than a short-lived trend. In other words, it should be 'normalised'. Normalising social science energy research in scientific curricula is essential for inculcating the core competencies comprising sustainability literacy (see Box 14.1). But most important, is it essential for tackling the energy challenges described in this book.

References

Alcamo, J. (2001). *Scenarios as tools for international environmental assessments. Experts' corner report.* Prospects and Scenarios No. 5. Environmental issue report No. 24. Copenhagen: European Environment Agency.

Alcott, B. (2008). The sufficiency strategy: Would rich-world frugality lower environmental impact? *Ecological Economics, 64*(4), 770–786.

Anderies, J. M., Janssen, M. A., & Ostrom, E. (2004). A framework to analyze the robustness of social-ecological systems from an institutional perspective. *Ecology and Society, 9*(1), 18. All web links accessed November 16, 2011, http://www.ecologyandsociety.org/vol9/iss1/art18

Annan, K. (2002a). Implementation of the United Nations Millennium Declaration. Report of the Secretary-General. Fifty-seventh session, Item 44 of the provisional agenda (A/57/150). Follow-up to the outcome of the Millennium Summit. A/57/270. General Assembly, 31 July. New York: United Nations.

Annan, K. (2002b). *An achievable agenda.* World Summit on Sustainable Development. New York: United Nations.

Argyris, C., & Schön, D. (1978). *Organizational learning: A theory of action perspective.* Reading, MA: Addison-Wesley.

Banfi Frost, S., Berg, M., Dupont, J.-F., Gysler, M., Kiener, E., Minsch, J., et al. (2011). *Erneuerbare Energien. Herausforderungen auf dem Weg zur Vollversorgung.* SATW Schrift Nr. 42. Bern: SATW Schweizerische Akademie der Technischen Wissenschaften.

Barkindo, M. (2006, February). *Energy supply and demand security.* Paper presented at EUROPIA Conference, London. http://www.opec.org/opec_web/en/1097.htm

Barry, J., Geraint, E., & Robinson, C. (2008). Cool rationalities and hot air: A rhetorical approach to understanding debates on renewable energy. *Global Environmental Politics, 8*(2), 67–98.

Beck, U. (1992). *Risk society: Towards a new modernity.* London: Sage Publications.

Beck, U. (1994). The reinvention of politics: Towards a theory of reflexive modernization. In U. Beck, A. Giddens, & S. Lash (Eds.), *Reflexive modernization* (pp. 1–55). Cambridge: Polity Press.

14 Lessons for Problem-Solving Energy Research in the Social Sciences 311

Bellucci, S., & Joss, S. (Ed.). (2002). *Participatory technology assessment: European perspectives.* Centre for the Study of Democracy. London: University of Westminster Publication.

Berlin-brandenburgische Akademie der Wissenschaften, Deutsche Akademie der Technikwissenschaften, & Leopoldina (2011). *Die Bedeutung der Gesellschafts- und Kulturwissenschaften für eine integrierte und systemisch ausgerichtete Energieforschung.* Berlin: Berlin-brandenburgische Akademie der Wissenschaften.

Biermann, F., & Pattberg, P. (2008). Global environmental governance: Taking stock, moving forward. *Annual Review of Environment and Natural Resources, 33*, 277–294.

Bijker, W. (1995). *Of bicycles, bakelites, and bulbs: Toward a theory of sociotechnical change.* Cambridge, MA: MIT Press.

Bijker, W. E., Hughes, T. P., & Pinch, T. (Eds.). (1987). *The social construction of technological systems: New directions in the sociology and history of technology.* Cambridge, MA: MIT Press.

Bloor, D. (1991). The strong program in the sociology of knowledge. In D. Bloor (Ed.), *Knowledge and social imagery* (pp. 3–23). Chicago: University of Chicago Press.

Blumer, Y. (forthcoming). Vulnerability and potential analysis (VPA) of the Swiss energy system. In *Cooperation project with the Swiss Federal Office of Energy (BFE).* PhD project. Natural and Social Science Interface. Zurich: ETH.

BP, British Petroleum. (2010). *Deepwater horizon: Accident investigation report. 8 September 2010.* London: BP. http://www.bp.com (BP global > Gulf of Mexico restoration > Investigating the accident > BP internal investigation).

Brand, R., & Karvonen, A. (2007). The ecosystem of expertise: Complementary knowledges for sustainable development. *Sustainability: Science, Practice & Policy, 3*(1), 21–31.

Brewer, G., & Stern, P. (Ed.). (2005). *Decision making for the environment: Social and behavioral science research priorities.* Washington, DC: National Academies Press.

Brown, M. A., & Sovacool, B. K. (2011). *Climate change and global energy security: Technology and policy options.* Cambridge: MIT Press.

BUND & Misereor (Eds.). (1996). *Zukunftsfähiges Deutschland. Ein Beitrag zu einer global nachhaltigen Entwicklung.* Studie des Wuppertal-Instituts für Klima, Umwelt, Energie. Basel: Birkhäuser.

Caldwell, L. K. (1976). Energy and the structure of social institutions. *Human Ecology, 4*(1), 31–45.

Callon, M., Law, J., & Rip, A., (Eds.) (1986). *Mapping the dynamics of science and technology: Sociology of science in the real world.* London: Macmillan.

Cash, D. W. (2003). Knowledge systems for sustainable development. *PNAS, 100*(14): 8087–8091.

CASS/ProClim. (1997). Research on sustainability and global change – Visions in science policy by Swiss researchers. http://www.proclim.ch/4dcgi/proclim/en/media?1122

Ceccarelli, L. (2004). Rhetoric of science and technology. In C. Mitchem (Ed.), *Encyclopedia of science, technology, and ethics* (vol. 3: L-R, pp. 1625–29). Detroit: Macmillan Reference.

Ceccarelli, L. (2005). A hard look at ourselves: A reception study of rhetoric of science. *Technical Communication Quarterly, 14*(3), 257–265.

Ceccarelli, L., Doyle, R., & Selzer, J. (1996). Introduction to the special issue on rhetoric of science. *Rhetoric Society Quarterly, 26*(4), 7–12.

Chabay, I. et al. (2011, draft). *Knowledge, learning, and societal change: Finding paths to a sustainable future.* Science plan for a cross-cutting core project of the International Human Dimensions Programme on Global Environmental Change, IHDP. 67 pp.

Chubin, D. E., & Restivo, S. (1983). The 'mooting' of science studies: Research programmes and science policy. In K. D. Knorr-Cetina & M. Mulkay (Eds.), *Science observed: Perspectives on the social study of science* (pp. 53–84). London: Sage.

Clark, W. C., & Dickson, N. (2003). Sustainability science: The emerging research program. *PNAS, 100*(14): 8059–8061.

Cohen, M. (2011). Book review perspectives: The end of modernity: What the financial and environmental crisis is really telling us, Stuart Sim. *Sustainability: Science, Practice, & Policy,* 7(2), http://sspp.proquest.com/static_content/vol7iss2/book.sim-print.html

Collins, H. M. (1985). *Changing order: Replication and induction in scientific practice.* Chicago: University of Chicago Press.

Committee to Review the IPCC. (2010). *Climate change assessments: Review of the processes and procedures of the IPCC.* Amsterdam: Interacademy Council.

Costanza, R., Cleveland, C., Cooperstein, B., & Kubiszewski, I. (2011, April 5). Can nuclear power be part of the solution? Solutions for a sustainable and desirable future. *The Solutions Journal,* 2(3). http://www.thesolutionsjournal.com/print/918

Coutard, O. (Ed.). (1999). *The governance of large technical systems.* New York: Routledge.

Craig, P. P., Gadgil, A., & Koomey, J. G. (2002). What can history teach us? A retrospective examination of long-term energy forecasts for the United States. *Annual Review of Energy and Environment, 27,* 83–118.

Crosbie, T. (2006). Household energy studies: The gap between theory and method. *Energy & Environment, 17*(5), 735–753.

Davison, A. (2001). *Technology and the contested meanings of sustainability.* New York: State University of New York Press.

de Carvalho, P. V. R., dos Santos, I. L., Gomes, J. O., da Silva Borges, M. R., & Huber, G. J. (2006). *The role of nuclear power plant operators' communications in providing resilience and stability in system operation.* Paper presented at the 2nd Symposium on Resilience Engineering, Juan-les-Pins, France, 8–10 November. http://www.resilience-engineering.org

de Haan, G., Kamp, G., Lerch, A., Martignon, L., Müller-Christ, G., Nutzinger, H. G. (Eds.). (2008). *Nachhaltigkeit und Gerechtigkeit. Grundlagen und schulpraktische Konsequenzen.* Ethics of Science and Technology Assessment, 33. Berlin: Springer.

De-Shalit, A. (1995). *Why posterity matters: Environmental policies and future generations.* New York: Routledge.

Diamond, R., & Moezzi, M. (2002). Becoming allies: Combining social science and technological perspectives to improve energy research and policy making. LBNL-50704. Berkeley: Lawrence Berkeley National Laboratory, LBNL.

Dietz, T., Ostrom, E., & Stern, P. (2003). The struggle to govern the commons. *Science, 302*(5652), 1907–1912.

Dobson, A. (Ed.). (1999). *Fairness and futurity: Essays on environmental sustainability and social justice.* Oxford: Oxford University Press.

Driessen, P. J., Leroy, P., & Van Vierssen, W. (2010). *From climate change to social change. Perspectives on science-policy interactions.* Utrecht: International Books.

EC, European Commission (2009). *The world in 2025 – Rising Asia and socio-ecological transition.* Brussels: Directorate-General for Research.

Edenhofer, O., Wallacher, J., Reder, M., & Lotze-Campen, H. (2010). *Global aber gerecht. Klimawandel bekämpfen, Entwicklung ermöglichen. Ein Report des Potsdam-Instituts für Klimafolgenforschung und des Instituts für Gesellschaftspolitik.* Munich: C. H. Beck.

EEA, European Environment Agency. (2009). *Looking back on looking forward: A review of evaluative scenario literature.* EEA Technical report, no. 3/2009. Copenhagen: EEA.

Egan, C. (2001). *The application of social science to energy conservation: Realizations, models, and findings.* Report No. E002. Washington, DC: American Council for an Energy-Efficient Economy.

Einarsson, S., & Rausand, M. (1998). An approach to vulnerability analysis of complex industrial systems. *Risk Analysis, 18*(5), 535.

Ekins, P. (2003, June). *Prospects and policies for step changes in the energy system: Developing an agenda for social science research.* Final report to the Economic and Social Research Council. London: Policy Studies Institute at the University of Westminster.

Elzen, B. E., Geels, F. W., & Green, K. (Eds.). (2004). *System innovation and the transition to sustainability: Theory, evidence and policy.* Cheltenham: Edward Elgar.

14 Lessons for Problem-Solving Energy Research in the Social Sciences

Emery, F. E., & Trist, E.L. (1960). *Toward a social ecology: Contextual appreciations of the future in the present.* New York: Plenum Books.

EMF, Energy Modeling Forum. (2011). *Energy efficiency and climate change mitigation.* EMF Report 25, 1, March 2011. Stanford, CA: Stanford University

Esser, J. K., & Lindoerfer, J. S. (1988). Groupthink and the space shuttle Challenger accident: Toward a quantitative case analysis. *Behavioral Decision Making, 2*, 167–177.

ETH Zurich, Eidg. Technische Hochschule Zürich (2008). *Energy strategy for ETH Zurich.* K. Boulouchos et al. (Eds.). Zurich: Energy Science Center. http://www.esc.ethz.ch/publications/energy

Eucken, W. (1968). *Grundsätze der Wirtschaftspolitik, 4. Auflage.* Tübingen: J.C.B. Mohr, Zürich; Polygraphischer Verlag.

Fachausschuss 'Nachhaltiges Energiesystem 2050' des ForschungsVerbunds Erneuerbare Energien (2010). *Energiekonzept 2050. Eine Vision für ein nachhaltiges Energiekonzept auf Basis von Energieeffizienz und 100% erneuerbaren Energien.* Berlin: Fraunhofer IBP, Fraunhofer ISE, Fraunhofer IWES, ForschungsVerbund Erneuerbare Energien.

Farrell, A. E., Zerriffi, H., & Dowlatabadi, H. (2004). Energy infrastructure and security. *Annual Review of Environment and Resources, 29*, 421–469.

Farwell, J., & Rohozinski, R. (2011). Stuxnet and the future of cyber war. *Survival, 53*(1), 23–40.

Fischhoff, B., Slovic, P., Lichtenstein, S., Read, S. & Combs, B. (1978). How safe is safe enough? A psychometric study of attitudes towards technological risks and benefits. *Policy Sciences, 9*, 127–152.

Florini, A. E., & Sovacool, B. K. (2009). Who governs energy? The challenges facing global energy governance. *Energy Policy, 37*(12), 5239–5248.

Flüeler, T. (2001). Options in radioactive waste management revisited: A proposed framework for robust decision-making. *Risk Analysis, 21*(4), 787–799.

Flüeler, T. (2006). *Decision making for complex socio-technical systems. Robustness from lessons learned in long-term radioactive waste governance. Vol. 42: Series Environment & Policy.* Dordrecht: Springer.

Flüeler, T. (2007). *Energy forecasts in search of today's society – Some insights from social science to bridge the gap.* Paper presented at the International Energy Workshop 2007, Stanford, CA.

Flüeler, T., Goldblatt, D., Minsch, J., & Spreng, D. (2007). *Meeting global energy challenges: Towards an agenda for social science research.* Final Report for EFDA and BP, Contract EFDA/05-1255. Zürich: ETH. http://www.esc.ethz.ch/box_feeder/ASRELEO-Projekt

Folke, C., Carpenter, S., Elmqvist, T., Gunderson, L., Holling, C. S., & Walker, B. (2002). Resilience and sustainable development: Building adaptive capacity in a world of transformations. *Ambio, 31*(5), 437–440.

German Federal Environment Agency. (2010). *Energy goal 2050: 100% renewable electricity supply by 2050.* Press release, no. 39/2010. Conference on July 7, 2010, Dessau-Rosslau: Umweltbundesamt.

Girod, B., de Haan, P., & Scholz, R. W. (2011). Consumption-as-usual instead of ceteris paribus assumption for demand: Integration of potential rebound effects into LCA. *The International Journal of Life Cycle Assessment, 16*(1), 3–11.

Girod, B., Wiek, A., Mieg, H., & Hulme, M. (2009). The evolution of the IPCC's emissions scenarios. *Environmental Science & Policy, 12*, 103–118.

Goldblatt, D. (2005a). Combining interviewing and modeling for end-user energy conservation. *Energy Policy, 33*(2), 257–271.

Goldblatt, D. (2005b). *Sustainable energy consumption and society: Personal, technological, or social change?* Dordrecht: Springer.

Goldblatt, D. (2007). Book review perspectives: The logic of sufficiency, Thomas Princen. *Sustainability: Science, Practice, & Policy, 3*(1). http://sspp.proquest.com/static_content/vol3iss1/SSPP-v3.1.pdf

Graham, B., Reilly, W. K., Beinecke, F., Boesch, D. F., Garcia, T. D., Murray, C. A., & Ulmer, F. (2011). *Deep water: The Gulf oil disaster and the future of offshore drilling. Report to the*

President. National Commission on the BP Deepwater Horizon Oil Spill and Offshore Drilling. http://www.oilspillcommission.gov/final-report

Grin, J., Rotmans, J., Schot, J. W., Geels, F. W., & Loorbach, D. (2010). *Transitions to sustainable development: New directions in the study of long term transformative change*. London: Routledge.

Gross, A. G. (2006). *Starring the text: The place of rhetoric in science studies*. Carbondale: Southern Illinois University Press.

Gunderson, L. H., & Holling, C. S. (Eds.). (2002). *Panarchy: Understanding transformations in human and natural systems*. Washington, DC: Island Press.

Hansson, A., & Bryngelsson, M. (2009). Expert opinions on carbon dioxide capture and storage – a framing of uncertainties and possibilities. *Energy Policy, 37*, 2273–2282.

Hård, M., & Jamison, A. (2005). *Hubris and hybrids. A cultural history of technology and science*. New York/London: Routledge.

Härtel, C., & Pearman, G. (2010). Understanding and responding to the climate change issue: Towards a whole-of-science research agenda. *Journal of Management & Organization, 16*(1).

Helmholtz Gemeinschaft (2011). *Zukünftige Infrastrukturen der Energieversorgung. Auf dem Weg zur Nachhaltigkeit und Sozialverträglichkeit. Proposal for the Establishment of a Helmholtz Alliance*. Karlsruhe: Karlsruhe Institute of Technology (KIT).

Herring, H., & Sorrell, S. (2009). *Energy efficiency and sustainable consumption: The rebound effect*. Basingstoke: Palgrave Macmillan.

Hessels, L. K., & van Lente, H. (2008). Re-thinking new knowledge production: A literature review and a research agenda. *Research Policy, 37*, 740–760.

Hobsbawm, E. (1962). *The age of revolution: Europe 1789–1848*. London: Weidenfeld and Nicolson.

Hofman, P. S., & Elzen, B. (2010). Exploring system innovation in the electricity system through sociotechnical scenarios. *Technology Analysis and Strategic Management, 22*(6), 653–670.

Holling, C. S. (1973). Resilience and stability of ecological systems. *Annual Review of Ecology and Systematics, 4*, 1–23. See also http://www.resalliance.org

Holling, C. S. (1996). Engineering resilience versus ecological resilience. In P. C. Schulze (Ed.), *Engineering within ecological constraints* (pp. 31–43). Washington, DC: National Academy Press.

Hughes, T. P. (1969). Technological momentum in history: Hydrogenation in Germany 1898–1933. *Past and Present, 44*(1), 106–132.

Hughes, T. P. (1983). *Networks of power: Electrification in Western Society, 1880–1930*. Baltimore, MD: Johns Hopkins University Press.

Hughes, T. P. (1987). The evolution of large technological systems. In W. E. Bijker et al. (Eds.), *The social construction of technological systems: New directions in the sociology and history of technology* (pp. 51–82). Cambridge, MA: MIT Press.

IAEA, International Atomic Energy Agency. (1991). *Safety culture. A report by the International Nuclear Safety Advisory Group*. Safety Series, 75, INSAG-4. Vienna: IAEA.

IBRD, The International Bank for Reconstruction and Development, & The World Bank. (2006). *A decade of measuring the quality of governance. Governance matters 2006. Worldwide governance indicators*. Washington, DC: IBRD/The World Bank.

IEA, International Energy Agency, UNDP, UNIDO. (2010). *Energy poverty: How to make modern energy access universal? Special early excerpt of the World Energy Outlook 2010 for the UN General Assembly on the Millennium Development Goals*. Paris: OECD/IEA.

IPCC, International Panel on Climate Change. (2007a). Summary for policymakers. In IPCC (Ed.), *Climate change 2007. Synthesis report*. Cambridge: Cambridge University Press.

IPCC. (2007b). Summary for policymakers. In IPCC (Ed.), *Climate change 2007. Impacts, adaptation and vulnerability*. Cambridge: Cambridge University Press.

Jackson, T. (2006). *An agenda for social science research in energy*. Summary of a Research Council Workshop held on 6th April 2006, University of Surrey.

14 Lessons for Problem-Solving Energy Research in the Social Sciences

Jaeger, C., & Jaeger, J. (2010). *Three views of two degrees.* EFC Working Paper. Potsdam: European Climate Forum.

Janda, K. (2009). *Exploring the social dimensions of energy use: A review of recent research initiatives. ECEEE 2009 Summer Study. Act! Innovate! Deliver! Reducing Energy Demand Sustainably.* Proceedings, vol. 4, pp. 1841–1852. Paper presented at the European Council for an Energy-Efficient Economy, ECEEE Summer Study, Colle Sur Loop, France, June 1–6. http://www.eci.ox.ac.uk/publications/downloads/janda09exploring.pdf

Janda, K. B. (2011). Buildings don't use energy: people do. *Architectural Science Review, 54,* 15–22.

Jochem, E. (Ed.). (2004). *Steps towards a sustainable development. A white book for R&D of energy-efficient technologies.* Novatlantis – Sustainability at the ETH domain. Dübendorf: Novatlantis. (pre-study 2002: Steps towards a 2000 Watt-Society. Developing a white paper on research & development of energy-efficient technologies)

Jochem, E., Sathaye, J., & Bouille, D. (Eds.). (2000). *Society, behaviour, and climate change mitigation.* Dordrecht: Kluwer.

Joerges, B. (1996). Large technical systems and the discource of complexity. In L. Ingelstam (Ed.), *Complex technical systems* (pp. 55–72). FRN, NUTEK, Tema T. Stockholm: Swedish Council for Planning and Coordination of Research, FRN.

Kates, R. W. (2010, ed.). *Readings in sustainability science and technology.* CID Working Paper, 213. Center for International Development. Cambridge, MA: Harvard University Press. http://www.hks.harvard.edu/centers/cid/publications/faculty-working-papers/cid-working-paper-no.-213

Kates, R. W. (2011). *From the unity of nature to sustainability science: Ideas and practice.* CID Working Paper, 218. Center for International Development (CID). Cambridge, MA: Harvard University. http://www.hks.harvard.edu/centers/cid/publications/faculty-working-papers/cid-working-paper-no.-218

Kemp, R. (1992): *The politics of radioactive waste disposal.* Manchester: Manchester University Press.

Kemp, R. (1997). *Environmental policy and technical change. A comparison of the technological impact of policy instruments.* Cheltenham: Edward Elgar.

Kesselring, P., & Winter, C.-J. (1994). *World energy scenarios: A two-kilowatt society – plausible future or illusion?* Proceedings. Villigen: PSI, pp. 103–116. Paper presented at the Conference 'Energietage 94', Villigen, Switzerland, 10–12 November.

Knorr Cetina, K. D. (1999). *Epistemic cultures. How the sciences make knowledge.* Cambridge: Harvard University Press.

Kopolow, D. (2011). *Nuclear power: Still not viable without subsidies.* Cambridge, MA: Earth Track.

Kuhn, T. S. (1962). *The structure of scientific revolutions.* Chicago: University of Chicago Press.

Kuhn, T. S. (1977). *The essential tension: Selected studies in scientific tradition and change.* Chicago: University of Chicago Press.

La Porte, T. (Ed.). (1991). *Social responses to large technical systems: Control or adaptation.* London: Kluwer.

Latour, B. (1987). *Science in action: How to follow scientists and engineers through society.* Cambridge: Harvard University Press.

Latour, B., & Woolgar, S. (1979). *Laboratory life: The construction of scientific facts.* Princeton, NJ: Princeton University Press.

Le Billon, P. (2005). *Fuelling war: Natural resources and armed conflict.* London: Adelphi Papers.

Le Coze, J. C., & Dupré, M. (2006). *How to prevent a normal Accident in a high reliable organisation? The art of resilience, a case study in the chemical industry.* Paper presented at the 2nd Symposium on Resilience Engineering, Juan-les-Pins, France, 8–10 November. http://www.resilience-engineering.org

Lenzen, M. (2008). Life cycle energy and greenhouse gas emissions of nuclear energy. A review. *Energy Conversion and Management, 49*(8), 2178–2199.

Lichtenberg, G. Chr. (1789–1793). Sudelbücher. In W. Promies (1971, ed.), *Schriften und Briefe. Band 1, Heft J (860)*. München: Carl Hanser.

Lidskog, R., & Elander, I. (1992). Reinterpreting locational conflicts: NIMBY and nuclear waste management in Sweden. *Policy & Politics, 20*(4), 249–264.

Lochard, J., & Prêtre, S. (1995). Return to normality after a radiological emergency. *Health Physics, 68*(1): 21–26.

Lohmann, L. (2010). Climate crisis: Social science crisis. In M. Voss. (Ed.), *Der Klimawandel: Sozialwissenschaftliche Perspektiven*. Wiesbaden: VS Verlag für Sozialwissenschaften.

Maclean, D. (1980). Benefit-cost analysis, future generations and energy policy: A survey of the moral issues. *Science, Technology, & Human Values, 5*(31), 3–10.

MacKenzie, D. (1993). *Inventing accuracy: A historical sociology of nuclear missile guidance*. Cambridge, MA: MIT Press.

Madlener, R., & Alcott, B. (2009). Energy rebound and economic growth: A review of the main issues and research needs. *Energy, 34*(3), 370–376.

Mayntz, R., & Hughes, T. P. (Eds.). (1988). *The development of large technical systems*. Boulder, CO: Westview Press.

Mazmanian, D., & Morell, D. (1990). The 'NIMBY' syndrome: Facility siting and the failure of democratic discourse. In N. J. Vig & M. E. Kraft (Eds.), *Environmental policy in the 1990s. Toward a new agenda* (pp. 125–143). Washington, DC: CQ-Press.

Merton, R. K. (1973). *The sociology of science*. Chicago: University of Chicago Press.

Mileti, D. S. (1999). *Disasters by design. A reassessment of natural hazards in the United States. National Academy of Sciences*. Washington, DC: Joseph Henry Press.

Modi, V., McDade, S., Lallement, D., & Saghir, J. (2005). *Energy services for the Millennium Development Goals*. Foreword by J. D. Sachs. Washington, DC: The International Bank for Reconstruction and Development/The World Bank and the United Nations Development Programme.

Möller, N. (2009). Should we follow the experts' advice? On epistemic uncertainty and asymmetries of safety. *International Journal of Risk Assessment and Management, 11*(4), 219–236.

Möller, N., & Hansson, S. O. (2008). Principles of engineering safety: Risk and uncertainty reduction. *Reliability Engineering and System Safety, 93*, 776–783.

Möller, N., Hansson, S. O., & Peterson, M. (2006). Safety is more than the antonym of risk. *Journal of Applied Philosophy, 23*(4), 419–432.

Mourik, R. M., Breukers, S., Heiskanen, E., Bauknecht, D., Hodson, M., Barabanova, Y., et al. (2009). *Conceptual framework and model: Synthesis report tailored for policy makers as target group. A practical and conceptual framework of intermediary demand-side practice*. European Commission Seventh Framework Programme (Theme: Energy).

NAS-BSD, National Academy of Sciences, Board on Sustainable Development. (1999). *Our common journey: A transition toward sustainability*. Washington, DC: National Academy Press.

Novatlantis. (2011). *Smarter living. Moving forward to a sustainable energy future with the 2000 watt society*. Novatlantis – Sustainability at the ETH Domain, with the support of Swiss Federal Office of Energy (SFOE) and the Swiss Engineers and Architects Association (SIA). Berne: Swiss Federal Office for Buildings and Logistics (BBL).

Orttung, R. W., & Perovic, J. (2009). Energy security. In M. D. Cavelty, et al. (Eds.), *The Routledge handbook of security studies*. London: Routledge.

Ostrom, E. (2009). *A polycentric approach for coping with climate change*. Report prepared for the WDR2010 Core Team, Development and Economics Research Group, World Bank. Bloomington, IN: Indiana University.

Ostrom, E. (2010). Polycentric systems for coping with collective action and global environmental change. *Global Environmental Change, 20*, 550–557.

Painuly, J. P. (2001). Barriers to renewable energy penetration. A framework for analysis. *Renewable Energy, 24*(1), 73–89.

14 Lessons for Problem-Solving Energy Research in the Social Sciences 317

Parkin, S., Johnston, A., Buckland, H., Brookes, F., & White, E. (2004). *Learning and skills for sustainable development. Developing a sustainability literate society. Guidance for Higher Education institutions.* London: Forum for the Future.

Perrow, C. (1982). The President's Commission and the normal accident. In D. L. Sills, C. P. Wolf, & V. B. Shelanski (Eds.), *Accident at Three Mile Island: The human dimensions* (pp. 173–184). Boulder, CO: Westview.

Perrow, C. (1984). *Normal accidents. Living with high-risk technologies.* New York: Basic Books.

Polanyi, M. (1966, reprint 2009). *The tacit dimension.* London: Routledge (Chicago: University of Chicago Press).

Polimeni, J. M., Mayumi, K., Giampietro, M., & Alcott, B. (2008). *The Jevons paradox and the myth of resource efficiency improvements.* London: Earthscan.

Porter, T. (1995). *Trust in numbers: The pursuit of objectivity in science and public life.* Princeton, NJ: Princeton University Press.

Poteete, A., Janssen, M., & Ostrom, E. (2010). *Working together: Collective action, the commons, and multiple methods in practice.* Princeton, NJ: Princeton Universtiy Press.

Price, D. de S., & Beaver, D. (1966). Collaboration in an invisible college. *American Psychologist, 21*, 1011–1018.

Princen, T. (2005). *The logic of sufficiency.* Cambridge: MIT Press.

Prins, G., Galiana, I., Green, C., Grundmann, R., Hulme, M., Korhola, A., et al. (2010). *The Hartwell Paper: A new direction for climate policy after the crash of 2009.* Oxford: Institute for Science, Innovation, and Society, University of Oxford.

Rammert, W., & Schulz-Schaeffer, I. (Eds.). (2002). *Können Maschinen handeln? Soziologische Beiträge zum Verhältnis von Mensch und Technik.* Frankfurt/M., New York: Campus.

Raven, R. P. J. M., Mourik, R. M., Feenstra, C. F. J., & Heiskanen, E. (2009). Modulating societal acceptance in new energy projects: Towards a toolkit methodology for project managers. *Energy, 34*(5), 564–574.

Rayner, S. (2010). Trust and the transformation of energy systems. *Energy Policy, 38*, 2617–2623.

Rayner, S., & Malone, E. L. (1998a). *Human choice and climate change: The societal framework, 1.* Columbus, OH: Battelle Press.

Rayner, S., & Malone, E. L. (1998b). *Human choice and climate change: The tools for policy analysis, 3.* Columbus, OH: Battelle Press.

Rayner, S., & Malone, E. L. (1998c). *Human choice and climate change: What have we learned?, 4.* Columbus, OH: Battelle Press.

Reason, J. (1987). The Chernobyl errors. *Bulletin of the British Psychological Society, 40*, 201–226.

Renn, O., & Levine, D. (1991). Credibility and trust in risk communication. In R. E. Kasperson & P. J. M. Stallén (Eds.), *Communicating risks to the public: International perspectives.* Dordrecht: Kluwer.

Research Council of Norway (2010). *Energy research gets infusion of social science.* http://www.forskningsradet.no/en/Newsarticle/Energy_research_gets_infusion_of_social_science/1253961272887

Rip, A. (1987). Controversies as informal technology assessment. *Knowledge: Creation, Diffusion, Utilization, 8*(2), 349–371.

Rittel, H., & Webber, M. (1973). Dilemmas in a general theory of planning. *Policy Sciences, 4*(2), 155–169.

Rychen, D. S., & Salganik, L. H. (Eds.). (2003). *Key competencies for a successful life and well-functioning society.* Cambridge, MA: Hogrefe und Huber.

Scholz, R. W., & Tietje, O. (2002). *Embedded case study methods: Integrating quantitative and qualitative knowledge.* Thousand Oaks, CA: Sage.

Schot, J. W., & Geels, F. W. (2007). Niches in evolutionary theories of technical change: A critical survey of the literature. *Journal of Evolutionary Economics, 17*(5), 605–622.

Schreurs, M. (2010). *A 100% renewable electricity supply by 2050: Climate-friendly, reliable, and affordable.* Berlin: German Advisory Council on the Environment (SRU).

Sen, A. (1993). Capability and well-being. In M. C. Nussbaum & A. Sen (Eds.), *The quality of life* (pp. 30–53). Oxford: Oxford University Press.

Seville, E., Fenwick, T., Brunsdon, D., Myburgh, D., Giovinazzi, S., & Vargo, J. (2009). *Resilience retreat. Current and future resilience issues*. Resilient Organisations Research Report 2009/05. Christchurch, NZ: Resilient Organisations Programme. http://www.resorgs.org.nz

Shell. (2008). *Energy scenarios to 2050*. The Hague: Shell International BV.

Shove, E. (2004). Efficiency and consumption: Technology and practice. *Energy & Environment, 15*(6), 1053–1065.

Singer, C. (2008). *Energy and international war: From Babylon to Baghdad and beyond*. Singapore: World Scientific.

Southernton, D., Cappells, H., & Van Vliet, B. (Eds.). (2004). *Sustainable consumption: The implications of changing infrastructures of provision*. Cheltenham and Northampton, MA: Edward Elgar.

Sovacool, B. (2006). Reactors, missiles, X-rays, and solar panels: Using SCOT, Technological Frame, Epistemic Culture, and Actor Network Theory to investigate technology. *Journal of Technology Studies 32*(1), 4–14.

Sovacool, B., & Brown, M. A. (2010). Competing dimensions of energy security: An international perspective. *Annual Review of Environment and Resources, 35*, 77–108.

Spaargaren, G., Mol, A. P. J., & Bruyninckx, H. (2006). Introduction: Governing environmental flows in global modernity. In G. Spaargaren, A. P. J. Mol, & F. Buttel (Eds.), *Governing environmental flows: Global challenges to social theory* (pp. 1–36). Cambridge: MIT Press.

Spreng, D. (2005). Distribution of energy consumption and the 2000 W/capita target. *Energy Policy, 3*, 1905–1911.

Stern, N. (2009). *The global deal: Climate change and the creation of a new era of progress and prosperity*. New York: PublicAffairs.

Stern, P. (2006). *Why social and behavioral science research is critical to meeting California's climate challenges*. The California Energy Commission Web site: http://www.energy.ca.gov/2006publications/CEC-999-2006-027/CEC-999-2006-027.pdf

Stirling, A. (2010). Keep it complex. *Nature, 468*, 1029–1031.

Summerton, J. (Ed.). (1994). *Changing large technical systems*. San Francisco: Westview Press.

Sweeney, J. L., & Weyant, J. P. (1979). *The Energy Modeling Forum: Past, present and future*. EMF PP6.1. Energy Modeling Forum, Stanford University. Stanford: Stanford University.

Tassey, G. (2007). *The technology imperative*. Northampton, MA: Edward Elgar.

Taylor, C. A. (1996). *Defining science: A rhetoric of demarcation*. Madison, WI: University of Wisconsin Press.

Travis, W. R. (2010). Going to the extremes: Propositions on the social response to severe climate change. *Climatic Change, 98*, 1–19.

Trutnevyte, E., Stauffacher, M., & Scholz, R. W. (2010a). *From visions to actions. Novel approach to linking energy visions with energy scenarios and assessing consequences*. Institute für Environmental Decisions. Natural and Social Science Interface. Zurich: ETH.

Trutnevyte, E., Stauffacher, M., & Scholz, R. W. (2010b). *Visions of stakeholders, engineering expertise and multicriteria assessment for energy strategies of a small community*. Paper presented at the 11th Biennial Conference 'Advancing sustainability in a time of crisis' of the International Society of Ecological Economics in Oldenburg and Bremen, Germany.

Tvedt, T., Chapman, G., & Hagen, R. (Eds.). (2010). *A history of water, vol. 3: Water and geopolitics in the new world order*. London: I. B. Tauris.

UN AGECC, The Secretary-General's Advisory Group on Energy and Climate Change (AGECC). (2010). *Energy for a sustainable future. Summary report and recommendations*. New York: United Nations.

UNDP, United Nations Development Programme, & GEF, Global Environment Facility. (2011). *Adapting to climate change*. UNDP-GEF Initiatives financed by the Least Developed Countries Fund, Special Climate Change Fund and Strategic Priority on Adaptation. New York: UNDP.

14 Lessons for Problem-Solving Energy Research in the Social Sciences

UNFCCC. (1992). *The United Nations Framework Convention on Climate Change.* http://unfccc. int (> The Convention)

Verbong, G. P. J., Christiaens, W. G. J., Raven, R. P. J. M., & Balkema, A. J. (2010). Strategic niche management in an unstable regime: Biomass gasification in India. *Environmental Science and Policy, 13*(4), 272–281.

Verbong, G. P. J., Geels, F. W., & Raven, R. P. J. M. (2008). Multi-niche analysis of dynamics and policies in Dutch renewable energy innovation journeys (1970–2006): Hype-cycles, closed networks and technology-focused learning. *Technology Analysis & Strategic Management, 20*(5), 555–573.

Vogl, J. (2010). *Das Gespenst des Kapitals.* Zürich: Diaphenes.

von Hayek, A. F. (1968). Der Wettbewerb als Entdeckungsverfahren. In A. F. von Hayek (1969, 2nd ed. 1994), *Freiburger Studien. Gesammelte Aufsätze.* Tübingen: J.C.B. Mohr/P. Siebeck.

Voss, J.-P., Bauknecht, D., & Kemp, R. (2006). *Reflexive governance for sustainable development.* Cheltenham: Edward Elgar.

Walker, B., Holling, C. S., Carpenter, S. R., & Kinzig. A. (2004). Resilience, adaptability and transformability in social-ecological systems. *Ecology and Society 9*(2): 5. http://www. ecologyandsociety.org/vol9/iss2/art5

WBGU, German Advisory Council on Global Change. (2011a). *Welt im Wandel. Gesellschaftsvertrag für eine grosse Transformation. Zusammenfassung für Entscheidungsträger.* Berlin: WBGU, Wissenschaftlicher Beirat der Bundesregierung für Umweltfragen.

WBGU. (2011b). *World in transition. A social contract for sustainability. Summary for policy-makers.* Berlin: WBGU. http://www.wbgu.de

WCED, World Commission on Environment and Development (Brundtland Commission). (1987). *Our common future.* Oxford: Oxford University Press.

Webler, T., & Tuler, S. P. (2010). Getting the engineering right is not always enough: Researching the human dimensions of the new energy technologies. *Energy Policy, 38,* 2690–2691.

WEC, World Energy Council. (2007). *Deciding the future: Energy policy scenarios to 2050.* London: WEC.

Weingart, P. (2008). How robust is 'socially robust knowledge'? In M. Carrier, D. Howard, & J. Kourany (Eds.), *The challenge of the social and the pressure of practice: Science and values revisited* (pp. 131–145). Pittsburgh, PA: University of Pittsburgh Press.

Westrum, R. (2006). *All coherence gone: New Orleans as a resilience failure.* Paper presented at the 2nd symposium on resilience engineering, Juan-les-Pins, France, November 8–10, 2006). http://www.resilience-engineering.org

Wilhite, H. (1996). A cross-cultural analysis of household energy use behaviour in Japan and Norway. *Energy Policy, 24*(9), 795–803.

Wilhite, H. (2008). New thinking on the agentive relationship between end-use technologies and energy-using practices. *Energy Efficiency, 1,* 121–130.

Wilhite, H., & Norgard, J. (2004). Equating efficiency with reduction: A self-deception in energy policy. *Energy and Environment 15*(3), 991–1011.

Wilhite, H., Shove, E., Lutzenhiser, L., & Kempton, W. (2000). The legacy of twenty years of energy demand management: We know more about individual behaviour but next to nothing about demand. In E. Jochem, et al. (Eds.), *Society, behaviour, and climate change mitigation* (pp. 109–126). Dordrecht: Kluwer.

Woolgar, S. (1988). Reflexivity is the ethnographer of the text. In S. Woolgar (Ed.), *Knowledge and reflexivity: New frontiers in the sociology of scientific knowledge* (pp. 1–13). London: Sage.

Wüstenhagen, R., Wolsink, M., & Bürer, M. J. (2007). Social acceptance of renewable energy innovation: An introduction to the concept. *Energy Policy, 35*(5), 2683–2691.

Name Index

A
Abelson, R. P., 8, 10
Adger, W. N., 242
AkEnd, 210, 218
Akrich, M., 109, 112
Alcamo, J., 224, 297, 310
Alcott, B., 304–305, 310, 316
Aldy, J., 146, 160
Altvater, E., 147, 160
Anderies, J. M., 277, 310
Anderson, J., 136–137, 160, 217–218, 225
Andsager, J. L., 235, 242
Anger, N., 138, 142, 145, 160
Annan, K., 300, 310
Archer, D., 49, 71, 202, 217–218
Argarwal, A., 99, 112
Argyris, C., 299, 310
Arneson, R., 148, 160
Arnstein, S. R., 228, 231, 234, 242
Ashworth, P., 217–218
Auld, G., 10

B
Bachram, H., 147, 161
Balkema, A. J., 298, 318
Bammer, G., 21, 40–43
Banfi Frost, S., 291, 310
Barkindo, M., 290, 310
Barnes, D., 79, 82, 84, 94
Barry, J., 299, 310
Bauknecht, D., 286, 316, 318
Beck, U., 216, 285, 299, 310
Beierle, T. C., 228–229, 231, 240, 242
Bellaby, P., 228, 243
Bellucci, S., 295, 311
Benson, S. M., 202, 207, 219
Berg, M., 310
Berlin-brandenburgische Akademie der Wissenschaften, 277, 311

Bernstein, S., 10
Biermann, F., 285, 311
Bijker, W. E., 168–169, 184, 188, 197, 207, 219, 294, 311, 314
Bloor, D., 279, 311
Blowers, A., 197, 219
Böhringer, C., 138, 144, 160–161
Bouille, D., 301, 315
Bourdieu, P., 108, 112
Boyd, E., 142, 147, 161, 164
Boyd, P. W., 211, 219
Bradley, D., 233, 243
Bradshaw, J., 198, 219
Brand, R., 41–42, 297, 311
Brewer, G., 277, 286, 311
British Petroleum (BP), 7, 118, 122, 128, 216, 293, 311
Bromley, D., 152, 161
Brown, K., 232, 241–242
Brown, M. A., 281, 285, 311, 318
Brunnschweiler, C. N., 116, 131
Bruyninckx, H., 285, 318
Bryngelsson, M., 285, 314
Bürer, M. J., 307, 319

C
Calderón, C., 129, 131
Caldwell, L. K., 279, 311
Callon, M., 169, 188, 294, 311
Caney, S., 146, 153, 158, 161–162
Carter, L. J., 197, 219
Cash, D. W., 300, 311
Cashore, B., 10
CASS, 31, 35, 287, 311
Ceccarelli, L., 299, 311
Chabay, I., 277, 311
Chilvers, J., 229, 241–242
Christiaens, W. G. J., 298, 318
Chubin, D. E., 279, 311

322 Name Index

Clarke, L., 191, 219, 259, 261
Clark, W. C., 299, 311
Cleveland, C., 291, 312
Colglazier, E. W., 197, 219
Collier, P., 12, 21, 94, 118, 131
Collins, H. M., 197, 219, 230, 242, 299, 311
Combs, B., 230, 243, 279, 313
Commission of the European Communities (CEC), 196–197, 207, 219
Committee on Radioactive Waste Management (CoRWM), 207, 210, 219
Commoner, B., 248, 261
Corbridge, S., 105, 112
Costanza, R., 291, 312
Coutard, O., 292, 312
Craig, P. P., 295, 312
Crosbie, T., 303, 312

D

D'Agostino, A. L., 47–71
Daly, H. E., 43
Davison, A., 279, 312
de Carvalho, P. V. R., 283, 312
de Coninck, H., 196, 200, 217, 220, 225
de Haan, G., 275, 312
de Haan, P., 313
den Elzen, M., 139, 142, 161
De-Shalit, A., 279, 312
Deutsche Akademie der Technikwissenschaften, 311
Diamond, R., 277, 312
Dickson, N., 299, 311
Dietz, T., 285, 312
Dobson, A., 279, 312
Driessen, P. J., 277, 312
Dupont, J.-F., 310

E

Edenhofer, O., 41–42, 145, 162, 214, 221, 288–289, 312
Edwards, P. N., 31–33, 42
Egan, C., 301, 312
Ekins, P., 277, 298, 312
Elander, I., 306, 315
Ellerman, D., 136–137, 140, 143, 161–162
Elzen, B. E., 169, 188, 298, 312, 314
Emery, F. E., 292, 312
Energy Modeling Forum (EMF), 295, 318
Erhard, L., 25, 42, 290
Esser, J. K., 293, 312
Eucken, W., 290, 312

European Commission (EC), 12, 14, 136–138, 161, 196, 274, 290, 297, 312, 316
European Environment Agency (EEA), 136, 161, 295, 312
European Union (EU), 12, 55, 62, 136–140, 161–162
Evans, A., 150, 162
Evans, R., 230, 242
Ezzati, M., 15, 21

F

Feenstra, C. F. J., 306, 317
Fineberg, V., 228, 245
Fiorino, D. J., 228, 231, 233, 240, 243
Fischedick, M., 211, 220, 226
Fischer, C., 144–145, 162
Fischhoff, B., 230, 243, 279, 313
Fitoussi, J., 257, 261
Florini, A. E., 285, 313
Flüeler, T., 3–10, 11–21, 23–42, 47–71, 73–94, 97–112, 115–131, 133–160, 167–188, 191–218, 227–242, 247–261, 263–269, 313
Flynn, R., 228, 243
Folke, C., 284, 313
Forester, J., 241, 243
Freudenburg, W. R., 197, 221
Frewer, L. L., 228, 231, 234, 244
Funtowicz, S. O., 232, 243

G

Gale, J., 210–211, 221
Gardiner, S., 153, 162
Garg, A., 216, 221
Geels, F. W., 169–170, 181, 183, 188–189, 298, 312–313, 317–318
Geraint, E., 299, 310
Gerard, D., 211, 214, 221, 226
German Advisory Council on Global Change (WBGU), 31, 35, 274, 276, 319
Gilbert, J., 148, 162
Girod, B., 297, 304, 313
Goldblatt, D. L., 3–10, 11–21, 23–42, 47–71, 73–94, 97–112, 115–131, 133–160, 167–188, 191–218, 227–242, 247–261, 263–269, 313
Goldstein, G., 193, 221
Goodin, R., 146, 149, 162
Gough, C., 197–198, 221
Green, K., 169, 188, 298, 312
Gregory, R., 229, 243
Grin, J., 298, 313
Gross, A. G., 299, 313
Grünwald, R., 200, 210, 221

Name Index

Gunderson, L. H., 284, 314
Gysler, M., 310

H
Hansson, A., 297, 314
Hansson, S. O., 279, 299, 316
Hård, M., 295, 307, 314
Harriss, J., 105, 112
Härtel, C., 276, 314
Heckscher, E. F., 25, 42
Heiskanen, E., 306, 316–317
Held, H., 41–42, 214, 221
Helmholtz Gesellschaft, 277, 314
Herring, H., 19, 22, 304, 314
Hessels, L. K., 297, 300, 314
Hirsch Hadorn, G., 40, 43, 152, 163, 243
Hobsbawm, E., 279, 314
Hofman, P. S., 298, 314
Holling, C. S., 284, 314, 319
Hourcade, J.-C., 136, 139, 163
Hughes, T. P., 168–169, 186, 188, 197, 219, 292–293, 311, 314, 316

I
Intergovernmental Panel on Climate Change (IPCC), 16, 31, 33–34, 92, 191–192, 194–195, 198–199, 206, 222, 286–289, 295, 297, 301
International Atomic Energy Agency (IAEA), 197, 204, 207, 213, 286, 293, 314
International Energy Agency (IEA), 13, 63, 92, 192, 196, 198, 200, 210–211, 214, 216, 274, 280–281, 314
International Monetary Fund (IMF), 105, 122, 126–127, 132
International Risk Governance Council (IRGC), 213, 222
Issing, O., 25, 42

J
Jackson, T., 277, 297, 314
Jaeger, C., 287–289, 314
Jaeger, J., 287–289, 314
Jamieson, D., 153, 162–163
Jamison, A., 295, 307, 314
Janda, K., 277, 302, 307, 314
Janssen, M. A., 285, 317
Jochem, E., 301, 304, 314–315
Joerges, B., 293, 315
Jørgensen, U., 167–188, 249–250, 252, 254–255, 259–260, 267–268
Joss, S., 295, 311
Jungk, R., 173, 188

K
Kammen, D. M., 224
Karnøe, P., 169, 172, 174, 176, 188
Karvonen, A., 41–42, 297, 311
Kasperson, R. E., 24, 197, 222, 317
Kates, R. W., 300, 315
Kemmler, A., 14, 21, 74, 87, 90, 95, 247, 261
Kemp, R., 110, 113, 286, 298, 306, 315, 318
Kesselring, P., 304, 315
Kiener, E., 310
Knorr Cetina, K. D., 294, 315
Koornneef, J., 200, 222
Krütli, P., 227–244
Kuhn, T. S., 299, 315

L
Lange, A., 144, 161
La Porte, T., 292, 315
Latour, B., 197, 222, 294, 299, 315
Leach, G., 82, 95
Lengwiler, M., 240, 243
Lenzen, M., 93, 95, 291, 315
Leopoldina, 311
Leroy, P., 277, 312
Levi, A., 8, 10
Levin, K., 8, 10
Levine, D., 307, 317
Lichtenberg, G. Chr, 278, 315
Lichtenstein, S., 230, 243, 279, 313
Lidskog, R., 306, 315
Lindoerfer, J. S., 293, 312
Liverman, D., 146, 154, 161, 163
Lohmann, L., 136, 146–147, 153, 164, 166, 276, 315
Loorbach, D., 298, 313
Luhmann, H.-J., 211, 220
Luhmann, N., 35, 42
Lundvall, B-Å, 169, 189
Lutzenhiser, L., 5, 47–49, 71, 276, 319

M
Maclean, D., 279, 316
Madlener, R., 304, 316
Malone, E. L., 217, 276, 288, 303, 317
Maul, P. R., 202, 210, 223–224
Max-Neef, M. A., 39–42, 73–77, 86, 88, 90, 95, 261, 263, 265, 268–269, 278
Mayntz, R., 35, 43, 292, 316
Mazmanian, D., 306, 316
McDaniels, T. L., 229, 232, 243
McKinsey, 213–215, 243
Meadows, D., 12, 21
Meinshausen, M., 139, 161, 191, 223

Name Index

Meitner, M., 232, 241, 244, 256, 261
Merton, R. K., 299, 316
Michaelowa, A., 143, 164
Midgley, G., 242–243
Minsch, J., 3–10, 11–21, 23–42, 47–71, 73–94, 97–112, 115–131, 133–160, 167–188, 191–218, 227–242, 247–261, 263–269, 273–310, 312–316, 318
Moezzi, M., 277, 312
Mol, A. P. J., 285, 318
Möller, N., 279, 297, 299, 316
Morell, D., 306, 316
Morone, J. G., 197, 223
Mourik, R. M., 297, 306, 316–317

N

Neuhoff, K., 140, 144, 163–164
Nordhaus, W., 152, 164
Nørgård, J. S., 19, 21
North, D. W., 196, 223
Nuclear Energy Agency (NEA), 197–198, 204, 207–208, 210–211, 223
Nuclear Waste Management Organization (NWMO), 210, 223
Nussbaum, M., 149, 164

O

Oberndorfer, U., 138, 160
Organization for Economic Cooperation and Development (OECD), 55, 63, 98, 119, 123, 136, 140, 164
Ostrom, E., 285, 312, 316–317
Ott, H., 151, 165
Otway, H., 229, 243

P

Paavola, J., 152, 161
Pacala, S., 224
Pachauri, R. K., 191–192, 198, 224
Pachauri, S., 73–95, 247–248, 251–252, 256–257, 266
Page, E., 153, 165
Page, S. C., 200, 224
Pahl-Wostl, C., 229, 231, 244
Painuly, J. P., 291, 316
Parkin, S., 275, 316
Parry, I., 144–146, 162, 165
Pattberg, P., 285, 311
Pearman, G., 276, 314
Perrow, C., 283, 293, 316
Peterson, M., 279, 316
Petts, J., 232, 240, 244
Pinch, T., 188, 197, 294, 311

Plato, 3–5, 10
Podobnik, B., 93, 95
Pohl, C., 40, 43, 233, 243
Polanyi, K., 98, 113
Polanyi, M., 300, 316
Polimeni, J. M., 304, 316
Porter, T., 145, 299, 316
Poteete, A., 285, 316
Pretty, J. N., 231, 244
Princen, T., 38, 43, 304, 317
Prins, G., 287–288, 317

R

Ramírez, A., 211, 217, 224
Rammert, W., 303, 317
Randers, J., 12, 21
Raven, R. P. J. M., 169–170, 181, 183, 188–189, 298, 306, 317–318
Ravetz, J. R., 197, 224
Ravetz, J. T., 232, 243
Rawls, J., 93–95
Rayner, S., 276, 285, 288, 303, 307, 317
Read, S., 230, 243, 279, 313
Reason, J., 293, 317
Reitman, W. R., 8, 10
Renn, O., 229, 244, 307, 317
Research Council of Norway, 277, 317
Restivo, S., 279, 311
Richardson, K., 191–192, 224
Rijkens-Klomp, N., 228, 231, 245
Rip, A., 110, 113, 169, 171, 189, 286, 298, 306, 311, 317
Rist, S., 233, 243
Rittel, H., 8, 10–11, 21, 23, 43, 273, 287, 317
Robinson, C., 299, 310
Ropohl, G., 197, 224
Rotmans, J., 298, 313
Rowe, G., 228, 231, 234, 244
Royal Society, 192, 224
Rychen, D. S., 275, 317

S

Sachs, J. D., 116, 132
Sachs, W., 151, 165
Salganik, L. H., 275, 317
Sathaye, J., 301, 315
Savage, D., 202, 210, 223–224
Scharpf, F. W., 35, 43
Schneider, D. 112
Schneider, F., 140, 165
Schneider, S. H., 192, 224
Scholz, R. W., 8, 10, 207, 222, 227–245, 296–297, 313, 317–318

Name Index

Schön, D., 299, 310
Schot, J. W., 298, 313, 317
Schulz-Schaeffer, I., 303, 317
Seifritz, W., 195, 224
Sen, A., 14, 21, 88, 95, 120, 132, 257–258, 261, 281, 306, 317
Sen, K., 104, 113
Servén, L., 129, 131
Shackley, S., 33, 197, 216–217, 221, 225
Shankleman, J., 115–132, 248–249, 253, 258–259, 267
Sheppard, S. R. J., 232, 241, 244, 256, 261
Shove, E., 5, 10, 47–49, 71, 276, 302, 317, 319
Shukla, P. R., 216, 221
Sieferle, R. P., 26, 43
Siegrist, M., 217, 226
Singleton, G., 217, 225
Slovic, P., 230, 243, 279, 313
Smil, V., 15, 18–21, 78, 95
Smith, A., 15, 25, 43, 187, 189, 195, 223
Smith, K. R., 82, 85, 95, 147, 151, 165
Socolow, R., 224
Solomon, B. D., 17, 21, 140, 162, 197, 219
Solomon, S., 203, 225
Sorrell, S., 19, 22, 304, 314
Sovacool, B. K., 47–71, 247, 274, 277, 281, 285, 294, 311, 313, 318
Spaargaren, G., 285–286, 318
Spreng, D., 3–10, 11–21, 23–42, 47–71, 73–94, 96, 97–112, 115–131, 133–160, 167–188, 191–218, 225, 227–242, 247–261, 263–269, 313, 318
Stallén, P.-J., 197, 226
Staudenmaier, J. M., 171, 189
Stauffacher, M., 207, 222, 227–245, 251, 256, 260–261, 268, 296, 318
Stavins, R., 135–136, 140, 142, 145–146, 160, 163, 165–166
Stern, N., 146, 152, 166, 191, 213, 215, 225, 288, 318
Stern, P., 24, 43, 277, 285–286, 302, 311–312, 318
Stern, P. C., 48, 50, 71, 228, 245
Sterner, T., 136, 140–141, 143–145, 148, 152, 164–166
Stevens, P., 116, 132
Stiglitz, J., 257, 261
Stirling, A., 187, 189, 240, 245, 297, 318
Stivens, M., 104, 113
Stringer, L. C., 229, 245
Stucki, S., 195, 225
Summerton, J., 292, 318

T

Tassey, G., 318
Taylor, C. A., 299, 318
Thurber, M. C., 260–261
Tietenberg, T., 136, 140, 166
Tietje, O., 8, 10, 234–236, 244–245, 297, 317
Tosato, G. C., 193, 221
Toth, F. L., 197, 225
Travis, W. R., 285, 318
Trist, E. L., 292, 312
Trutnevyte, E., 240, 245, 296, 318
Tuler, S. P., 197, 222, 277, 319

U

UN AGECC, 13, 22, 281, 318
United Nations Development Programme (UNDP), 13, 22, 100, 118, 132, 274, 318
Unruh, G. C., 216, 225

V

van Asselt, M. B. A., 228, 231, 245
Vanderheiden, S., 153, 166
van der Vleuten, E. B. A., 170, 189
van Kerkhoff, L., 40, 43
van Lente, H., 297, 300, 314
Van Vierssen, W., 277, 312
van Vuuren, D., 139, 161, 193, 225
Verbong, G. P. J., 298, 318
Victor, D. G., 192, 225, 260–261
Visschers, V. H. M., 217, 226
Vlek, C., 197, 226
Vogl, J., 299, 318
von Hayek, A. F., 290, 318
Voss, J.-P., 285–286, 318

W

Walker, B., 284, 319
Walker, G., 228, 245
Wallquist, L., 217, 226
Warde, A., 108, 113
WBGU German Advisory Council on Global change, 319
Webber, M., 8, 10–11, 21, 23, 43, 273, 287, 317
Webler, T., 228, 231, 245, 277, 319
Weinberg, A. M., 196, 225–226
Weingart, P., 286, 319
Weitzman, W., 144, 152, 166
Wiek, A., 232, 236, 239, 242–245, 313
Wiesmann, U., 21–22, 39–40, 43, 233, 243

Wilhite, H., 48, 97–112, 248, 252–253, 257–258, 267, 276, 303–304, 319
Wilson, E., 211, 213–214, 226
Winter, C.-J., 304, 315
Winters, M. S., 120, 131
Wolsink, M., 307, 319
Woodhouse, E. J., 197, 223
Woolgar, S., 279, 294, 315, 319

World Bank, 81, 96, 116, 118, 120–125, 128–129, 132, 249, 286, 314, 316
World Health Organization (WHO), 13, 22, 84, 96
World Resource Institute (WRI), 218, 226
Wuppertal Institute, 200, 225–226
Wüstenhagen, R., 307, 319
Wynne, B., 197, 226, 229, 245

Subject Index

A

Abatement, 134, 137–145, 151, 198, 213–214
Acceptability, 240, 278, 306–307
Acceptance, 4–5, 17, 24, 27, 38, 40–41, 127, 137, 207, 217–218, 260, 268, 274, 288, 306–307
Access, 7, 12–14, 20, 28–29, 45, 54, 78–83, 87–91, 99, 105–106, 122, 125, 128–129, 180, 184–185, 247–251, 253, 257–258, 261, 266, 274, 280–281, 284, 288, 291–292, 302–303
Accountability, 125, 127, 158, 196, 258, 284, 286, 290
Actor
  non-state, 285, 289
  state, 35, 285, 289
Adaptability, 284, 295
Adaptation, 94, 254, 261, 263, 282, 288, 297, 299
Adaptive capacity, 284, 299
Agenda, 5, 7, 15, 125, 173, 185, 210, 248, 273, 276, 288, 300, 304, 308–309
Allocation, 26, 120, 125, 136–140, 142, 144, 146, 150, 153, 155, 157–159, 209, 230, 254
Alternative future, 297
Ambiguity, 11, 308
Ambivalence, 285
Analysis
  data, 303
  flow, 294
  network, 298
  reflective, 253, 267
Anticipation, 174, 275, 286
Approach
  national, 285
  systemic, 278, 292
  technocratic, 173, 254
  technoeconomic, 302

Appropriation, cultural, 307
Artefact, 34, 292–294
Assessment, 5, 19, 33–34, 69, 88, 93, 110, 133–160, 168, 183, 197, 201, 205, 207–208, 210–211, 217, 230, 232, 237, 240, 255–256, 284–285, 289, 295–296, 299, 304–305
Asymmetry, 15, 138, 216, 282, 291
Attitudes, 36, 159, 294, 301–302
Availability, 12, 14, 17, 79, 81, 104, 209–210, 281, 285
  See also Energy, availability
Awareness, 79, 84, 251, 297, 305–306

B

Backcasting, 298
Back end, 197
Barriers, 17, 36, 56, 64, 67, 174, 176, 196, 204–205, 207, 209, 253, 282, 291
Basic needs, 88, 148, 153–158, 160, 281
Behaviour, 19–20, 23–24, 30–31, 35, 45, 50, 78, 84, 108, 129, 140, 149, 156, 174, 184–185, 196, 207, 287, 294–295, 301–305
Benefit, 67, 78–79, 81, 84, 121, 124–125, 139, 147, 149, 152, 154, 179, 185, 197, 211, 216, 265, 274, 280, 288, 292, 294, 297
Best practice, 286
Bonus, 291
Bottom-up strategy, 296
Bridging function, 200, 255
Business, 4, 27–29, 35, 48, 53–54, 57, 73, 75–76, 97, 118, 121, 137, 143, 214, 268, 288–289, 297, 304
Business model, 27–28, 288

327

C

Capability, 14, 37, 88
Capacity, 13–14, 36, 88, 99, 122, 124, 129–130, 147–149, 156, 158, 178–179, 181–182, 187, 193, 202, 233, 250, 284, 295, 299
  adaptive, 284, 299
Capital
  natural, 251
  social, 100, 285
Cap and trade, 68, 136, 259
Carbon capture and storage (CCS), 16–17, 191–218, 250–251, 255, 260, 285
Carrying capacity, 14, 285
Catastrophic event, 306
2° Celsius target, 191, 287
Challenger space shuttle, 293
Challenges, 6–9, 11–21, 28, 30–31, 34, 36, 38–39, 50, 82, 134–135, 147, 179–180, 196, 233, 247, 274, 277, 279, 281–282, 297, 300, 309
Change
  climate, 7–8, 12, 15–16, 20, 31–32, 36, 55, 64, 67, 70, 94, 111, 143, 145, 147, 152–153, 187, 192–193, 196, 211, 213, 215, 254, 259, 267, 274, 276, 282, 287–289, 295, 297, 300–301, 307–308
  social, 17, 97, 104, 106, 111, 168, 170–171, 184, 305
  systemic, 292, 297–298
Chernobyl, 15, 293, 296
Choice, 8, 20, 24, 48, 82–84, 98, 110, 141–142, 144, 150, 152–153, 171, 184–185, 200, 208–209, 279
Civil society, 8, 124, 256, 274, 309
Climate policy, 153, 201, 217, 259, 285–289
Closed issue, 207
Closure, 119, 138, 140, 202, 204, 207, 217
Cluster, 7–8, 41, 111–112, 232, 257
CO_2, 12, 15–18, 85, 91–93, 133, 135–136, 139, 141, 155, 159, 168, 179–180, 184–187, 191–196, 198–200, 202–207, 211, 213–218, 249–251, 255, 259–260, 288, 302, 305
Co-construction, 186
Co-evolution, 31, 34
Collaboration, 9, 39, 70, 122, 201, 227–242, 263, 265, 268, 275–276, 310
Commodity, 29, 81, 108, 111, 204
Communication, 23, 32, 36, 39, 50, 52–53, 57–58, 70, 81, 129, 228, 230, 234, 250, 257, 260, 278, 280, 283, 285, 299, 306–308

Community
  local, 176, 280
  See also Scientific community
Competition, 12, 20, 126, 129, 140, 170, 184, 198, 201, 268, 284, 290, 292
Complementarity, 6
Complexity, 4, 21, 29, 41, 152–153, 159, 175, 256, 275, 281, 283, 285, 294, 297, 308
Compliance, 137, 139, 141, 154, 197, 204, 210, 249, 260, 285, 293
Concealment, 4, 279
Conceptualisation, 111
Conflict, 14, 28–29, 36–38, 84, 91–92, 116–117, 119, 121, 124–125, 130, 153, 157, 160, 168–171, 173, 180, 186–188, 201, 205, 209, 229, 233, 239, 274–275, 289–290, 296–297, 301–302
Consensus, 8, 17, 20, 88, 116–117, 121–130, 192, 207, 233, 253, 258–259, 308
Consequences, 4, 7–8, 16, 20, 25, 28–30, 78, 116–117, 144, 151, 168, 201, 208–210, 252, 258, 282, 285, 290, 296
  unintended, 20, 30, 116, 201, 285
Consistency, 197, 297
Constellation, 177, 179, 184, 186, 258
Constraint, 50, 78, 88, 133, 168, 197, 200, 255–256
Construct, 172, 234, 236, 238–239, 241
Consumption, *see* Energy, consumption
Context, 4, 26–27, 31–32, 34, 37, 50–51, 84, 90, 106, 109, 115–131, 136–137, 139–140, 142, 144–145, 148, 150, 153, 155, 158, 169, 171, 177, 180, 187, 192, 207, 231–233, 241, 251, 254, 257–259, 267–268, 284–285, 294, 303
Contextualisation, 253
Contingency, 288, 296
Contradiction, 40, 253, 267, 269, 293
Control, 35, 76, 83, 119–121, 149–150, 178, 184, 186, 197, 201–202, 210–213, 231, 239, 248, 250, 286
Controversy, 33, 143, 169, 173–174, 176
Cooking fuels, 13, 74, 77 85, 281
Cooperation, 27, 31, 35, 39, 55, 130, 234, 236, 238–239, 261, 275, 285, 302
Corruption, 16, 116, 118–119, 123, 127
Cost-benefit ratio, 292
Costs, 213–215
  social, 215, 289
Coverage, 80, 124, 134–135, 139, 140–143, 155, 159, 213, 254
Creativity, 38, 274
Credibility, 34, 288–289, 297, 299, 307

Subject Index

Credit, 16, 29, 55, 65, 67, 128, 136, 151, 154, 202
Crisis, 20, 26, 29, 48, 117, 138, 173–174, 213, 218, 229, 260, 288
  financial, 138, 213, 218
Cross-disciplinarity, 275, 301
Cultural practice, 17, 98, 101, 253
Culture
  risk, 280
  safety, 250, 293
  scientific, 34
  traditional, 280

D

Damage, 13, 17, 74, 121, 131, 139, 143–144, 151–153, 193, 208, 213, 215, 252–253, 267, 280
Danger, 36, 138, 149, 275, 287, 296
Decarbonisation, 298
Decentralisation, 79, 138, 201, 282, 292, 296, 301
Decision making, 35, 41, 125, 149, 187, 216–217, 228–232, 234, 240–241, 251, 256, 261, 275, 282, 285, 296, 301, 306
  pluralistic, 285
Decommissioning, 201, 291
Deepwater horizon drilling rig, 293
Deficiency, 88, 288
Deficit, 31, 35–37, 105, 124, 258, 260
Deliberation, 39, 141, 242, 287
Demand security, 290
Democracy, 38, 148, 274, 300
Democratic conditions, 28
Dependence, 13, 17, 25, 82, 87, 93, 119–120, 204, 248–249, 277, 281, 285, 290–291, 302
  See also Energy
Design, 4, 7, 16, 34, 38–39, 41, 49–50, 110, 112, 123–124, 137–138, 140, 143–144, 146–147, 157, 159, 172, 174–176, 178, 181, 187, 195, 201, 206–207, 209, 227, 239–242, 248, 252, 254–255, 277, 282–283, 286, 289, 302
Developing countries, 14, 16–18, 26–27, 55, 63, 67, 70, 73–94, 128, 142–143, 150–151, 153–154, 158, 196, 251–252, 254, 256–259, 266–267, 281–282, 303, 305, 307
Development, 7, 11–13, 16, 18–21, 30–31, 55, 81, 91, 93, 118, 129, 142, 213, 232–240, 251–252, 256–261, 280
Dialogue, 3, 5, 91, 131, 135, 297, 299, 304
Dilemma, 8, 48, 106, 275

Disciplinarity, 37, 40–42, 49, 56–58, 76–77, 135, 171, 184, 236, 241, 263–264, 269, 275, 278, 296, 299, 301, 309–310
Discipline, 5, 25, 36, 40, 48, 54, 57–58, 61, 77, 116, 183, 235, 253, 264–265, 267
Disclosure, 119
Discount, 152–153, 279
Discourse, 36, 104, 171, 207, 261, 296, 299–300, 306–308
Disposal of waste, 201, 204–205, 212, 250, 260, 291, 306–307
Distribution, 7, 12, 62, 90, 93–94, 100, 122, 124, 134–135, 140–142, 144, 146, 148, 154–155, 157–158, 169–170, 180, 185, 200, 202, 216, 279, 283
Diversification, 14, 292
Diversity, 58, 99, 159, 233, 300, 307
Dominance, 4
Driver, 74, 82, 100, 121, 200–201, 247, 253, 258, 260, 280, 302

E

Economic aspects, 4, 141, 197, 213–215
Education, 14, 24, 27, 37, 39, 41, 57–58, 84–85, 87–88, 90–91, 99–100, 102, 172, 251, 257, 276, 289, 291
Effectiveness, 16, 35, 39, 83, 147, 176, 187, 196, 201, 218, 233, 253–255, 260, 277, 284, 291
Efficiency
  energy, 18–19, 49, 55–56, 66, 168, 171, 175–176, 182–183, 194, 200–201, 240, 279, 281
  environmental, 16, 253–255, 277
  system, 286
Egalitarian, 249, 282
Elites, 120, 125, 130, 158
Emissions, 16, 91–94, 133–160, 179–180, 249
Emissions trading, 133, 144
Empathy, 275
Empowerment, 231, 234, 238–239, 260
Endowment, 116, 134, 253, 267
End-use, 20, 77, 88, 280
End-user, 276, 298, 304
Energy
  availability, 281
  access, 13–14, 28–29, 83, 89–91, 247–248, 250, 258, 274, 281, 288, 302, 314
  conservation, 17, 19–20, 28, 296, 298
  consumption, 14, 18–20, 28, 85–88, 90, 92–93, 99, 108, 111, 178, 248, 252, 257, 267, 281, 284, 294, 300, 302–305, 308

330   Subject Index

Energy (*cont.*)
   contracting, 291
   demand, 13, 18–19, 99, 196, 248, 258, 276,
      282, 297
   efficiency, 18–19, 49, 55–56, 66, 168, 171,
      175–176, 182–183, 194, 200–201, 240,
      279, 281
   embodied, 88
   foreign policy, 14
   independence, 290
   intensity, 19, 110, 112, 257, 284
   model, 20, 86, 295
   planning, 183–185
   practice, 112, 253, 257
   research, 5–7, 9–10, 23–42, 48, 111, 176,
      228, 256, 264, 273–310
   resources, 12, 14–15, 17, 25, 38, 116,
      248–249, 251, 279, 284
   scenario, 92, 297
   service, 28, 74, 78, 83, 89, 248, 280, 291,
      302
   system, decentralised, 79, 292, 296
   transition, 73–94, 251, 266
Enforcement, 35, 134, 145, 196, 201, 213, 249,
      260, 285–286
Environment, 8, 12, 15–18, 85, 123, 174, 182,
      233, 240, 251–256, 285, 296, 308
Epistemic community, 48
Equal footing, 7, 30–31, 235
Equality, 277, 280–283, 285
Equity, 7–8, 91, 93, 127, 135, 252, 257, 259,
      279–280, 292, 299
Ethical, 4, 91, 127, 135–136, 139, 146, 149,
      152–153, 156, 158–159, 259, 275,
      278–279
Evaluation, 36, 54, 62, 81, 197, 230–231, 237,
      288–289
Experiment, 5, 124, 184–185, 287
   thought, 85–94
Expert, 33–34, 41, 207, 231, 235–238, 251,
      295–297
Expertise, 4, 6, 15, 39, 41, 159, 182, 184, 210,
      241, 296–297
Exploitation, 29, 94, 117, 130, 150, 153
Expropriation, 288
External effects, 291
Externalities, 26, 135, 149, 284, 290

F
Failure, 19, 24, 30, 39, 48, 55, 73, 78, 84, 150,
      174–175, 181, 217, 251, 260, 283, 288,
      290, 293, 298, 308–309
Fairness, 17, 93, 240, 307

Feed-in tariffs, 17–18, 55, 65, 67, 176, 182,
      291
First mover, 196
Fix, technical, 5, 23, 191–218, 268,
      284
Forecasting, 54, 278, 295–297
Foresight, 38, 295
Framework, 7, 9, 16, 23, 28, 31–32, 39–40,
      73–84, 121, 123, 145, 180, 211,
      213, 232, 235, 242, 250–251,
      256, 260–261, 266, 274, 277,
      285–288, 290–291, 298, 301,
      309
Framing, 169, 171, 183–184, 287, 299, 309
Freedom, 34, 38, 88, 102, 106, 150, 155, 280,
      306
Fukushima, 15, 284, 293, 296
Fund, 105, 122–124, 127, 129, 218
Future, 143, 300
   alternative, 297
Futurity, 277, 278–280

G
Gap, 48, 70, 139, 142, 217, 232, 256
Gender, 49, 53–54, 56, 60, 70, 77, 83, 91, 104,
      257, 281
Generation, 12, 19–20, 31–33, 40, 62, 66, 109,
      139, 152–153, 168, 181, 187, 204, 216,
      232, 250, 252, 260, 275, 279–281, 285,
      302
   future, 12, 152–153, 216, 279–280,
      285
Geoengineering, 192
Globalisation, 26, 31, 36, 105, 107–108,
      285–286
Goal, 9–10, 17, 19, 25, 31, 34, 39, 81, 91, 93,
      100, 134–136, 138–140, 142, 144, 158,
      178, 187, 201–202, 227, 229, 232–233,
      235, 241–242, 251, 256, 259, 261, 265,
      276, 280, 286–287, 296, 298–299, 304,
      306, 308
Goal setting, 276, 296, 299
Good life, 30, 88, 278, 280
Goods, 14, 18, 25, 50, 88–89, 93, 98–100,
      105–107, 108, 117, 146–150, 153, 157,
      281, 283–284, 289
Good science, 279
Governance
   adaptive, 286
   model, 286
   polycentric, 285
   reflexive, 285

Subject Index

Greenhouse gas, 14–15, 18, 28, 137, 147, 191–192, 198, 200, 203–205, 210, 213–214, 250, 252, 259, 291

Growth, 12, 19, 25–26, 85–86, 97–98, 100, 105–106, 107–110, 116–119, 120, 122, 125, 127, 129–130, 139, 143, 167, 171–173, 176–177, 248, 252, 280, 282, 300, 304–305

H

Hardware, 30, 254, 278

Harmonisation, 16, 137, 210, 254

Hazard potential, 13, 15, 27, 279, 293

Household, 14, 73–94, 97–112, 247–248, 252, 256–257, 267, 302–304, 308

Human dimension, 48, 308

Human rights, 28, 121–123, 125, 127, 274, 280, 300

I

Identity, 100, 105, 120

Impact, 7, 12–13, 15–18, 20, 29, 45, 52, 75, 77, 80–81, 83–85, 91, 121, 123–124, 126–131, 152, 168, 170–171, 173, 177, 180, 184, 187, 195, 200, 202, 210, 217, 229, 236, 238–239, 251–256, 259, 267, 274, 277, 279–280, 282, 287–288, 297, 301, 304–305, 308

Implication, 20, 79, 81, 84–85, 111, 116, 148–149, 253, 268–269, 293, 297

Incentive, 40, 50, 110, 119, 139, 145–146, 196–197, 204, 211, 213, 216, 249, 253, 260, 284

Indicator, 25, 87–88, 90, 100, 118, 205, 210, 255, 257–258, 284, 295, 308

Indulgence, 146, 151, 153, 155–158, 259

Inequality, 14, 91–93, 118, 129, 149, 154

Inequity, 91, 93, 154

Information
asymmetry, 15, 138, 282, 291
complex, 275

Infrastructure, 11, 13, 15, 26, 28, 30–33, 78, 87, 90, 97–98, 124, 128–130, 137, 169, 200, 216, 233, 247, 252–253, 256–258, 267, 278, 283–284, 287, 291, 303, 307
critical, 278, 283–284

Innovation, 6, 12, 20, 27–29, 34, 36–38, 55, 78, 145, 169, 173–174, 176, 180–181, 184–186, 197, 215, 249, 259–260, 282, 289–292, 296, 298
performance, 290

Instability, 119, 280, 284

Institution, 11–12, 15, 17–20, 24, 28, 30, 34–36, 48–49, 51, 53–54, 56, 81, 116, 121, 137, 169–171, 173, 182, 196, 213, 229, 250, 255, 278, 283–291, 297, 300–301, 304, 306–307, 310

Instrument
planning, 150
support, 16–17

Instrumentalisation, 5, 27, 268

Integration, 9, 26, 37–42, 90, 135, 168, 185, 200, 202, 217, 234, 241, 247, 261, 276, 288, 292

Interaction, 12, 20, 28, 30, 48, 70, 78, 108, 117, 120, 138, 143, 148, 159, 257, 260, 275, 278, 284–285, 287–288, 292, 298

Interdependency, 24, 27, 130, 277, 302–303

Interdisciplinarity, 39, 47–71, 77

Interests, 4, 8, 13, 15, 17, 28, 36, 126–128, 138, 155, 168–169, 171, 180, 187, 216, 228–230, 235, 239, 255–256, 274, 279–280, 286, 289, 293, 297, 302, 306–307

Internalisation, 18, 26, 134, 213, 291

Interplay, 24, 32, 34, 82, 200, 239–240, 284, 289–290

Interpretative flexibility, 197

Intervention, 26, 50, 186, 209, 290–291, 298, 302–304

Investment, 19, 26, 29, 55, 64, 67, 70, 78–79, 105, 117, 127–130, 168, 174, 176, 179–182, 184–187, 249, 252, 260, 280, 283, 290–291, 298

Invisibility, 3–5, 279

Invisible hand, 25

Involvement, 24, 33, 45, 77, 85, 87, 127, 129–130, 177–178, 184–185, 209, 228–232, 234, 236–241, 256, 259–261, 268

J

Justice, 5, 34, 93, 127, 134–135, 143, 146, 148, 153–158, 240, 253, 259, 275, 282

K

Knowledge
disciplinary, 42
infrastructure, 31–33, 287
integration, 38–42, 227, 234
local, 112
management, 9, 38–42, 261
production, 33, 40, 233, 286, 296
systems, 33, 35
tacit, 108, 300

Subject Index

Knowledge (*cont.*)
  target, 31, 35–36, 39, 274, 287, 300
  transdisciplinary, 39
  transformation, 31, 37–39, 274, 276, 287–288, 300

L
Landgrabbing, 301
Land use, 7, 15, 55, 67, 168, 254, 276
Large technical systems (LTS), 32, 293
Layperson, 282, 306
Legacy, 102, 276
Legitimacy, 35, 187, 229–230, 235, 241, 256, 261, 286, 288, 297
Level, 25, 28, 30, 35, 40, 50, 61, 77, 86–89, 91, 120, 122, 124, 127–128, 134, 142–144, 150, 152–153, 178, 180, 184, 192, 202–203, 205, 208, 210, 212, 215, 229, 232, 235–240, 247, 250, 261, 263, 267, 274–276, 278, 281, 287, 289, 292, 296, 298–299, 305, 307–309
Levies, 292
Liability, 17, 26, 211, 260, 291
Liberalisation, 64, 81, 138–139, 145, 180, 182, 184, 186, 255, 291–292
Life cycle analysis/assessment (LCA), 183, 199–200, 256, 284, 294, 304
Lifelines, 283
Life pattern, 111
Lifestyle, 14, 19, 28, 75, 93, 107, 149, 196, 248, 267, 285, 298, 300–303, 307
Limitation, 28, 36, 42, 51, 99, 135, 159, 267, 295
Loans, 67, 105, 125–126, 128, 132, 182, 258, 291
Lock-in, 146, 216, 260, 296
Longevity, 202, 205, 255, 295
Long term, 3, 6–9, 16–18, 23, 27–29, 31, 39, 50, 93, 130, 137, 142, 146, 186–187, 191–218, 268, 278, 283, 286, 288, 291–292, 295–298, 302, 308–310

M
Mainstream, 5, 98, 135, 298
Market
  barriers, 56, 64, 67, 291
  crash, 290
  failure, 48, 55, 150, 290
Materialism, 280
Maturity, 199–200, 202, 207, 291
Media, 15, 50, 84, 98, 105, 111, 124, 235–236, 238–239, 257, 275, 306–307

Mediation, 37–38
Mercantilism, 24–30, 280, 291
Metabolism, 295
Method, 16, 19, 39–40, 48–52, 54, 61, 69–70, 159, 172, 175, 195, 208, 228, 231–233, 236–239, 241, 252, 256, 258, 261, 275, 294–296, 298, 303
Millennium Development Goal (MDGs), 13, 81, 91, 93, 251, 280
Mindset, 151, 302
Mitigation, 20, 94, 144–145, 153, 180, 185, 192–193, 198, 201, 211, 213, 251, 255, 288
Model, 25, 27–29, 32–34, 48, 86, 98–100, 105, 107, 130, 210, 282, 301
Modernisation/modernization, 78, 106, 257, 280, 285
Monetary policy, 289
Money, 12, 26, 48, 50, 88, 98, 108, 117, 120, 140, 144, 147–148, 150, 153, 155, 158, 204, 249, 257, 291
Monitoring, 31, 122, 134, 137, 139, 145, 181, 204–205, 207, 209, 211–214, 217, 240, 285, 303
Multi-criteria analysis (MCA), 152–153, 234, 237
Multidisciplinarity, 264
Multiplicity, 7–8

N
Nation, 93, 137, 155, 286, 290
Negligence, 290, 296
Negotiation, 34–35, 121, 123–124, 126, 187, 230, 233, 242, 289, 293
Network, 13, 23, 35, 75, 106, 108, 169, 176, 179, 182, 185–186, 200, 216, 240, 254, 289, 293–294, 298–300
NIMBY (Not In My Back Yard), 17, 306
Normalisation, 28, 310
Norms, 35, 41, 77, 97, 303

O
Objectives
  conflicting, 36, 168, 275
Obstacle, 29, 118, 186, 259, 284, 291, 306
Oil companies
  national, 115, 125, 290, 292
Opponent, 120, 149, 307
Option, 153, 159, 193–194, 209, 218, 250–251
Oversight, 15, 17, 125, 181–182, 260, 284, 293
Ownership, 91, 102–104, 106, 108–109, 118, 147–148, 184, 303

Subject Index

P

Paradigm, 3, 6, 23, 41, 48, 182, 187, 204, 276, 309

Participation, 37, 143, 150, 158, 173, 196, 213, 227–232, 235, 240–241, 252, 256, 258–261, 263, 265, 275, 286, 306–307

Path dependency, 169, 186, 255, 273, 278, 285, 295–296

to pay, 148, 150

Peace, 34, 282–283, 287, 300

Peer review, 6, 210, 239

Perception, 35–37, 149, 177–178, 217, 240, 250, 302

Performance, 17, 79, 97, 116, 123, 137–138, 167, 186, 198–199, 201–202, 205, 207, 218, 251, 253, 255, 257

Perspective

business, 75

cost, 74

economic, 5, 90

pluralistic, 15, 250, 285

resource, 73–74

sociotechnical, 171

systems, 255, 298

technical, 73–75, 171

time, 75, 80, 83, 85

top-down, 174, 186–187, 251, 255, 268, 306

user, 75

Pilot project, 291

Planning, 6–8, 12, 24, 40–41, 53, 57, 78, 100, 178, 183–185, 208, 228, 235, 241, 251, 255, 256–257, 261, 288, 297–298

Plurality, 88, 288, 306

Policy

climate, 153, 201, 217, 259, 285–289

decision, 29, 134, 229

energy, 7–8, 14, 27, 29, 37, 45, 47–71, 109–112, 172, 176, 182, 197, 228, 286–287, 309

environmental, 27, 133, 145–146, 149, 288

implementation, 276

instrument, 28, 133–134, 141–142, 144–147, 149, 151–153, 158–160

integration, 23

intervention, 50, 298, 303–304

monetary, 289

strategy, 38, 50, 173, 183, 200, 240, 255, 286, 292, 305

Pollution, 15, 55, 67, 76, 84, 95, 99, 115, 134–135, 141, 145, 147, 152, 158, 173, 182, 185, 252, 255, 279

Poverty, 90–94, 116, 119, 122, 129, 150, 160, 248, 266, 273, 277, 280–282, 285, 299–300

Power

asymmetry, 143, 150, 155–160, 254, 259

Practice

cluster, 111–112

cultural, 17, 98, 101, 253

theory, 108, 111

Practitioner, 6, 9, 49, 101, 116, 184, 236, 263, 267, 294, 299

Precursor, 293

Prediction, 34, 74, 167, 267

Problem

complex, 39, 159, 214, 242

ill-defined, 8, 287

ownership, 303

real-world, 9, 31, 69, 233

solving, 8, 34–37, 176, 231, 233, 273–310

space, 286

'super-wicked', 8

'wicked', 8, 287

Procedure, 51, 137–138, 152, 204, 207, 210–212, 215, 230, 234–237, 241, 250, 284–285, 290, 295

Process, 9, 16, 27, 31–35, 39, 41, 81, 91, 118–119, 122–123, 128, 137, 139, 141, 143, 147, 169, 174, 180–183, 185–187, 194–195, 199, 201, 209–211, 228–234, 237, 239–242, 259, 261, 274, 277, 282, 290, 297, 306, 308

Production, 4–5, 7, 13–14, 15, 18, 25–26, 33, 35, 70, 74–76, 99, 117–118, 119, 121, 124–126, 128, 130, 136, 144, 171–173, 175–180, 182–183, 185, 194, 201, 213, 233, 251–252, 255, 280–282, 286, 288, 294, 296, 300, 303

Progress, 19, 31, 38–40, 78, 80, 100, 116, 125, 201–202, 217, 234, 237–239, 248, 260, 288, 304, 308

Projection, 47, 50, 55, 64, 70, 138, 252

Promotion, 24, 26, 170, 213, 235, 240, 291–292

Property rights, 3, 145, 147

Proponent, 135–136, 249, 288, 290, 299, 307

Prosperity, 25, 34, 38, 280, 282, 304

Protectionism, 26

Public

resistance, 306

Public goods, 284

Q

Quality of life, 117, 285, 305
Quota, 148, 179–180, 184, 186, 291

R

Radioactive waste, 17, 191, 196–197, 201–208,
    211–213, 216–217, 232, 239, 250, 255,
    268, 291, 307
Rational actor/individual, 98, 276
Rationality, 24, 36, 306–308
Rebound, 28–29, 38, 303–305
Reflection, 5, 9, 23, 35–38, 45, 108, 160, 168,
    240, 253, 267, 274–275, 288
Reflexivity, 45, 108, 273–274, 277–280
Reform, 27, 75, 99, 101, 122, 288–289
Regime, 110–111, 169, 173, 176, 185–187,
    201, 213, 216, 218, 254–256, 260, 268,
    286
Region, 13–14, 58, 64, 88, 120, 135, 208–209,
    233, 235–238, 240–241, 244, 248, 281,
    284
Regulation, 7, 18, 20, 30, 94, 100, 136,
    140–142, 145–147, 149, 155–156,
    158–159, 182, 187, 197, 204, 211–213,
    218, 249, 283, 286, 291, 293
Relevance, 3, 6, 9, 50, 77, 139–140, 150, 153,
    155, 157, 267, 300
Reliability, 13, 51, 56–57, 258, 281
Renewable energy, 16–17, 19, 27, 29, 67, 145,
    147, 167–168, 173, 176, 178–182, 184,
    186, 240, 250, 254, 260, 267, 282,
    289, 292, 296, 298, 301, 306–307,
    310
Representation, 33, 52, 69, 157
Reputation, 3, 124, 213, 280, 307
Research
    agenda, 3, 5, 308
    analytic, 263
    applied, 263
    area, 278, 300
    basic, 211, 309
    community, 17, 39, 310
    design, 39, 41, 49, 277
    field, 54, 61, 70
    interdisciplinary, 49, 57, 241
    long-term, 308
    management, 48
    method, 48–51, 227–228, 230, 232–233,
        236–239, 303
    policy, 26, 309
    programme, 175–176, 276, 308–310
    question, 5, 10, 37, 241, 263–266,
        277–278

synthetic, 263–264
    transdisciplinary, 39–40, 227–228, 265,
        275, 278–279
Research and Development (R&D), 9, 67,
    145–146, 151, 198, 201, 204, 210, 214,
    216, 277
Resilience, 283–284
Resource, 26, 117–119, 121–125, 194
Resource curse, 115–118, 121–122, 124–126,
    128–130, 158, 248–249, 253, 267, 282,
    292, 301
Response capacity, 285
Responsibility, 5, 7–9, 27, 78, 91, 93–94, 102,
    104, 127, 129, 151, 158–159, 173, 202,
    258, 290
Revolution
    'dual', 279
Rhetoric, 4, 70, 187, 299, 306
Risk
    ambiguous, 199, 307
    analysis/assessment, 197, 204, 213, 217,
        230, 299
    communication, 250, 260, 306–307
    creeping, 307
    society, 285
Robustness
    social, 286
Roles, 83–84, 104, 169, 177, 240, 257, 281,
    285, 294
Rule, 4, 34, 93, 102, 120, 124, 134, 136, 140,
    159, 196, 213, 218, 230, 237, 255,
    283–285, 287, 296

S

Safety
    assessment, 191, 207–210, 217
    culture, 250, 293
Sanction, 250, 285
Scales, 15, 202, 217, 273, 285, 309
Scarcity, scarcities, 25–27, 29, 31, 37–38, 118,
    134, 137–138, 143, 172, 252, 274
Scenario
    analysis, 198, 236, 241, 243, 297
    participatory, 297
    sociotechnical, 7, 70, 191, 197, 254, 287,
        292, 298
    techniques, 295, 298
School of thought, 298, 310
Science
    'good', 209, 279
    'hard', 70, 276
    pluralistic, 285
    'soft', 6

Subject Index

Science and technology studies (STS), 9, 20, 197, 254, 292
Scientific community, 273, 278, 292, 297–300
Seamless web, 168, 293
Security
  demand, 290
  financial, 275
  institutional, 185
  supply, 13–14, 249–250, 281, 284, 292
Self-organisation, 37–38, 274
Self-reflection, 5, 263
Self sufficiency, 89
Shortcoming, 202, 249, 258, 276, 297
Short term, 20, 76, 107, 137, 167, 179, 191, 202, 251, 255–256, 276, 309
Side-effect, 211
Social learning, 16, 273, 278, 292, 297–300
Society
  civil, 8, 124, 256, 274, 282, 309
  learning, 23, 31–35
  risk, 123, 285
Sociotechnical, 7, 70, 191, 197, 254, 287, 292, 298
Software, 30, 196, 254, 278
Solution
  long-term, 187, 197, 202, 208–210, 214, 259, 286
  short-term, 20, 76, 107, 137, 167, 179, 191, 202, 251, 255–256, 276, 309
Source, 8, 14, 18, 49, 54–56, 61, 65, 73, 77–78, 82, 89, 99, 168–169, 171, 176, 179, 185, 187, 192, 194–195, 201, 204, 216, 236, 241, 248, 250, 252, 257, 277, 281, 289, 291, 301
Spill-over effects, 288
Stability, 11–12, 18, 115, 119, 209, 280, 284, 299
Stagnation, 276
Stakeholder, 39, 77, 128, 157, 170, 209–211, 217, 228–232, 234–235, 237–239, 250–251, 256, 265, 268, 285, 295–296, 298–299
  See also Actor
Stewardship, 281
Stopping rules, 287
Storyline, 170–171, 297
Strategy, 26, 38, 50, 173, 183, 208, 240, 255, 286, 292, 305
Subsidy, 74, 82
Sufficiency, 89, 304–305
Supervision, 102, 129, 286, 293
Supplier, 13, 290–292
Supply security, *see* Security

Survey, 54, 61, 70, 87, 99–100, 103–104, 108–109, 146, 237–240, 266, 295, 303
Sustainability
  analysis/asssessment, 295
  literacy, 310
Sustainable development, 31, 34–35, 37, 93, 135, 196, 232, 242, 257, 275, 285–286, 288, 299–300
Symptom, 8, 27, 207, 288
System
  analysis, 197, 198–201, 232, 234, 236, 238–239
  boundary, 33
  human, 285
  inertia, 216
  knowledge, 31–34, 39, 265, 287, 300
  large technical, 32, 293
  limit, 277
  longevity, 202, 205, 255, 295
  natural/physical, 32
  open, 11, 23, 202
  scientific, 275, 287
  social/political, 32–34
  sociotechnical, 7, 70, 191, 197, 254, 287, 292, 298
  technical, 7, 12, 32–34, 70, 197, 210, 254, 287, 292–293, 298, 303
  technoeconomic, 3, 5–6, 18, 302, 306
  theory, 197

T

Target knowledge, *see* Knowledge
Taxes, 18, 55, 65, 67, 118, 120, 122, 145–146, 149, 292
Technical fix, 5, 191–217, 268, 284
Technology
  assessment, 191, 197, 256, 285, 295
  development, 30–31
  discourse, 308
  niche, 298
  policy, 183, 191, 197, 297
  portfolio, 296
Tensions, 18, 28, 173, 254
Theory, 6, 34–35, 48, 93, 98, 108, 111, 144, 150, 169–170, 186, 197, 249, 267, 297, 301
Thought experiment, 85–94
Three Mile Island, 293
Time
  frame, 17, 214, 217, 296, 309
  horizon, 289
1 Ton CO_2 society, 305

Tool, 6, 14, 16, 27, 61, 70, 120, 184, 200, 276, 284, 294–295
Top-down perspective, 174, 186–187, 251, 255, 268, 306
Traceability, 205, 297
Traditional culture, 280
Transaction costs, 145, 285
Transdisciplinarity, 23, 39–41, 73–77, 86, 233, 261, 265–266, 268
Transformation, 23, 31, 35, 37–39, 174, 187, 257, 274, 276, 280, 282, 287–288, 297, 300
Transition, 27, 35, 73, 79, 82, 84, 90, 102, 169–170, 186–188, 199–200, 227–228, 260, 276, 299
Transparency, 121–127, 201, 230, 241–242, 258, 261, 286, 290, 297
True-cost pricing, 290
Trust, 17, 36, 102, 214, 229–230, 240, 255–256, 260, 290, 307

U
Uncertainty, 32–34, 38, 141, 144, 182, 191, 207, 210, 232, 285, 291, 297, 308

V
Validation, 34, 215
Value, 6, 8, 24, 31, 37, 56, 86–88, 93, 119, 124, 129, 148–149, 152, 156, 159, 229, 253, 260, 265, 273, 278–279, 289, 300–305, 309
Verification, 34, 137, 139, 211
Viability, 19, 187, 298
Vision, 100, 173–174, 181–185, 255, 260, 273, 298, 304, 306
Vulnerability, 283–284

W
2000 Watt society, 304–305
Wealth, 18, 24–25, 93–94, 101, 115, 117–125, 129–130, 136, 140, 148, 150, 154–155, 157–158, 233, 247, 258
Welfare, 25, 75, 134, 145, 257–258
Well-being, 12, 19, 88, 90, 127, 257, 300
Willingness
    to accept, 288
    to experiment, 185